Edited by Robert Puers, Livio Baldi,
Marcel Van de Voorde, and
Sebastiaan E. van Nooten

Nanoelectronics

Further Volumes of the Series "Nanotechnology Innovation & Applications"

Axelos, M. A. V. and Van de Voorde, M. (eds.)

Nanotechnology in Agriculture and Food Science

2017
Print ISBN: 9783527339891

Cornier, J., Kwade, A., Owen, A., Van de Voorde, M. (eds.)

Pharmaceutical Nanotechnology

Innovation and Production

2017
Print ISBN: 9783527340545

Fermon, C. and Van de Voorde, M. (eds.)

Nanomagnetism

Applications and Perspectives

2017
Print ISBN: 9783527339853

Mansfield, E., Kaiser, D. L;, Fujita, D., Van de Voorde, M. (eds.)

Metrology and Standardization for Nanotechnology

Protocols and Industrial Innovations

2017
Print ISBN: 9783527340392

Meyrueis, P., Sakoda, K., Van de Voorde, M. (eds.)

Micro- and Nanophotonic Technologies

2017
Print ISBN: 9783527340378

Müller, B. and Van de Voorde, M. (eds.)

Nanoscience and Nanotechnology for Human Health

2017
Print ISBN: 978-3-527-33860-3

Raj, B., Van de Voorde, M., Mahajan, Y. (eds.)

Nanotechnology for Energy Sustainability

2017
Print ISBN: 9783527340149

Sels, B. and Van de Voorde, M. (eds.)

Nanotechnology in Catalysis

Applications in the Chemical Industry, Energy Development, and Environment Protection

2017
Print ISBN: 9783527339143

*Edited by Robert Puers, Livio Baldi, Marcel Van de Voorde,
and Sebastiaan E. van Nooten*

Nanoelectronics

Materials, Devices, Applications

Volume 2

WILEY-VCH
Verlag GmbH & Co. KGaA

Volume Editors

Prof. Robert Puers
Catholic University Leuven
Elektrotechniek - ESAT-MICAS
De Croylaan
Kasteelpark Arenberg 10
3001 Leuven
Belgium

Dr. Livio Baldi
Micron Semiconductor
via C. Olivetti 2
20864 Agrate Brianza
Italy

Prof. Dr. Dr. h.c. Marcel H. Van de Voorde
Member of the Science Council
of the French Senate and National
Assembly, Paris
Rue du Rhodania, 5
BRISTOL A, Appartement 31
3963 Crans-Montana
Switzerland

Dr. Sebastiaan E. van Nooten
Semi Consulting
Bosuillaan 303
3722 XM Bilthoven
Netherlands

Series Editor

Prof. Dr. Dr. h.c. Marcel H. Van de Voorde
Member of the Science Council
of the French Senate and National
Assembly, Paris
Rue du Rhodania, 5
BRISTOL A, Appartement 31
3963 Crans-Montana
Switzerland

Covercredit: Circuit: © fotolia, Petrovich 12
Central Processor: © fotolia, Maksym
Yemelyanov

Library of Congress Card No.: applied for

British Library Cataloguing-in-Publication Data
A catalogue record for this book is available from the
British Library.

**Bibliographic information published by the
Deutsche Nationalbibliothek**
The Deutsche Nationalbibliothek lists this
publication in the Deutsche Nationalbibliografie;
detailed bibliographic data are available on the
Internet at http://dnb.d-nb.de.

© 2017 Wiley-VCH Verlag GmbH & Co. KGaA,
Boschstr. 12, 69469 Weinheim, Germany

Print ISBN: 978-3-527-34053-8
ePDF ISBN: 978-3-527-80071-1
ePub ISBN: 978-3-527-80073-5
Mobi ISBN: 978-3-527-80074-2
oBook ISBN: 978-3-527-80072-8

Cover Design Adam Design
Typesetting Thomson Digital, Noida, India
Printing and Binding Strauss GmbH, Mörlenbach,
Germany

Printed on acid-free paper

*Thanks to my wife for her patience with me spending
many hours working on the book series through
the nights and over weekends.
The assistance of my son Marc Philip related to the complex
and large computer files with many sophisticated scientific
figures is also greatly appreciated.*

Marcel Van de Voorde

Series Editor Preface

Since years, nanoscience and nanotechnology have become particularly an important technology areas worldwide. As a result, there are many universities that offer courses as well as degrees in nanotechnology. Many governments including European institutions and research agencies have vast nanotechnology programmes and many companies file nanotechnology-related patents to protect their innovations. In short, nanoscience is a hot topic!

Nanoscience started in the physics field with electronics as a forerunner, quickly followed by the chemical and pharmacy industries. Today, nanotechnology finds interests in all branches of research and industry worldwide. In addition, governments and consumers are also keen to follow the developments, particularly from a safety and security point of view.

This books series fills the gap between books that are available on various specific topics and the encyclopedias on nanoscience. This well-selected series of books consists of volumes that are all edited by experts in the field from all over the world and assemble top-class contributions. The topical scope of the book is broad, ranging from nanoelectronics and nanocatalysis to nanometrology. Common to all the books in the series is that they represent top-notch research and are highly application-oriented, innovative, and relevant for industry. Finally they collect a valuable source of information on safety aspects for governments, consumer agencies and the society.

The titles of the volumes in the series are as follows:

Human-related nanoscience and nanotechnology

- *Nanoscience and Nanotechnology for Human Health*
- *Pharmaceutical Nanotechnology*
- *Nanotechnology in Agriculture and Food Science*

Nanoscience and nanotechnology in information and communication

- *Nanoelectronics*
- *Micro- and Nanophotonic Technologies*
- *Nanomagnetism: Perspectives and Applications*

Nanoscience and nanotechnology in industry

- Nanotechnology for Energy Sustainability
- Metrology and Standardization of Nanomaterials
- Nanotechnology in Catalysis: Applications in the Chemical Industry, Energy Development, and Environmental Protection

The book series appeals to a wide range of readers with backgrounds in physics, chemistry, biology, and medicine, from students at universities to scientists at institutes, in industrial companies and government agencies and ministries.

Ever since nanoscience was introduced many years ago, it has greatly changed our lives – and will continue to do so!

March 2016   *Marcel Van de Voorde*

About the Series Editor

Marcel Van de Voorde, Prof. Dr. ir. Ing. Dr. h.c., has 40 years' experience in European Research Organisations, including CERN-Geneva and the European Commission, with 10 years at the Max Planck Institute for Metals Research, Stuttgart. For many years, he was involved in research and research strategies, policy, and management, especially in European research institutions.

He has been a member of many Research Councils and Governing Boards of research institutions across Europe, the United States, and Japan. In addition to his Professorship at the University of Technology in Delft, the Netherlands, he holds multiple visiting professorships in Europe and worldwide. He holds a doctor honoris causa and various honorary professorships.

He is a senator of the European Academy for Sciences and Arts, Salzburg, and Fellow of the World Academy for Sciences. He is a member of the Science Council of the French Senate/National Assembly in Paris. He has also provided executive advisory services to presidents, ministers of science policy, rectors of Universities, and CEOs of technology institutions, for example, to the president and CEO of IMEC, Technology Centre in Leuven, Belgium. He is also a Fellow of various scientific societies. He has been honored by the Belgian King and European authorities, for example, he received an award for European merits in Luxemburg given by the former President of the European Commission. He is author of multiple scientific and technical publications and has coedited multiple books, especially in the field of nanoscience and nanotechnology.

Contents

Foreword *by Andreas Wild* XXV
Nanoelectronics for Digital Agenda *by Paul Rübig*
and Livio Baldi XXXVII
Electronics on the EU's Political Agenda *by Carl-Christian Buhr* XLI
Preface *by Livio Baldi and Marcel H. van de Voorde* XLVII

Volume 1

Part One   Fundamentals on Nanoelectronics *1*

1   **A Brief History of the Semiconductor Industry** *3*
             Paolo A. Gargini
1.1   From Microelectronics to Nanoelectronics and Beyond *3*
1.1.1   You Got to Have Science, Genius! *3*
1.1.2   What Would Science Be Without Technology? *5*
1.1.3   The Magic of Economics *11*
1.1.4   Back to the MOS *14*
1.1.5   Technology Innovation Must Go On! *15*
1.1.6   Bipolar against MOS! *16*
1.1.7   Finally It All Comes Together *20*
1.2   The Growth of the Semiconductor Industry: An Eyewitness Report *22*
1.2.1   The Making of the PC Industry *23*
1.2.2   The DRAM Wars *26*
1.2.3   The Introduction of New Materials *30*
1.2.4   Microprocessors Introduction Cycle Goes from 4 to 2 Year *31*
1.2.5   The 300 mm Wafer Size Conversion *31*
1.2.6   The 1990s: Scaling, Scaling, Scaling *33*
1.2.7   Equivalent Scaling: Designers Will Never Know What We
             Have Done *34*
1.2.8   Is There Life Beyond the Limits of CMOS and of Von Neumann
             Architecture? *39*
1.2.9   Nanoelectronics to the Rescue *41*
1.2.10   The New Manhattan Project *45*

1.2.11   System Requirements and Heterogeneous Integration *48*
1.2.12   Evolve or Become Irrelevant *49*
1.2.13   Bringing It all Together *51*
          Acknowledgments *52*

2   **More-than-Moore Technologies and Applications** *53*
          Joachim Pelka and Livio Baldi
2.1   Introduction *53*
2.2   "More Moore" and "More-than-Moore" *54*
2.3   From Applications to Technology *56*
2.4   More-than-Moore Devices *58*
2.4.1   Interacting with the Outside World *58*
2.4.2   Powering *59*
2.4.3   More-than-Moore Technologies *60*
2.5   Application Domains *61*
2.5.1   Automotive *61*
2.5.2   Health Care *62*
2.5.2.1   Wearable Health Care *62*
2.5.2.2   Biochips and Lab-on-Chips *63*
2.5.3   Safety and Security *65*
2.5.4   Industrial Applications *67*
2.5.4.1   Integrated Power *67*
2.5.4.2   Lighting *69*
2.6   Conclusions *70*
          Acknowledgement *71*
          References *71*

3   **Logic Devices Challenges and Opportunities in the Nano Era** *73*
          Frédéric Boeuf
3.1   Introduction: Dennard's Scaling and Moore's Law Trends and
          Limits *73*
3.2   Power Performance Trade-Off for 10 nm, 7 nm, and Below *75*
3.2.1   Electrostatics of Advanced CMOS Devices *75*
3.2.2   Speed Performance Metrics of CMOS Technologies *78*
3.2.2.1   Switching Delay Formulation *78*
3.2.2.2   Effective Current and MOSFET Electrostatics *80*
3.2.3   Parasitics Capacitance in Logic Devices *81*
3.2.3.1   Effective Capacitance of an Inverter Switch *81*
3.2.3.2   Parasitic Capacitance Calculation Method *83*
3.2.4   Power Dissipation in Transistor Devices *84*
3.2.4.1   Static Power Dissipation *84*
3.2.4.2   Dynamic Power Dissipation *85*
3.2.4.3   Limitation of the Minimum Voltage Supply: The V_{th} Variability *87*
3.2.5   Summary of the Key Points of CMOS Devices *88*
3.3   Device Structures and Materials in Advanced CMOS Nodes *89*

3.3.1   SCE Immune MOSFET Architectures   *89*
3.3.1.1   Fully Depleted SOI, UTB, and UTBB Structures   *90*
3.3.1.2   FinFET and Double-Gate Devices   *93*
3.3.1.3   Gate-All-Around Transistors and Nanowires   *96*
3.3.2   Parasitic Capacitances in Advanced Device Structures   *97*
3.3.3   High-Mobility Materials and Devices   *100*
3.3.3.1   Transistor Current in Ultrashort Devices   *100*
3.3.3.2   Material Engineering for Transport Enhancement   *101*
3.3.3.3   Choice of Materials for Advanced CMOS   *103*
           References   *105*

4   Memory Technologies   *113*
           Barbara De Salvo and Livio Baldi
4.1   Introduction   *113*
4.2   Mainstream Memories (DRAM and NAND): Evolution and Scaling
           Limits   *115*
4.3   Emerging Memories Technologies   *120*
4.3.1   Ferroelectric Memories   *120*
4.3.2   Magnetic Memories   *122*
4.3.3   Phase Change Memories   *124*
4.3.4   Resistive RAMs: OxRAM and CBRAM   *126*
4.3.5   Other Memory Concepts   *129*
4.4   Emerging Memories Architectures   *130*
4.4.1   From Cell to Arrays   *130*
4.4.2   3D RRAM Architectures   *132*
4.5   Opportunities for Emerging Memories   *133*
4.5.1   Storage Class Memory   *133*
4.5.2   Embedded Memories   *133*
4.6   Conclusions   *134*
           References   *135*

Part Two   Devices in the Nano Era   *137*

**5   Beyond-CMOS Low-Power Devices: Steep-Slope Switches for
           Computation and Sensing**   *139*
           Adrian M. Ionescu
5.1   Digital Computing in Post-Dennard Nanoelectronics Era   *139*
5.2   Beyond CMOS Steep-Slope Switches   *143*
5.3   Convergence of Requirements for Energy-Efficient Computing and
           Sensing Technologies: Enabling Smart Autonomous Systems
           for IoE   *148*
5.4   Conclusions and Perspectives   *149*
           References   *151*

6   **RF CMOS**   *153*
    Patrick Reynaert, Wouter Steyaert and Marco Vigilante
6.1   Introduction   *153*
6.2   Toward 5G and Beyond   *153*
6.3   CMOS @ Millimeter-Wave: Challenges and Opportunities   *156*
6.4   Terahertz in CMOS   *159*
6.5   Conclusions   *161*
    References   *162*

7   **Smart Power Devices Nanotechnology**   *163*
    Gaudenzio Meneghesso, Peter Moens, Mikael Östling, Jan Sonsky, and Steve Stoffels
7.1   Introduction   *163*
7.2   Si Power Devices   *164*
7.2.1   Discrete versus Integrated Power Devices   *164*
7.2.2   Low-Voltage MOSFETs   *166*
7.2.3   High-Voltage MOSFETs   *170*
7.2.4   IGBTs   *173*
7.2.5   Device versus Application Landscape   *175*
7.3   SiC Power Semiconductor Devices   *176*
7.3.1   High-Voltage Blocking   *178*
7.3.2   SiC Diodes/Rectifiers   *179*
7.3.3   Switch Devices   *180*
7.3.4   JFETs and MOSFETs   *180*
7.3.5   Bipolar Junction Transistors   *182*
7.3.6   Ultrahigh Voltage–High-Injection Devices   *183*
7.3.7   Concluding Remarks and Issues of Concerns for SiC Power Devices   *183*
7.4   Power GaN Device Technology   *184*
7.4.1   GaN Material and Device Physics   *184*
7.4.2   Device Architectures   *187*
7.4.2.1   HEMT (Schottky)   *187*
7.4.2.2   MISHEMT   *188*
7.4.2.8   Vertical Devices   *188*
7.4.3   Ohmic Contacts   *190*
7.4.4   E-MODE Devices   *191*
7.4.4.1   Thin AlGaN Gate Barrier   *191*
7.4.4.2   Charge Incorporation   *191*
7.4.4.3   P-GaN or P-AlGaN Gate Structure   *192*
7.4.4.4   HEMT/FET Hybrid   *192*
7.4.4.5   Cascode   *192*
7.4.5   Breakdown Voltage Engineering and Limitations   *193*
7.4.5.1   Buffer Engineering   *193*
7.4.5.2   Substrate Implantation   *194*

7.4.5.3   Substrate Removal   *194*
7.4.6   Dispersion Phenomena   *195*
7.4.6.1   Surface-Induced Dispersion   *195*
7.4.6.2   Buffer-Induced Dispersion   *197*
7.4.7   Conclusion   *197*
7.5   New Materials and Substrates for WBG Power Devices   *198*
References   *201*

8   Integrated Sensors and Actuators: Their Nano-Enabled Evolution into the Twenty-First Century   *205*
Frederik Ceyssens and Robert Puers
8.1   Introduction   *205*
8.2   Sensors   *208*
8.2.1   Mechanical Sensors   *208*
8.2.1.1   Pressure Sensors and Microphones   *208*
8.2.1.2   Gyroscopes and Accelerometers   *209*
8.2.1.3   Resonators   *210*
8.2.2   Vision/IR   *210*
8.2.3   Terahertz (Thz) Imaging   *211*
8.2.4   Radar/Lidar   *212*
8.2.5   Gas Sensors   *212*
8.2.6   Biosensors   *213*
8.3   Actuators   *214*
8.3.1   Electrostatic, Electromagnetic, and Piezoelectric   *214*
8.3.2   Pneumatic, Phase Change, and Thermal Actuators   *216*
8.3.3   Artificial Muscles   *216*
8.4   Molecular Motors   *217*
8.5   Transducer Integration and Connectivity   *218*
8.6   Conclusion   *219*
References   *220*

Part Three   Advanced Materials and Materials Combinations   *223*

9   Silicon Wafers as a Foundation for Growth   *225*
Peter Stallhofer
9.1   Introduction   *225*
9.2   Si Availability and Technologies to Produce Hyperpure Silicon in Large Quantities   *226*
9.2.1   Metallurgical Silicon Production   *226*
9.2.2   Purification of Metallurgical Silicon via Trichlorosilane   *227*
9.2.3   Production of Electronic Grade Polysilicon   *228*
9.2.4   Monocrystalline Silicon Production   *229*
9.2.4.1   CZ Growth Method   *229*
9.2.4.2   FZ Growth Method   *232*

9.2.5   Process Sequence of Silicon Wafer Production   *232*
9.2.5.1   Mechanical Treatment   *233*
9.2.5.2   Chemical Treatment   *234*
9.2.5.3   Chemical–Mechanical Polishing   *234*
9.2.5.4   Final Cleaning and Packaging   *235*
9.2.5.5   Epitaxy   *236*
9.3   The Exceptional Physical and Technological Properties of Monocrystalline Silicon for Device Manufacturing   *237*
9.3.1   Doping   *237*
9.3.2   Crystal Structure   *237*
9.3.3   Silicon Dioxide   *238*
9.3.4   Intrinsic Defect Categories   *239*
9.3.5   Defect Kinetic Behavior   *240*
9.4   Silicon and New Materials   *241*
9.5   Example of Actual Advanced 300 mm Wafer Specification for Key Parameters   *242*
            Acknowledgments   *242*
            References   *242*

10   **Nanoanalysis**   *245*
            Narciso Gambacorti
10.1   Three-Dimensional Analysis   *246*
10.1.1   X-Ray Tomography for the Analysis of TSV   *247*
10.1.2   Progress in Atom Probe Tomography for Semiconductor Analysis   *249*
10.2   Strain Analysis   *250*
10.2.1   State-of-the-Art Strain Analysis by Precession Electron Diffraction   *252*
10.2.2   X-Ray for Strain Measurements   *253*
10.3   Compositional and Chemical Analysis   *256*
10.3.1   Advanced Characterization of HKMG Stacks for Sub-14 nm Technology Nodes   *256*
10.3.2   TEM Composition Analysis of NMOS Device   *259*
10.4   Conclusions   *260*
            Glossary   *261*
            Acknowledgments   *262*
            References   *262*

Part Four   Semiconductor Smart Manufacturing   *265*

11   **Front-End Processes**   *267*
            Marcello Mariani and Nicolas Possémé
11.1   A Standard MOS FEOL Process Flow   *267*
11.2   Cleaning   *268*

11.2.1   Wet Cleaning   *268*
11.2.2   Advanced Aqueous Cleaning   *268*
11.2.3   Nonaqueous Advanced Cleaning Approaches   *269*
11.2.4   Advanced Drying Techniques   *270*
11.3   Silicon Oxidation   *271*
11.4   Doping and Dopant Activation   *272*
11.4.1   Coimplantation   *273*
11.4.2   Defect Engineering and Surface Treatment   *273*
11.4.3   Flash Anneal, Laser Annealing, and Nonthermal Activation
           Techniques   *274*
11.4.4   Plasma Doping   *274*
11.4.5   Molecular Monolayers Doping   *275*
11.5   Deposition   *275*
11.5.1   Thin Film Deposition   *275*
11.5.2   Atomic Layer Deposition   *277*
11.5.3   Other Monolayer Deposition Techniques   *279*
11.6   Etching   *279*
11.6.1   Wet Etching   *279*
11.6.2   Dry Etching   *280*
11.6.3   Limitation of Plasma Etching for Critical Dimension Control at the
           Atomic Scale   *281*
11.6.4   Existing Solutions   *284*
11.6.5   Plasma Etch Challenges for Nanotechnologies: ALE Wishes or
           Reality?   *285*
           References   *285*
           Bibliography   *288*

12   **Lithography for Nanoelectronics**   *289*
           Kurt Ronse
12.1   Historical Perspective of Lithography for Nanoelectronics   *289*
12.1.1   Traditional "Geometrical Scaling" by Optical Lithography   *289*
12.1.2   From Lithography to Patterning as Driver for Geometrical Scaling   *291*
12.1.3   Layout Optimization for Improved Printability   *292*
12.2   Challenges for Lithography in Future Technology Nodes   *292*
12.2.1   193 nm Immersion Lithography with Multiple Patterning   *292*
12.2.2   Insertion of Extreme UV Lithography   *294*
12.2.2.1   EUVL Progress in Source   *295*
12.2.2.2   EUVL Progress in Masks   *295*
12.2.2.3   EUVL Progress in Resist   *297*
12.2.2.4   EUV Insertion into N7   *298*
12.2.2.5   EUV Lithography Extendibility toward N5 and Beyond   *301*
12.2.3   Directed Self-Assembly (DSA)   *302*
12.2.3.1   DSA Principles and Some DSA Flows   *302*
12.2.3.2   DSA Challenges and Progress   *303*
12.2.3.3   DSA Insertion into N7   *307*

12.2.3.4   DSA Extendibility   *309*
12.2.4   Alternative Lithographies: E-Beam Maskless, Nanoimprint   *309*
12.2.4.1   Parallel E-Beam Direct Write Status and Challenges   *309*
12.2.4.2   Nanoimprint Lithography Status and Challenges   *311*
12.3   Pattern Roughness: The Biggest Challenge for Geometrical
            Scaling   *311*
12.4   Lithography Options in Previous and Future Technology Nodes   *313*
            References   *315*

13   Reliability of Nanoelectronic Devices   *317*
            Anthony S. Oates and K.P. Cheung
13.1   Introduction   *317*
13.2   Interconnect Reliability Issues   *318*
13.2.1   Reliability of Porous Inter-Metal-Level Dielectrics (ILD)   *318*
13.2.2   Reliability of Cu Conductors   *320*
13.3   Transistor Reliability Issues   *322*
13.4   Radiation-Induced Soft Errors in Silicon Circuits   *325*
13.5   Conclusions   *327*
            Acknowledgments   *328*
            References   *328*

Volume 2

Part Five   Circuit Design in Emerging Nanotechnologies   *331*

14   Logic Synthesis of CMOS Circuits and Beyond   *333*
            Enrico Macii, Andreas Calimera, Alberto Macii, and Massimo Poncino
14.1   Context and Motivation   *333*
14.2   The Origin: Area and Delay Optimization   *335*
14.2.1   Two-Level Optimization   *336*
14.2.2   Multilevel Optimization   *337*
14.2.3   Sequential Synthesis   *339*
14.3   The Power Wall   *340*
14.3.1   Dynamic Power   *340*
14.3.2   Leakage Power   *343*
14.4   Synthesis in the Nanometer Era: Variation-Aware   *345*
14.4.1   Logic Synthesis for Manufacturability and PV Compensation   *346*
14.4.2   Thermal-Aware Logic Synthesis   *347*
14.4.3   Aging-Aware Logic Synthesis   *348*
14.5   Emerging Trends in Logic Synthesis and Optimization   *350*
14.5.1   Logic Synthesis for Approximate Computing   *351*
14.5.2   Approximate Logic Synthesis (ALS)   *352*
14.5.3   Design of Approximate IPs   *353*
14.5.4   Post-CMOS and Beyond Silicon   *354*
14.5.4.1   Emerging Devices   *354*

14.5.4.2   New Logic Primitive and Possible Implementation Styles   *355*
14.6   Summary   *358*
References   *358*

15   **System Design in the Cyber-Physical Era**   *363*
Pierluigi Nuzzo and Alberto Sangiovanni-Vincentelli
15.1   From Nanodevices to Cyber-Physical Systems   *363*
15.2   Cyber-Physical System Design Challenges   *365*
15.2.1   Modeling Challenges   *365*
15.2.2   Specification Challenges   *367*
15.2.3   Integration Challenges   *368*
15.3   A Structured Methodology to Address the Design Challenges   *370*
15.3.1   Coping with Complexity in VLSI Systems: Lessons Learned   *370*
15.3.2   Platform-Based Design   *373*
15.3.3   Contracts: An Overview   *375*
15.3.3.1   Assume-Guarantee Contracts   *375*
15.3.3.2   Horizontal and Vertical Contracts   *378*
15.4   Platform-Based Design with Contracts and Related Tools   *380*
15.4.1   Requirement Formalization and Validation   *380*
15.4.2   Platform Component-Library Development   *384*
15.4.3   Mapping Specifications to Implementations   *386*
15.4.3.1   Architecture Design   *387*
15.4.3.2   Control Design   *388*
15.5   Conclusions   *390*
Acknowledgments   *390*
References   *390*

16   **Heterogeneous Systems**   *397*
Daniel Lapadatu
16.1   Introduction   *397*
16.2   Heterogeneous Systems Design   *400*
16.2.1   Design Considerations   *401*
16.2.2   Design Analysis   *402*
16.2.2.1   Mechanical Design   *404*
16.2.2.2   Electrical Design   *405*
16.2.2.3   Thermal Design   *409*
16.2.2.4   Reliability Design   *410*
16.2.3   Assembly and Testing Design   *412*
16.3   Heterogeneous Systems Integration   *414*
16.4   Testing the Performance and Reliability of Heterogeneous
Systems   *418*
16.5   Conclusions   *423*
Acknowledgments   *424*
References   *424*

17 **Nanotechnologies Testing** *427*
Ernesto Sanchez and Matteo Sonza Reorda
17.1 Introduction *427*
17.2 Background *428*
17.3 Current Challenges *433*
17.3.1 SoCs and Embedded Instruments *433*
17.3.2 Process Variations *435*
17.3.3 Combining End-of-Manufacturing and In-Field Test *436*
17.4 Testing Advanced Technologies *437*
17.4.1 Resonant Tunneling Diodes and Quantum-Dot Cellular Automata *438*
17.4.2 Crossbar Array Architectures *441*
17.4.3 Carbon Nanotubes *442*
17.4.4 Silicon Nanowires FETs *443*
17.5 Conclusions *444*
References *444*

Part Six **Nanoelectronics-Enabled Sectors and Societal Challenges** *447*

18 **Industrial Applications** *449*
L. Baldi and M. Van de Voorde
18.1 Introduction *449*
18.2 Health, Demographic Change, and Well-being *450*
18.3 Food Security, Sustainable Agriculture and Forestry, Marine and Maritime and Inland Water Research, and the Bioeconomy *450*
18.4 Secure, Clean, and Efficient Energy *451*
18.5 Smart, Green, and Integrated Transport *451*
18.6 Climate Action, Environment, Resource Efficiency, and Raw Materials *452*
18.7 Europe in a Changing World – Inclusive, Innovative, and Reflective Societies *452*
18.8 Secure Societies – Protecting Freedom and Security of Europe and Its Citizens *452*

19 **Health** *455*
Walter De Raedt and Chris Van Hoof
19.1 Introduction *455*
19.2 The Worldwide Context *455*
19.3 Requirements and Use Cases for Emerging Wearables *459*
19.3.1 Assisted Living *460*
19.3.2 Congestive Heart Failure (CHF) *461*
19.3.3 Cancer and Point of Care *462*
19.3.4 Sleep Monitoring – Sleep Apnea *463*
19.3.5 Presbyopia *464*

19.3.6   Fitness and Stress   *465*
19.3.7   Pregnancy   *466*
19.3.8   Advanced Computing Needs Only Grow   *466*
19.4   Conclusions   *467*
            References   *468*

20   Smart Energy   *471*
            Moritz Loske
20.1   Energy Revolution – Why Energy Does Have to
            Become Smart?   *471*
20.1.1   Smart Energy and Systems   *473*
20.1.2   Smart Energy Effect-Matrix   *474*
20.1.2.1   Smart Generation   *474*
20.1.2.2   Smart Storage   *475*
20.1.2.3   Smart Transmission and Distribution   *475*
20.1.2.4   Smart Consumption   *475*
20.1.2.5   Energy Management   *475*
20.2   Applications of Smart Energy Systems and their Societal
            Challenges   *476*
20.2.1   Multi-energy Smart Grid   *476*
20.2.2   High Voltage Transmission and Distribution Systems   *478*
20.2.3   Microenergy Grid   *480*
20.2.4   Energy Harvesting Systems   *481*
20.2.5   Mobility   *482*
20.3   Nanoelectronics as Key Enabler for Smart Energy
            Systems   *483*
20.3.1   Key Products for Smart Energy systems   *483*
20.3.2   Technological Requirements and Challenges   *484*
20.3.2.1   Requirements of Power-Electronics   *484*
20.3.2.2   Requirements of Micro-/Nanoelectronics   *485*
20.4   Summary and Outlook   *486*
            References   *487*

21   Validation of Highly Automated Safe and Secure Vehicles   *489*
            Michael Paulweber
21.1   Introduction   *489*
21.2   Societal Challenges   *490*
21.3   Automated Vehicles   *491*
21.4   Key Requirements to Automated Driving Systems   *493*
21.5   Validation Challenges   *496*
21.6   Validation Concepts   *497*
21.7   Challenges to Electronics Platform for Automated Driving
            Systems   *498*
21.8   Conclusion   *499*
            References   *499*

22 **Nanotechnology for Consumer Electronics** *501*
Hannah M. Gramling, Michail E. Kiziroglou, and Eric M. Yeatman
22.1 Introduction *501*
22.1.1 2D Materials and Flexible Electronics *502*
22.2 Communications *503*
22.3 Energy Storage *506*
22.4 Sensors *509*
22.4.1 Motion Processing Units *510*
22.4.2 Nanosensors for Biomedical Applications *511*
22.4.3 Optical Sensors *513*
22.5 Internet-of-Things Applications *514*
22.6 Display Technologies *515*
22.6.1 Self-Illuminating Displays *516*
22.6.2 Reflective Displays *517*
22.6.3 Transparent Conductors *518*
22.7 Conclusions *520*
References *520*

Part Seven **From Device to Systems** *527*

23 **Nanoelectronics for Smart Cities** *529*
Joachim Pelka
23.1 Why "Smart Cities"? *529*
23.2 Infrastructure: All You Need Is Information *531*
23.3 Nothing Will Work Without Energy *535*
23.4 Application: What Can Be Done with Information *537*
23.4.1 Smart Buildings *538*
23.4.2 Mobility and Transport *540*
23.4.3 Production and Logistics *543*
23.5 Trusted Hardware: Not Only for Data Security *546*
23.6 Closing Remarks *548*
Acknowledgement *548*

Part Eight **Industrialization: Economics/Markets – Business Values – European Visions – Technology Renewal and Extended Functionality** *551*

24 **Europe Positioning in Nanoelectronics** *553*
Andreas Wild
24.1 What is the "European" Industry *553*
24.2 European Strategic Initiatives *554*
24.2.1 The European Commission *554*
24.2.2 ECSEL Joint Undertaking *554*

24.2.3   Combining Instruments   *555*

24.3   Policy Implementation Instruments   *556*

24.3.1   In The World   *556*

24.3.2   In Europe   *557*

24.4   Europe's Market Position   *558*

24.4.1   European Market Share: Consumption   *559*

24.4.2   European Market Share: Supply   *560*

24.4.3   European Manufacturing Capacities   *563*

24.5   European Perspectives   *564*

25   **Thirty Years of Cooperative Research and Innovation in Europe: The Case for Micro- and Nanoelectronics and Smart Systems Integration**   *567*

Dirk Beernaert and Eric Fribourg-Blanc

25.1   Introduction   *567*

25.1.1   The European R&D Program in the European R&D Landscape   *569*

25.2   Nanoelectronics and Micro-Nanotechnology in the European Research Programs   *570*

25.3   A Bit of History Seen from an ICT: Nanoelectronics Integrated Hardware Perspective   *571*

25.4   ESPRIT I, II, III, and IV   *572*

25.5   The 5th Framework (1998–2002)   *574*

25.6   The 6th Framework (2002–2006)   *575*

25.7   The 7th Framework (2007–2013)   *576*

25.8   H2020 (2014–2020)   *579*

25.9   Some Results of FP7 and H2020   *581*

25.9.1   At Program Level   *581*

25.9.2   The ICT Research in FP7   *582*

25.9.3   Micro/Nanoelectronics and Smart Systems   *582*

25.10   Results of the JTI ENIAC and ARTEMIS   *583*

25.11   An Analysis of Beyond CMOS in FP7 and H2020   *584*

25.12   MEMS, Smart Sensors, and Devices Related to Internet of Things   *586*

25.13   From FP6 to FP7: An integrated approach for micro-nanoelectronics and micro-nanosystems   *587*

25.13.1   Research cooperation between the Framework and Eureka initiatives   *587*

25.14   Enabling the EU 2050+ Future: Superintelligence, Humanity, and the "Singularity"   *589*

25.15   EU 2050±: Driven by a Superintelligence Ambient   *590*

25.16   Conclusion   *592*

26   **The Education Challenge in Nanoelectronics**   *595*

Susanna M. Thon, Sean L. Evans, and Annastasiah Mudiwa Mhaka

26.1   Introduction   *595*

26.2   Traditional Programs in Nanoelectronics Education   *596*
26.2.1   Fields of Study   *596*
26.2.2   Topics of Study   *596*
26.2.3   Example Programs   *598*
26.3   Challenges in Nanoelectronics Education   *600*
26.3.1   Bridging the Disciplines   *600*
26.3.2   Theory versus Practice in Classwork   *601*
26.3.3   Resource Availability   *601*
26.3.4   New Applications   *602*
26.3.5   Industry and Translation   *602*
26.3.6   Degree Levels   *603*
26.3.7   Cultural Challenges   *604*
26.4   New Cross-Discipline Applications   *604*
26.5   Future Education Programs   *605*
26.5.1   Scenario A: Modification of Current University Approach   *608*
26.5.2   Scenario B: Comprehensive Nanoelectronics Education System   *608*
           Acknowledgments   *610*
           References   *610*

27   Conclusions   *613*
           Robert Puers, Livio Baldi, and Marcel Van de Voorde   *613*

           Index   *617*

Foreword

Motto: The future of integrated electronics is the future of electronics itself.

G.E. Moore[1]

1
The Nanoelectronics Industry

The electronic components industry, generically described as "nanoelectronics," is an industry with specificities that set it apart from almost all other industries. Its perimeter is expanding continuously; it started by relying on chemists and physicists handling semiconductor crystals; then added electrical engineers to build circuits and functional blocks; now it also employs considerable numbers of software and system engineers. Its customers achieve increased economic efficiency by allowing functionality to be integrated in components; this way, they allow their vendors to expand their competence and move up the value chain.

The nanoelectronics positioning in the global economy is often depicted as the reversed pyramid shown in Figure 1. At the tip of the pyramid, there is the nanoelectronics industry producing components – popularly known as "computer chips." At the next level, "original equipment manufacturers" (OEMs) use the components to build electronic products with a market value roughly five times higher than that of the components. The electronic equipment industry enables information and communications services with a market value about five times higher than that of the equipment they use. This way, it can be estimated that nanoelectronics enable economic activities with a total value around 25 times higher than its own market value: in 2014, they approached $9000 billions, or 11% of the approximately $80,000 billions gross domestic product of the world. Their weight continues increasing year after year.

The electronic components are used in almost any artifact produced by the industry: they can be found everywhere, from the lock on a hotel door to the space

---

1) G.E. Moore (1965) Cramming more components onto integrated circuits. *Electronics*, **19**, 114; reprinted in *Proceedings of the IEEE*, **86** (1), 82, 1998.

Figure 1 Nanoelectronics enabling products and services.

shuttle. They are manufactured under extreme cleanliness conditions on slices of monocrystalline silicon called "wafers" in dedicated facilities called "wafer fabs." A wafer fab operates highly sophisticated equipment using specialty materials to build hundreds or thousands of structures on each wafer. A structure can contain billions of devices, essentially transistors, but also resistors, capacitors, inductors, and so on; it is so complex that it can only be conceived using "electronic design automation" (EDA) tools, in fact computer programs that assemble predefined functionalities from a library containing blocks capable to perform arithmetic and logic calculations, memory blocks to store software and data, connectivity blocks, and so on. Before delivering them to the users, the structures are diced from the wafer, put in packages foreseen with electrical contacts, tested, and marked; these operations are performed in specialized "assembly lines."

The nanoelectronics industry consists essentially of all the entities that contribute toward delivering electronic components to the OEMs: they are primarily "integrated devices manufacturers" (IDM) and their suppliers, although the IDM denomination is not exactly correct. First, not all component providers build "integrated" devices; in fact, the "discrete" components (such as individual transistors, diodes, etc.) continue being an important part of the total production, with specific components showing significant growth, such as light-emitting diodes (LEDs) used as lamps, power devices, or micro-electromechanical systems (MEMS). Second, not all component providers are also "manufacturers"; an increasing part is represented by an "emerging" value chain consisting of "fabless" companies using contract manufacturing executed by third parties called "foundries." This trend started in 1987 with the establishment of the Taiwan Semiconductor Manufacturing Company (TSMC), the first "pure play" foundry, but became highly significant in the last 5 years since two fabless companies rank among the top 10 sales leaders. Third, a number of specialties (like equipment, materials, design automation or assembly and test) split off from the IDMs forming branches of a dedicated supply chain that must be also given proper consideration. Figure 2 illustrates the segmentation of the industry in different specialties and business models.

This overview of the nanoelectronics industry takes into account all types of discrete and/or integrated electronic components suppliers, together with their dedicated supply chains.

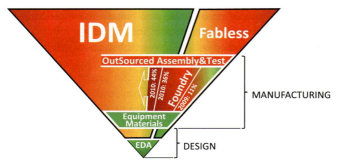

Figure 2 The segmentation of the nanoelectronics industry.

2
The Nanoelectronics Ecosystem

The nanoelectronics industry has one of the highest innovation rates in the economy, often ranking number 1 in terms of R&D expenditures as a percentage of sales. The industry capitalizes upon ingenuity from everywhere in the world, and from any sources, including commercial companies of all sizes, academic and institutional research, and individual investigators. It succeeded sustaining over more than half a century an unparalleled flux of innovation.

The extreme precision and cleanliness necessary to achieve reasonable manufacturing yields at nanometric scale results in unusually high fixed costs of the research and manufacturing infrastructure. It is actually quite impossible to confirm the value of an innovation at low technology readiness levels (TRLs)[2]: positive laboratory results are no more than a hope; successful implementations in realistic environments are no more than a possibility; any novel idea must be taken all the way to an operational environment before concluding on its viability. Since the operational environments are extremely costly, typically in the multibillion dollar range, the industry uses "lab–fabs," that is, facilities used both for research and for manufacturing of commercial products that can absorb the majority of the fixed costs. This approach is practically adopted across the board.

Around each company operating lab–fabs, there is a considerable number of small- and medium-sized companies, of research institutes, and university laboratories collaborating to maintain a technology pipeline filled with new ideas that are continuously scrutinized and moved toward higher TRLs to narrow the selection to the ones that can be included in future recipes. The metaphor of the industry is an ecosystem, relying on the large sequoia trees to withstand fires and tempests in the forest, on medium-sized trees and small bushes to provide a habitat bringing creative ideas to life, and on grass root innovation from university and institutional research to maintain a soil reach in nutrients.

---

2) http://ec.europa.eu/research/participants/data/ref/h2020/wp/2014_2015/annexes/h2020-wp1415-annex-g-trl_en.pdf

The industry makes effective use of project-oriented collaborative research; it is natural to find it well represented in programs carried out by alliances or consortia that naturally cross boundaries between geographic areas and between disciplines.

Also, its systemic and strategic significance attracts the attention of public entities; some of them get involved in setting directions and priorities, some other simply provide financial incentives to facilitate the progress or promote a particular location.

3
Miniaturization

The primary engine of progress in the industry is the "miniaturization." Unparalleled advances in equipment, materials, and manufacturing techniques enable a continuous reduction in size of the elementary function, the transistor. The peculiarity of the semiconductor technology consists in the fact that this improves simultaneously not only all performances parameter but also the unit costs. This trend was recognized already in 1965 (see footnote 1), being known as the "Moore's law"; it initially stated that the number of components per integrated function will double every year. Today, it is usually formulated in terms of the number of components per unit area doubling every (so many) month. In fact, the number of months is of secondary importance as long as this quasi-exponential progression continues, as it did since half a century, in spite of periodical warnings about insurmountable barriers – always overcome by the ingenuity of the researchers in the field. This is described as the "More Moore" progression.

Nanoelectronics follows since 1994 the "International Technology Roadmap for Semiconductors"[3] (ITRS) generated by hundreds of specialists from all around the world. It identifies the challenges to overcome and the timing of the industrial deployment of the successive technology generation called "nodes." Each node is characterized by a "feature size" expressed in nanometers, a rather generic identifier for a whole new set of technology capabilities that obviously depend on many more parameters than just one geometric dimension. Each feature size is smaller by the square root of 2 than the previous one, so that every new node appears to cut in half the silicon real estate needed for a function, in reference to the Moore's law. Companies try to beat the ITRS schedule and be first to market with the next node; in fact, the differences in time are small, and industry moves more or less in lockstep. This quasi-synchronization induced by ITRS guarantees the demand for the equipment and materials suppliers that could therefore invest in R&D at least 5 years before a new node was expected, enabling in due time the subsequent development of new manufacturing processes. Nowadays the industry is considerably widening its markets, serving numerous applications with technology needs that do not always evolve in

3) www.itrs.net/

synchronicity. It becomes increasingly difficult to define a unique, all-encompassing roadmap. ITRS is currently in a restructuring process. It remains to be seen to which extent its success in providing guidance for the industry will continue.

Making the devices smaller require high capital investments in advanced wafer fabs in order to keep the manufacturing yield close to 100%; today, a viable fab costs in excess of $10 billions. Surprisingly, the more expensive the fab, the lower the unit costs of the products it builds, thanks to an overproportional increase in productivity and the beneficial effects of the economy of scale. The decision whether to operate or not own fabs is essential for each company: If the business volume is not commensurate with the capacity of a commercially viable fab, it is preferable to rely on contract manufacturing that can aggregate the demand from several users to reach the needed economy of scale. In this case, the business model may be "fab-lite" when outsourcing most standard but maintaining some proprietary manufacturing generating market differentiation, or entirely "fabless" when relying on system and circuit design to compete. This drives down the number of the companies that participate in the miniaturization race.

As the number of devices per unit area increases, complex functions that were realized before by OEMs can now be integrated on a chip by the components suppliers. Advanced components enable electronic equipment with increased capabilities, better performance, lower power consumption, and smaller form factors. Applications can move from being stationary to becoming mobile, then portable, and eventually even wearable by a person – or go even further enabling autonomous functionality incorporated in communicating objects building the "Internet of Things."

New applications can be addressed at every stage on the road, fueling a continuous increase in demand that is yet far from saturation. Modern applications as high-performance computing, data centers, Internet routers, cloud computing, or big data primarily rely on the newest technology nodes. There is no doubt that nanoelectronics will continue on the miniaturization path that will fuel growth in the foreseeable future.

4
Functional Diversification

Although a new technology node is ready every 2 years or so, each node will be used in manufacturing for 10 or 20 years after introduction. As a technology generation matures, the cost diminishes and it becomes affordable to add new features in the manufacturing recipe to address specific application requirements. They usually include specific device architectures for nonvolatile memories, power, radio frequency, sensing, actuating using either electronic effects or micromachined structures. These enrichments prolong the life expectancy of a technology generation; increase the volume of the commercial production it enables; and improve the overall return on investments. Since they create value

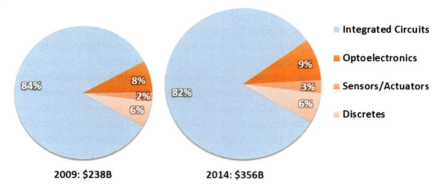

Figure 3 Semiconductor market split; in 2014: integrated circuits 82%; optoelectronics 9%; sensors/actuators 3%; discretes 6%.

through diversification rather than through miniaturization, they are referred to as the "More than Moore" progression.

The "More Moore" and "More than Moore" directions have been for some time depicted as orthogonal. In fact, diversification builds upon processing capabilities introduced in the miniaturization progress.

In market surveys, diversification products are partially reported together with the integrated circuits (ICs) and partially separated under the title opto-electronics – sensors/actuators – discretes (O–S–D). However, the distinction is not always sharp, for example, camera chips are classified together with the LED lamps among the optoelectronics, although they may be closer to the ICs and surely benefit from miniaturization. ITRS 2013 recognizes that there are more innovation streams in the industry, but represents them running on three parallel paths, highlighting the synergy between the "More Moore" mainstream evolution, the "More than Moore" enrichment of existing technologies on one side, and the "Beyond CMOS" exploration of new avenues on the other side.

The diversification has an essential role in enabling nanoelectronics to penetrate additional application areas. Over the last 5 years, the O–S–D products grew only marginally faster that the ICs, benefitting in the first place the progress in optoelectronics, and to some extent in sensors/actuators, while discretes grew as fast as the ICs (Figure 3). Nonetheless, the O–S–D TAM represented a business opportunity of about $65 billions in 2014. This is large enough to entice even companies ranking in the top 25 sales leaders to participate, or even to specialize in this segment.

5
Embedding Software

At the beginning of the digital revolution, hardware and software used to be often interrelated and therefore codeveloped; for example, it was desirable to design computer instructions that could be executed during a single turn of the hard disk. Today, complex computing structures are manufactured as an integrated circuit, and it is mandatory to colocate on the same chip the software defining its functionality and thereby build a system on chip (SoC).

For clarification, not all software encountered in the industry matters here; design software tools, either generated in-house or purchased from outside vendors, software systems for manufacturing control, scheduling, logistics, HR, and so on are not of interest for this overview. Likewise, operating systems, Internet-based businesses, or the plethora of applications ("apps") are usually considered as belonging to a separate industry.

Embedded software is a constitutive element of the products delivered to the customers of the industry and a major contributor to the value created in nanoelectronics. There is a commercial market for "embedded systems," consisting typically of subassemblies of hardware and software providing well-defined functionality that can be assembled by the OEMs in their end products. It is currently estimated at about $150 billions per year, the value being attributed to both hardware (88%) and software (12%). These numbers are quoted here only as an example. In fact, most embedded systems are captive, being generated inside the nanoelectronic companies and/or by their customers. The value of the embedded systems in the captive production surely exceeds by far the commercial market, being estimated in the range of billion dollars per year; the share between hardware and software may differ considerably from the quoted values.

Absent reliable quantitative data, it shall be noted here that software became an essential competence of the nanoelectronics industry, an essential enabler for the usability of the nanoelectronic products, and for sure one of the elements with an increasing significance and weight in the future.

6
Restructuring the Value Chain

The nanoelectronics value chain has continuously evolved since its beginning in 1956 with the Shockley Semiconductor Laboratory (a division of Beckman Instruments, Inc.), quickly followed next year by the split off of Fairchild Semiconductor (as a division of Fairchild Camera and Instrument Corporation), and then by further 65 start-ups launched in the following 20 years. The technology also diffused through numerous licenses, both for captive production and commercial activities.

6.1
Value Chain Fragmentation

The products of the industry evolved from individual diodes and transistors, to integrated circuits, and then to entire systems on a chip or in a package including embedded software. A growing number of disciplines got involved in the process, demanding frequent "make or buy" decisions and creating opportunities for externalization. Long ago, the components manufacturers stopped building equipment for processing, packaging, or testing; it is now a separate branch with yearly sales around $50 billions. The semiconductor materials are another separate branch with yearly sales around $30 billions since the chip makers decided to purchase high-purity fluids, slurries, and further special chemicals from outside suppliers, and stopped pulling silicon monocrystals, purifying, slicing, and polishing them to wafers. Although many IDMs operate own assembly lines, they use outsourced assembly and test (OSAT) for the vast majority of their volume production, another separate branch approaching $30 billions per year. Some of the IDMs still develop in-house specialty design automation tools, but the industry relies by and large on commercially available systems summing up yearly to about $3 billions. Many other activities are subcontracted, like building lithography masks, cleaning wafer fab gear, reclaiming nonyielding wafers or those used in trial runs, and so on.

This fragmentation of the value chain was taking place naturally when a specialist vendor could find numerous potential customers, that is, semiconductor companies with similar needs. This may not be the case in the future. Under the pressure of the economy of scale, the industry evolves toward a smaller number of increasingly larger fabs. This evolution is further accelerated by the foundry model: one company (the foundry) operates fabs, many other use it and go fabless reducing the number of companies running fabs.

Under these circumstances, the trend toward fragmentation may be reversed, at least in some cases. Wafer fabs operators may have to develop special relationships with their suppliers, or even to reintegrate some activities previously outsourced when the shrinking customer base would force some specialized suppliers out of business. In fact, Intel, Samsung, and TSMC coinvested billions of dollars and acquired some ownership in ASML to ensure the progress to the next lithography generations. This trend reversal will surely affect the European equipment and materials suppliers that currently have a higher market share than the European components suppliers. They will have to cope with the challenge posed by a shrinking customer basis.

6.2
Vertical Integration

Long ago, many semiconductor sales leaders used to be a segment of an OEM organization. In the meantime, many vertically integrated companies spun off their component departments, following the general belief that winning in the future economy requires moving up the value chain and closer to the end user, shifting the center of gravity from manufacturing to software to services. In Europe, Philips externalized NXP 9 years ago, Siemens separated Infineon

16 years ago. Thomson contributed its semiconductor department to the creation of STMicroelectronics 29 years ago.

Not all companies followed this path. Even now, some of the top-ranking positions have been taken up by the semiconductor divisions of vertically integrated companies. Even if some of them show profitable growth, they are rather in minority.

Recent evolutions seem to indicate that in some cases there may be a trend opposite to this conventional wisdom. Vertical integration may become on occasion attractive again for the same old reasons: exclusive access to a specific technology (including system on chip architecture) creating a competitive advantage; unrestricted availability of manufacturing capacity; security, better protection against hardware/software hacks by controlling the critical steps in the supply chain. This trend is illustrated by a fabless company like Qualcomm acquiring an IDM like NXP, a software specialist such as Microsoft building smartphones or by a software/equipment specialist such as Apple designing its own components and engaging directly the foundries. Apple already ranks among top 50 semiconductor suppliers, even if its production is captive.

The future evolution of the electronics industry is no more a one-way street. Some companies reconsider vertical integration or other types of privileged relations with their suppliers, similar to some extent to the convergence observed between chip manufacturers and some of their suppliers.

If such trends seem to appear on a global basis, they did not manifest yet in Europe. No European electronic system leader indicated at this time an interest in vertical integration or in a special relationship with its component suppliers beyond the conventional commercial interactions.

6.3
Emerging Value Chain

The strategy to move up the value chain was also embraced by component suppliers, in particular considering the natural evolution of integrated components that kept absorbing competencies previously exercised by their customers. This requires however caution, taking steps only in "win-win" situations to avoid entering into competition with the own customers. In this context, manufacturing was perceived as commodity, low-value, and low-profit – a rather unattractive – business. The fab-lite and even fabless strategies represent a valid approach that has been successfully demonstrated in all regions.

In fact, semiconductor manufacturing turned out to be a very good business when it could fully exploit the economy of scale. Indeed, the leading foundry manufactures chips that generate higher sales numbers at its fabless customers (combined) than those of the largest IDM; it became the pace setter for miniaturization; it operates with healthy profit margins; and the foundry business continues growing faster than the market, as indicated in Figure 2. The share of the contract manufacturing in the digital IC is already dominant, considering that memories are not build in foundries.

7
Opportunities and Perspectives

7.1
Emerging Market Opportunities

Often, the applications that created big surges in demand fueling nanoelectronics growth have been either underestimated or not foreseen at all. The last example is the explosion of smartphones, tablets, and other portable devices that blurred the boundaries between the computing, communication, and consumer market segments. It is therefore risky to state what the next big opportunity will be. Nonetheless, even if the details of the future products are yet to be defined, there are areas in which the growth is likely to accelerate.

A quick overview of the electronic systems market and the component consumption per market segment shown in Figure 4 indicates that in most markets the component penetration is in the range of 25%, except for the segment Industrial/Medical/Other for which it is less than 18% (government applications also show low penetration, but they are a segment too small to matter in this context). The last years have experienced an acceleration of the component consumption in automotive, and this trend is likely to continue under the impact of new technologies enabling various types of electric vehicles, highly automated or even autonomous driving, and on-board infotainment. The "Industrial/Medical/Other" sector however seems to present the biggest opportunity: It can increase its consumption of components by 50% only to be at a par with the other segments. This could well happen within the "Industry 4.0" concept put forward by a European initiative, paralleled by the "Industrial Internet" concept put forward

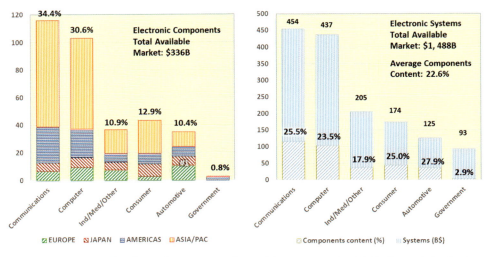

Figure 4 The electronics systems market and the component weight in the market value in different market segments.

in the Unites States of America. It is based on the observation that the industry historically moved from mechanization to electrification and to information technology, and now has reasons to expect that the next significant boost in productivity and capabilities will occur by merging Internet technologies in the industrial processes. There is almost a unanimous expectation that industry will strongly move in this direction, even if particular implementation examples are still in the process of taking shape.

Likewise, computers were initially intended for about 100 governments, then they became business machines addressing about 50 k corporation, and then they eventually became personal and could interest a billion people. The next step in growing consumption is foreseeable: It will consist in embedding computing capabilities in objects. The "Internet of Things" (IoT) will further increase the number of "users" by one or two orders of magnitude, boosting demand. The concrete implementation cases are still in the process of being defined, but there is quasi-unanimity that the IoT will occur, taking the industry to the next level.

Of course, the unforeseen products and services idea should not be forgotten. The nanoelectronic industry creates opportunities for anybody, located anywhere in the world, to change the world with the force of a good idea.

Independent consultant, former executive   *Dr. Andreas Wild*
director of ECSEL and ENIAC Joint Undertakings
Munich, Germany

Nanoelectronics for Digital Agenda

Paul Rübig and Livio Baldi

The digital economy is developing rapidly worldwide. It permeates now countless aspects of the world economy, impacting sectors as varied as banking, retail, energy, transportation, education, publishing, media, and health. Information and communication technologies are transforming the ways social interactions and personal relationships are conducted, with fixed, mobile, and broadcast networks converging, and devices and objects increasingly connected to form the Internet of Things. The volume of data traffic on Internet has grown by a factor of 20 between 2005 and 2013, reaching the astonishing amount of more than 51 exabytes per month in 2013.[1] The digital economy will reach EUR 3.2 trillion in the G-20 economies and already contributes up to 8% of GDP, powering growth and creating jobs. In addition, over 75% of the value-added created by the Internet is in traditional industries, due to higher productivity gains.[2] It is the single most important driver of innovation, competitiveness, and growth, and it holds huge potential for European entrepreneurs and small- and medium-sized enterprises (SMEs).

This evolutionary trend was recognized by Europe already in year 2000, when the Lisbon Agenda set the ambitious target to make the European Union "the most competitive and dynamic knowledge-based economy in the world capable of sustainable economic growth with more and better jobs and greater social cohesion", by 2010.

Unfortunately, the huge potential of the digital economy is still underexploited in Europe, with 41% of enterprises being nondigital, and only 2% taking full advantage of digital opportunities.

However, Europe is well positioned to succeed in the global digital economy, thanks to its world-class research organizations and regional ecosystems. Also, the European industry also has a strong position in several critical sectors, such as embedded digital system, with 30% world market share, and in building complex systems such as cars, trains, and planes. It is estimated that digital technologies "inside the car" determine more than 50% of the key selling features and

---

1) OECD data. 1 exabyte $= 10^{18}$ byte.
2) http://ec.europa.eu/growth/sectors/digital-economy/importance/index_en.htm

represents the key differentiator, and European companies are leaders in the market of automotive electronics. All product, services, and market sectors can profit by the digital revolution. The digitization of manufacturing can transform the entire industry, offering prospects for the relocation of industry in Europe. To capture these advantages, Europe needs both to establish a strong digital sector *and* to facilitate the adoption of digital technologies in all sectors in Europe. It has been estimated that if all EU countries mirrored the performance of the United States or the best-performing EU countries, 400 000 to 1.5 million new jobs could be created in the EU Internet economy.

To this purpose, the Commission has launched in the frame of the Horizon 2020 Programme the Digital Agenda as one of the seven pillars of the Europe 2020 strategy. It aims at improving the environment and infrastructure in Europe for the digital economy, providing the right regulatory and legal frameworks in place, removing national restrictions toward a real single market, building a digital economy, and promoting the e-society.

However, these actions would only make Europe into a more appetizing market for the ICT industry of other regions, if the strength of European industry is not properly reinforced, investing in world-class ICT research and innovation to boost growth and jobs.

In order to define a strategy to help Europe reap the advantages of the Digital Revolution, the Commission supported the formation of an Electronics Leaders Group (ELG) bringing together the leaders of Europe's 10 largest semiconductor and design companies, equipment and materials suppliers, and the three largest research technology organizations, with the task of establishing a strategy to reverse the downward trend of electronic industry in Europe.

The ELG proposed to the Commission to focus efforts in three areas for a stronger ICT industry:

- First, the ELG identified the emerging markets of smart connected objects and the Internet of Things where a leading position and growth can be captured. There is a lot of opportunity if Europe leads on the platforms on which IoT will develop.
- The ELG proposed a second line of action on vertical markets, such as the automotive, energy, and security sectors, where Europe is strong and where disruptions will probably occur much faster than expected due to the increasing importance of electronic content.
- The third area is the changing landscape of mobile convergence. Europe is to gain a leading capability in the future communication networks and devices. 5G offers opportunities in the years to come.

In the new Horizon 2020 Programme, instruments have been introduced to support innovations and prepare European industry to be at the forefront. In this programme, about €12 billion will be invested by the Union between 2014 and 2020 in ICT research and innovation. It is breaking new ground for delivering innovation and will mean that good ideas make the jump from the laboratory to the marketplace.

If ICT will be the engine of the economic growth of Europe, nanoelectronics will have to provide the fuel. Ms. Kroes, the European Commissioner for Digital Agenda in the second Barroso Commission, put forward the challenge "to double the economic value of the semiconductor component production in Europe by 2020–2025" and create an "Airbus of Chips" since the technology development, design, and manufacturing of electronic components and systems is of strategic importance for Europe. In order to ensure that Europe will be a key player in this area in the future, there is a need to put the sector on a steep growth path. This is essential for the electronics industry itself and for the whole of the industrial fabric in Europe.

In this sector, the main initiative has been the establishment of the Joint Undertaking ECSEL in 2014, a unique industry-led public–private partnership for "Electronic Components and Systems for European Leadership" to fund research and innovation actions on Nanoelectronics, Cyber-Physical, and Smart Systems. The Union contributes to it about €1.2 billions, to be matched by contributions from participating Member States and industry, in order to reach a total investment level of some €5 billions by 2020. Building on the successes of its predecessors ENIAC (a public–private partnership focusing on nanoelectronics) and ARTEMIS (a technology platform bringing together key players in the embedded computing arena), it aims at supporting the full industrial development chain, down to large-scale actions to close the gap to the market, including pilot lines preparing for first-time production and further production capacity increase in Europe.

ECSEL is a structuring instrument, aiming at helping the industry to coordinate itself across value chains, integrating the most advanced technologies of components, software, and architectures into highly innovative smart embedded cyber–physical systems. And this is driven to create growth and jobs and to address pressing societal needs for Europe in domains such as transport, energy, or health.

Of course, more basic research will continue to be funded in the regular Horizon 2020 calls of the Leadership in Enabling and Industrial Technologies (LEIT) section, and in the Excellent Science section, under the Future & Emerging Technologies (FET) actions and the continuation of FET Flagships initiative.

Investments will not only be delivered via Horizon 2020. Regions will also be active in mobilizing funds to scale up competence centers and infrastructures and further support industry, under the Smart Specialization Program.

In addition, President Juncker recently announced an investment plan of €315 billions to inject public and private funds into the economy over the next 3 years, which could contribute to cover the €35 billions investment that ELG identified as required in order to double the value of production in Europe.

Several actions are also needed to accelerate the demand and improve the regulatory environment and infrastructure in Europe. To this purpose, the industry is working very hard on an Important Project of Common European Interest

(IPCEI) in the area of electronics, building on the pilot lines sustained by ECSEL. It will bring together competences in Europe and will have a leverage effect on an extensive supply network throughout Europe and the economy. This discussion is taking place at a moment when the business scene in electronics is changing, as we can see from the recent acquisitions of US companies, International Rectifier and FREESCALE, by Infineon and NXP. This is probably not the end – the industrial landscape is expected to continue to change drastically.

Member of the European Parliament   *Dr. Paul Rübig*
Strasbourg/Brussel

Micron Semiconductor, Agrate Brianza, Italy   *Dr. Livio Baldi*

Electronics on the EU's Political Agenda

Carl-Christian Buhr[1]

1
Digital Action in the European Union

The Digital Agenda for Europe,[2] one of the flagship initiatives of the global "Europe 2020" Strategy,[3] was set out, in 2010, with one overarching goal: to bring the benefits of the digital revolution to everybody – and not just to every person but to everybody in their various roles such as student, patient, entrepreneur, researcher, innovator, and so on.

What can the European Union do to have any impact on this? Three things:

1) *Legislation:* for example, to pave the way for new technologies by doing away with outdated legislative constraints, to enable and promote the introduction of new technologies, to make investments in research more likely, and so on.
2) *Funding:* for example, to support research and innovation activities by academia and industry (the latest installment of this kind of funding is the "Horizon 2020"[4] programme, worth about €80 billions until 2020). Or, to support regions in investing in relevant installations and infrastructures, for example, via European structural and investment funds[5] such as the funds for regional and rural development.
3) *Convening discussions:* bringing topics to the European political agenda, inviting Member States to take positions, make proposals, exchange best practices, and so on.

---

1) The author, an advisor to European Commission Vice-President Neelie Kroes 2010–2014, reports in a purely personal capacity on some of the Vice-President's work during that time.
2) http://ec.europa.eu/digital-agenda/digital-agenda-europe
3) http://ec.europa.eu/europe2020/index_en.htm
4) http://ec.europa.eu/programmes/horizon2020/
5) http://ec.europa.eu/contracts_grants/funds_en.htm

2
A Focus on Micro- and Nanoelectronics

So this was the goal and these were the available instruments when European Commission Vice-President Neelie Kroes first looked at the electronics sector and in particular the micro- and nanoelectronics sectors. She made a number of visits to relevant facilities (Infineon and GlobalFoundries in Dresden, Intel in Leixlip near Dublin, STMicroelectronics and LETI in Grenoble, IMEC in Leuven, and ASML in Eindhoven) and met with industry leaders and sector experts.

3
Difficult Times Ahead

The European Union was in danger of losing its ability to master and to deliver the whole electronics value chain from development down to competitive production.

But would that be a problem? After all, many companies in the sector prided themselves on shedding production facilities and ordering their designs to be produced by others, outside of Europe.

And we had seen this pattern before, but with what results? Do we still have significant computer or hifi or TV manufacturers in the European Union? No, what we have are only niche players.

4
Why the Electronics Sector Matters

If electronics went that way, it would be a real problem. For even a largely automated microchip production facility has a much more important function for the overall economy than the few hundred direct jobs it represents at best:

- These are not just any jobs, but well-paid and sought-after ones. University education in the closer or farther vicinity will have to step up its game to supply highly qualified staff.
- Such a factory also depends on long and diverse supply chains, largely made up of SMEs in the same or neighboring fellow EU countries.
- For manufacturing to stay cutting-edge, continuous research and development is essential, with all the investments and spin-offs that it entails.
- Most important, unlike other technologies, digital will be everywhere; microchips and other nanoelectronic components will be ubiquitous.

This is a massive industrial opportunity and an economic bloc such as the European Union, one of the largest in the world, cannot afford not to fight for a leading role in it. Europe produces airplanes and cars that are globally renowned. Yet these machines rapidly morph into winged and wheeled computers. What future could a European car manufacturer really have if more and more of a car's value came from elsewhere?

5
Europe Has a Chance

Europe has well-qualified graduates, very good infrastructure, short distances to customers, a world-class and reliable legal system, a very large common market, and so on. There is no reason why the sector should not be able to thrive here in the same way as it thrives in places such as Taiwan, South Korea, or Israel.

At the same time, Europe still has a very strong presence in some parts of the value chain: for example, for research and development, in the machinery and equipment area, and for specialized production. Europe also aims at "reindustrialization" (defined at bringing the share of industry/manufacturing in the economy back to 20%[6]) and still provides extensive public support for research and innovation, at both national and EU levels.

6
Industrial Strategy for Micro- and Nanoelectronics in Europe

Vice-President Kroes thus started a process of extensive engagement with industrial stakeholders across the European Union at the end of which, in 2013, she convinced her fellow Commissioners to support and agree with her "Electronics Strategy"[7] for Europe. It set a goal of doubling the EU's weight in relevant manufacturing by mobilizing investments of up to €100 billions by 2020. And it set out a way to ensure better targeting and better pooling of available resources across the European Union.

7
Pooling Resources for Research and Development

Neelie Kroes also created ECSEL (Electronic Components and Systems for European Leadership), a European body to bring together the large-scale projects and investments needed to get ahead and reach critical mass in research and development. ECSEL was launched in June 2014[8] and large EU Member States

---

6) See page 23 of the European Commission's Communication "For a European Industrial Renaissance" (http://eur-lex.europa.eu/legal-content/EN/TXT/PDF/?uri=CELEX:52014DC0014&from=EN).
7) http://europa.eu/rapid/press-release_IP-13-455_en.htm
8) http://europa.eu/rapid/press-release_SPEECH-14-540_en.htm

confirmed their decision to contribute large amounts over and above the fore-seen EU funding.

8
Getting Industry to Act

At the same time, stakeholder engagement with the industry continued. An "Electronics Leaders Group"[9] involving CEOs of the large-sector companies was set up to work out an overarching industrial strategy for the sector and its future development, going much beyond research and innovation funding. A strategic roadmap was delivered in February 2014,[10] followed by an Implementation Plan in June 2014.[11]

9
State Aid or No State Aid?

But other fronts were also tackled. The Commission renewed its guidelines for regional state aid,[12] setting out what kind of state aid (i.e., state support) is acceptable. This is no free for all (and no country should give aid where no aid is needed) but it could allow, for example, support to a new chip factory on a green field, or even for a significant extension of existing facilities.

After all, it would not make sense to prevent, say, the Free State of Saxony to support a large electronics investment for fear of the aid preventing the company in question from locating elsewhere in the European Union if the result was that the factory is then built outside Europe. It is simple: in this case, we all lose.

10
The EU Cannot Give Aid But It Can Help

The European Union as such cannot give support to any business endeavor or market investment. It supports research and investment actions that can be relevant for the concerned businesses' decisions (hence, the setup of ECSEL) – but it does so only to the extent of 10% of public support in the European Union, the remainder coming from the Member States.

The European Investment Bank (EIB) Group can support projects, as can other promotional or private banks. But state aid needs to come from

---

9) http://ec.europa.eu/digital-agenda/en/news/european-electronics-companies-set-invest-%E2%82%AC100-billion-create-250000-jobs-and-double-european;   http://europa.eu/rapid/press-release_MEMO-13-903_en.htm
10) http://europa.eu/rapid/press-release_IP-14-148_en.htm
11) http://ec.europa.eu/information_society/newsroom/cf/dae/document.cfm?doc_id=6293
12) http://ec.europa.eu/competition/state_aid/regional_aid/regional_aid.html

governments. And governments normally do not go out and search projects to give money to them. So it is essential that it can work the other way round. It needs to be clear, and get repeated at every opportunity, that Europe is open for business. One idea supported by Vice-President Kroes was that a future EU agency or office could be tasked with ensuring this overseas, for example, in South-East Asia and in the United States. And when ideas and plans come to Europe in local, regional, and national governments, they need to find people with open arms and knowledge and willingness to make projects happen.

11
What Next? The EU Investment Plan

It is interesting to keep this in mind when looking at the new European Commission's Investment Plan.[13] The EIB and other lenders have received a boost of capital in order to attract more funds and invest in projects that benefit goals of societal importance, including topping up funding from other sources. A new European fund for strategic investments (EFSI), an important element of the plan, is up and running.

The electronics industry and the whole sector now has an opportunity to bring this development to the attention of the authorities in the regions and EU Member States where they have invested or want to invest. So that it can be ensured that funds, whether from the EFSI or from elsewhere (e.g., national promotional banks), stand ready to finance what undertaking entrepreneurs dream up for their own benefit – and for the benefit of the European economy and society as a whole.

---

13) http://ec.europa.eu/priorities/jobs-growth-investment/plan/index_en.htm

Preface

What is nanoelectronics? In the mind of most people, nanoelectronics is something highly technical, and matter for scientists, university professors, and highly specialized engineers. If pressed, they could perhaps recognize that, yes, smartphones and the Internet have probably something to do with it. What they probably do not realize is the fact, that nanoelectronics surrounds us, and it is the basis on which our everyday life is built upon. In the last 50 years, microelectronics at first and nanoelectronics afterward have experienced an exponential growth, not only in performance and volume production but also in entering into every single aspect of daily life and even completely influencing it. We do not have to look back 50 but only 10–15 years to realize that things we are now taking for granted are just a very recent addition to our life, and in many cases have completely modified our daily routines and social contacts in this short time span.

Nanoelectronics enables us to keep in touch with each other and to be social in a new sense, giving us instantaneous access to information and entertainment, is reducing energy consumption plus enabling "green" energy, driving our cars, taking care of our health and security, and making easy and improving our efficiency at work. There is hardly any field of technology that is not relying on nanoelectronics deep inside: from medicine to energy, from mobility to security. The Internet and mobile communication with the ever-increasing data speed and data storage (the "Cloud") is unthinkable without nanoelectronics.

Until now, nanoelectronics has been mostly a matter for specialists, each focusing on its own field, and without a global overview of challenges and potential. It is time that industry and society learn more about it, in order to understand its potential and to be prepared for the revolutionary changes it will introduce in economy and society in the near future. Nanoelectronics has already experienced an incredible development in the past years, but it is likely to see an even faster evolution in the future. As mentioned, it already had a large impact on our lives in the last decade, and it will be even more so in 2017.

To this purpose, this book has been composed in a way to be accessible to a large spectrum of educated persons, and is placed between fundamental science and dedicated applications. The contributors of this book are globally located experts from academia, research institutes, and industry. It aims to provide an overview, also to newcomers in the field, of the basics of the technology, the still unexploited growth potential of nanoelectronics, of its technical challenges, and of the ways in which it can and will play a continuing dominating role in industry and society, drastically changing our lives for the better.

Nonspecialists in the field will discover an outline of the main technical issues and challenges of the technology and will obtain an understanding of its trends. Specialists will find useful information in the field of applications of nanoelectronics in industry and society, but also the basis of the technology is covered. The final chapter provides an overview of the main economic factors behind nanoelectronics and of such issues such as European policies and education in this domain.

The book will be of interest to students in electronics, micromechanics, physics, chemistry, medicine, and biosciences. It will serve as background knowledge for those developing software applications. But it will certainly be of value to scientists, teachers, policymakers, and industrialists.

We have to bear in mind that nanoelectronics is already the key enabling factor in our society, almost all products and most services would not be feasible without it, supported by smart software solutions. However, its potential is still largely unexploited.

Nanoelectronics has a global dimension, of particular importance to Europe, the United States, and Asian countries such as Japan, Taiwan, and Korea; for highly populated countries such as China, India, and South America, its impact is quickly rising. "Nanoelectronics" is becoming a dominating force and this book aims at providing clarity and boosting confidence in this fascinating world.

The editors would like to thank everyone without whose help this book would not have become a reality. We express our sincere gratitude, especially to all the eminent authors for their outstanding contributions. Lastly, but not the least, we would like to take this opportunity to express our deepest gratitude and appreciation to all the experts who have painstakingly reviewed the manuscripts on highly specialized topics pertaining to their domain expertise.

Summer 2016

Livio Baldi
Marcel H. Van de Voorde

Part Five
Circuit Design in Emerging Nanotechnologies

14

Logic Synthesis of CMOS Circuits and Beyond

Enrico Macii, Andrea Calimera, Alberto Macii, and Massimo Poncino

Politecnico di Torino, Dip. di Automatica e Informatica, Corso Duca degli Abruzzi 24, 10129 Torino, Italy

14.1
Context and Motivation

In 1965, Gordon Moore was asked to predict what was going to happen in the semiconductor components industry over the next years. The answer was what today we all know as the Moore's Law [1]: *the number of devices per chip roughly doubles every 18 months, resulting in higher operating frequency and lower cost per transistor.*

More similar to an empirical observation and forecast than a rigorous law, there is no intrinsic, or scientific reason behind Moore's law. If so, why the VLSI community invested huge economical and research efforts to sustain this trend? Moore's law can be seen as a social contract between the semiconductor industry and its customers to keep technology growing at an exponential rate. In this sense, whatever the financial cost of keeping it up, the social cost of not doing so would be far greater. Hence, honoring this informal contract was, is, and will be the challenge for the entire VLSI community.

Advancements in the CMOS fabrication process is certainly one of the pillars of Moore's law, but it is not the only one. A less showy, yet equally important actor plays a key role: Design Automation. To manage the ever-increasing complexity of digital circuits with shorter and shorter time-to-market, people sought the support of computer-aided design (CAD) tools. This opened a new research branch in computer science, the electronic design automation (EDA).

To cope with the complexity wall dictated by more complex designs and more sophisticated technologies, EDA engineers faced the problem as typically humans do, namely, by following a divide-and-conquer strategy. Starting from a very abstract description of the system, implementation details are refined step-by-step by solving easier optimization problems. The result is what we find today in commercial integrated circuits (ICs) design frameworks, a vertical iteration of three basic steps, that is, modeling, synthesis, and optimization, repeated in sequence at

Nanoelectronics: Materials, Devices, Applications, First Edition. Edited by Robert Puers, Livio Baldi, Marcel Van de Voorde, and Sebastiaan E. van Nooten.
© 2017 Wiley-VCH Verlag GmbH & Co. KGaA. Published 2017 by Wiley-VCH Verlag GmbH & Co. KGaA.

different levels of abstraction, that is, from system and architectural level down to register transfer level (RTL), gate level, transistor level, and layout level.

Logic synthesis is the process that takes place at the RTL, where a behavioral description of a digital design, that is, the behavioral view, is automatically processed in order to generate a functionally equivalent gate-level description [2], that is, the structural view. The resulting netlist is then optimized in order to meet design constraints with minimum resources.

The evolution of logic synthesis tools has been driven by the scaling of CMOS technology. At the origin, when the semi-custom standard-cell based implementation was recognized as a universal standard, synthesis and optimization strategies differed in terms of (i) rules applied to manipulate the circuit, (ii) the sequence of execution of such rules, and (iii) the min-search heuristic implemented to explore the area-delay design space and find the cheapest implementation.

As soon as the CMOS technology entered the nanometric scale, the main differentiation factor shifted to the cost function rather than the algorithms to minimize it. Indeed, the basic optimization kernel remained unchanged, the same used for area-delay optimization, while new design techniques able to tackle different metrics, that is, dynamic/leakage power and delay uncertainty due to process-temperature-aging variation, started to be stacked one on top of the other. The results of this integration process is what big EDA vendors are marketing: a powerful, push-button multimetric optimization engine that can process multibillion transistor chips in a quantifiable portion of time.

Is that the end of the line? Can we mark logic-synthesis as a solved problem? The answer is no. Indeed, it is again the time to talk about logic synthesis. The reason is still the same: surf the technology wave and break the complexity wall.

Silicon and CMOS style are, and most probably will be for at least the next 10 years, the substrate for the industry of mainstream electronic goods. That said, today's trend is clearly highlighting the need of a radical shift in thinking on digital hardware. The electronics market is calling for new application-centric ICs that can sustain the Internet-of-Things (IoT) revolution. On one hand, data-intensive applications, like, for instance, sensor fusion on wireless nodes, need new design paradigms that guarantee a better energy-performance tradeoff; the most representing one being the recently introduced approximate computing, where circuits barter accuracy for energy. On the other hand, new technologies with enhanced electrical and mechanical properties suitable for wearable devices are pushing to replace standard semiconductors: ambipolar silicon-nanowires, graphene p–n junctions, graphene nanoribbons, magnetic tunnel junctions, and domain-wall nanowiress, are just a few examples. Such technologies will enable new logic primitives with higher expressive power and less power consumption. At this preliminary stage, it is hard to predict which emerging design paradigm or technology will reach first massive production. However, similar to what it was done in the past, logic synthesis will serve as catalyst during the selection process. That is why logic synthesis is, again, a hot research topic.

The aim of this chapter is not that of providing an exhaustive survey of existing logic synthesis algorithms and tools; given the vastness of the topic, that would

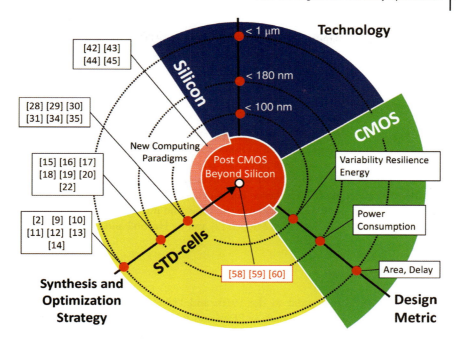

Figure 14.1 Taxonomy of logic synthesis strategies.

require an entire book *per se* and probably it would not add much to readers. Following the chronological evolution of electronics technologies, we introduce and describe the main challenges/issues that CAD researcher involved with logic synthesis had to face. For each of them we briefly describe the problem, we analyze the main knobs that can be used to eliminate and/or mitigate it, and we present a selected pool of solutions; some of them belong to our direct experience in this field. This would serve as the starting point for those who are not familiar with the topic and a reference for experts in the sector. From the dawn of multi-level synthesis we move up to space age strategies for post-CMOS and beyond silicon emerging technologies, passing through state-of-the-art power-, thermal-, and aging-aware optimization. The flow of information contained in the chapter follows the technology-centered taxonomy depicted in Figure 14.1.

14.2
The Origin: Area and Delay Optimization

Historically, circuit area and speed represented the main points of concern in logic synthesis. Designers were faced with the issue of finding the smallest possible implementation fitting a given timing constraint; or, vice-versa, finding the fastest possible design under a given area constraint.

The capability of synthesis tools to explore complex trade-off spaces in short time is what makes automated design preferable to manual design, especially in view of the number of circuit elements featured by modern electronic systems. Clearly, the tools used for optimization and the underlying methods and algorithms strongly depend on the chosen design style, which, in turn, is very much dependent on the technology to be used for the physical implementation of the circuit.

In this section, we consider two main design styles for synthesizing combinational circuits: two-level logic and multilevel logic. We then take a quick look at sequential systems, which represent the large majority of circuits utilized in practice. For the sake of brevity, we will only discuss the basic principles of combinational and sequential synthesis. For a deeper coverage of modeling, design and optimization methods, algorithms and tools, the interested reader may refer to the excellent textbooks by De Micheli [2] and by Hachtel and Somenzi [3].

14.2.1
Two-Level Optimization

Two-level logic was popular in the 1970s and 1980s, thanks to the advent of programmable logic devices, such as PLAs and PALs, which offered flexibility in the implementation of different functions, as well as regularity of the physical design structure. PLAs and PALs have lost interest in subsequent years, due to the appearance of more sophisticated technologies; however, the basic principles of two-level synthesis remain a relevant background for today's digital designers; therefore, we briefly review them here.

We limit our discussion to the case of area versus delay minimization, leaving to other sections in this chapter the treatment of broader kinds of optimizations, such as those targeting power consumption. More specifically, in two-level circuits implemented as PLAs, delay can be considered as constant. Therefore, the only parameter which matters for optimization purposes is area.

The simplest way to specify a logic function is via a truth table, which can be translated into a sum-of-product algebraic representation; the latter is in one-to-one correspondence to a PLA implementation.

Figure 14.2 shows the concept at the basis of two-level logic minimization. In particular, the left-hand side of the figure represents the truth table of a two-output logic function; its corresponding representation as sum-of-minterms and the PLA structure that implements it are shown in the top part of the figure. Finally, the bottom part of the figure presents one possible optimization of the function in the form of sum-of-products and the corresponding PLA structure.

Each product term is implemented by one row in the AND plane of the PLA, while each output of the function is implemented by one column in the OR plane of the PLA. Therefore, a first-order metric for PLA minimization consists of synthesizing logic functions with minimum number of product terms. A further area reduction can be achieved by limiting the number of transistors connected to each row; transistors correspond to literals in the product terms;

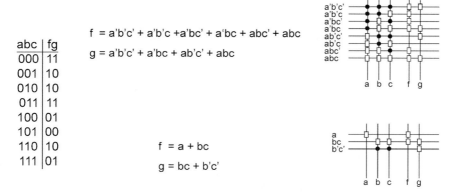

abc	fg
000	11
001	10
010	10
011	11
100	01
101	00
110	10
111	01

f = a'b'c' + a'b'c + a'bc' + a'bc + abc' + abc

g = a'b'c' + a'bc + ab'c' + abc

$f = a + bc$

$g = bc + b'c'$

Figure 14.2 Truth table, algebraic expressions, and PLA implementations of a logic function.

therefore, a second-order optimization metric relates to the total literal count of the function.

In a nutshell, the objective of two-level minimization is that of finding the algebraic expression of the function, which uses the minimum number of product terms covering all the ones in the output columns of the truth table; and, since multiple solutions are possible, to select the list of product terms such that the total number of literals in the algebraic expression (normally called, cover) is minimized.

Determining the best possible cover is computationally difficult; therefore, exact minimization methods are often replaced by heuristic ones, where computation time and computer storage space are traded for approximate, suboptimal implementations.

Heuristic two-level minimization has been the subject of intensive research and investigation. The most successful approach ever is named ESPRESSO [4]. It is based on an algorithm that starts with an initial cover of the function, and that iterates the application of some basic transformations to the product terms (expand, reduce, reshape, irredundant) until a convergence criterion is met.

Common to all such transformations is an efficient manipulation of the logic function, which happens thanks to the exploitation of Boole's Expansion Theorem and recursive function decomposition based on cofactoring.

14.2.2
Multilevel Optimization

The appeal of PLAs and PALs has diminished with technology evolution; therefore, in the vast majority of the cases, two-level logic has been replaced by multilevel logic for combinational circuits of practical interest.

The preferred model for the representation of a multilevel logic network is the so-called Boolean network. This is a directed graph whose nodes consist of simple logic functions, and whose edges represent inputs and outputs of the nodes.

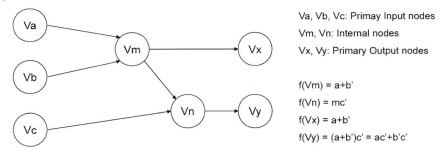

Va, Vb, Vc: Primay Input nodes

Vm, Vn: Internal nodes

Vx, Vy: Primary Output nodes

f(Vm) = a+b'

f(Vn) = mc'

f(Vx) = a+b'

f(Vy) = (a+b')c' = ac'+b'c'

Figure 14.3 Boolean network and algebraic expressions of a logic function.

As a hole, the Boolean network represents a Boolean function for which a multi-level circuit needs to be implemented. Figure 14.3 below shows a simple, three-input, two-output Boolean network; its algebraic expressions are represented in the bottom part of the figure.

In multilevel networks, delay can no longer be considered constant, as in the case of two-level circuits; designers (and design tools) can then explore the area-delay trade-off to determine an implementation of the circuit that satisfies the given constraints.

The size of the space to be investigated to come up with an optimal design makes exact optimization infeasible even for small circuits. Multilevel synthesis then relies on scripts, that is, sets of network transformations that try to move the existing solution toward an optimal point of the design space [5]. There are two classes of transformations: algebraic and Boolean. Algebraic transformations are used to manipulate the algebraic expressions of the functions of the nodes of the Boolean network as they were purely mathematical expressions. Boolean transformations are more powerful than the algebraic ones, as they exploit the Boolean features of the nodes' functions. In particular, Boolean transformations exploit the concept of don't care, that is, conditions for which the values of a Boolean function are irrelevant, thus they can be used for optimization purposes.

Starting with the Boolean network model of a logic function, algebraic and Boolean transformations are applied heuristically in order to obtain a new Boolean network whose topology is optimal with respect to the design constraints (area or delay) imposed as specification. The so-obtained technology-independent model of the circuit (i.e., not yet composed of logic gates from a technology library) is then translated into an interconnection of real gates by a final transformation named technology mapping (or library binding).

A cell library is a set of primitive logic gates to be used for implementing a logic circuit. Each primitive (called cell) is characterized by logic function, area, delay, capacitive load, and so on. The objective of technology mapping is that of selecting a set of cells from the library and to properly interconnect them to obtain a minimum size (or minimum delay) circuit that implements the logic function represented by the Boolean network model. Different modifications can be done by the technology-mapping algorithm to the Boolean network model to determine the solution (i.e., the list of cells and their interconnection topology) that minimizes the cost function of interest. As for the case of Boolean network optimization, also technology mapping is normally done heuristically, as the complexity of the space of the solutions that must be explored makes exact algorithms not applicable for computation time and storage space reasons.

14.2.3
Sequential Synthesis

Most real-life circuits are sequential by nature. In addition, they are synchronous, that is, their operation is regulated by a clock signal, which is ticking at a given frequency. Synthesizing this kind of devices is thus a key need for modern VLSI design. A popular way for modeling a sequential system is through a finite state machine (FSM), that is, a graph representation of the behavior of the circuit, where nodes indicate circuit states and edges indicate transitions between states.

Figure 14.4 shows the state diagram of a FSM and the corresponding circuit implementation.

Given the FSM model, objective of sequential synthesis is that of generating the gate-level netlist of a circuit including combinational gates and sequential registers, properly interconnected [6]. Obviously, this is done by taking into account the area and delay constraints provided at specification time. The first stage of the synthesis flow consists of minimizing the number of states in the FSM graph. In fact, a smaller number of states implies a smaller number of registers in the final circuit implementation. Binary codes are then assigned to the states of the FSM; this operation implies determining the total number of registers to be included in the final design. Also, the combinational logic functions that generate the circuit outputs and next state outputs are computed; being

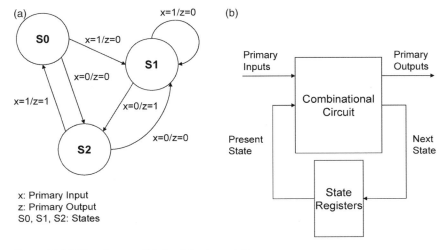

Figure 14.4 FSM and sequential circuit implementation.

them combinational, they can be synthesized using the algorithms for multilevel synthesis discussed earlier in this chapter. A final stage of technology mapping, using a technology library featuring also sequential elements (i.e., flip-flops, latches, registers), completes the synthesis process, thus creating an interconnected netlist of cells that is ready for the back-end part of the design flow (i.e., placement, routing, clock tree synthesis).

14.3
The Power Wall

CMOS technologies have dominated the landscape of digital (and not only) circuits of the last three decades. Power consumption for this kind of circuits has increased as the feature size of the transistors has shrunk.

Since power efficiency is one of the most important figures of merit of modern electronic systems, the ability of designing low power circuits has emerged as a key need. New methods that enable tight power consumption control during design have been the subject of extensive research. In this section, we review the basic contributions in the field of low power logic synthesis.

There are three major sources of power dissipation in a CMOS circuit [7]

$$P = P_{\mathrm{Dyn}} + P_{\mathrm{SC}} + P_{\mathrm{Leak}},$$

where P_{Dyn}, called dynamic or switching power, is due to charging and discharging capacitors driven by the gates in the circuit, P_{SC}, called short-circuit power, is caused by the short circuit currents that arise when pairs of PMOS/NMOS transistors are conducting simultaneously, and P_{Leak}, called leakage or static or stand-by power, originates mainly from subthreshold currents caused by transistors with low threshold voltages and from gate currents caused by reduced thickness of the gate oxide.

For older technologies (e.g., 0.25 μm and above), P_{Dyn} was dominant. For deep-submicron processes, P_{Leak} has become much more important.

In the sequel, we discuss separately solutions for dynamic and leakage power minimization that are applicable at the logic level.

14.3.1
Dynamic Power

Dynamic power for a CMOS gate working in a synchronous environment is modeled as

$$P_{\mathrm{Dyn}} = \frac{1}{2} G_{\mathrm{L}} V_{\mathrm{dd}}^2 f_{\mathrm{Ck}} E_{\mathrm{Sw}}$$

where C_{L} is the output load of the gate, V_{dd} is the supply voltage, f_{Ck} is the clock frequency, and E_{Sw} is the switching activity of the gate, defined as the probability of the gate's output to make a logic transition during one clock cycle.

In general, reductions of P_{Dyn} are achievable by combining the minimization of the four parameters in the formula above. However, at the logic level, since supply voltage, clock frequency and, to some extent, output load are not defined, power optimization translates to the problem of minimizing circuit switching activity. In other words, synthesis for low power adopts the following, simplified dynamic power model, based on the zero-delay assumption:

$$P \propto \sum_g p_g(1 - p_g) = E_{SW}$$

where $p_g = p(g)$ is the signal probability of gate g.

The target of logic synthesis for low power is then to obtain a gate-level netlist for which the outputs of the gates have signal probabilities highly skewed toward 0 or 1, since this condition minimizes E_{SW}.

In the case of two-level logic, power minimization is tackled, simply, by changing the usual cost functions (i.e., total number of product terms and total number of literals in the cover) to account for signal probabilities [8]; however, the algorithms for cover manipulation remain the same as for the case of area minimization.

For multilevel logic, some notion of gate output load can be considered in the power model by taking into account the fan-out of the gates. Then, the objective of the logic synthesis algorithm is that of minimizing the total switched capacitance of the circuit, the latter being defined as the sum over all the gates of the product of switching activity and fan-out. The heuristic transformations used for area and delay optimization can be applied to this purpose, although the cost function to be minimized should include the switched capacitance as the main driver. Algebraic techniques based on extraction [9] and Boolean techniques based on exploitation of do not cares [10] are examples of successful reuse of known concepts to achieve low power multilevel circuits.

Technology mapping can also be made power aware; the multilevel network generated by the synthesis algorithm should be annotated with switching activity information, which can then be exploited by the mapping algorithm to hide high switching nodes inside the gates, as shown in Figure 14.5, taken from Ref. [11].

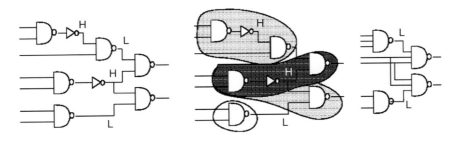

H = high switching activity node
L = low switching activity node

Figure 14.5 Technology mapping for low power.

Since power consumption depends, strongly, on the technology used for physical implementation of the circuit, a more accurate view of the power requirements of a gate-level netlist is available after the mapping step; in fact, the standard cells in the libraries are characterized by parameters such as area, timing, load capacitance, internal power consumption, supply voltage, operating frequency, and so forth. Power reductions can then be achieved by applying to the mapped circuit some further, technology-dependent transformations.

Among others, a very effective one is called gate resizing; it relies on the fact that cell libraries have several instances of cells with different sizes. The idea is to replace some cells of the circuit with others having the same functionality, but smaller area and smaller capacitive load [12]. Gate replacement is a difficult task, because smaller gates are also slower. Then, only gates not belonging to critical paths can be replaced. Obviously, the application of this technique entails the availability of a timing analysis tool for identifying the critical gates in the circuit [13].

Other post-mapping transformations do exist, either functional (e.g., redundancy addition and removal), or topological (e.g., buffer insertion, pin swapping, phase assignment). Although they are normally supported by many commercial synthesis tools, we do not cover them here for the sake of brevity.

As in the case of area and delay minimization, *ad hoc* techniques have been developed to address power optimization in sequential circuits. As an example, we discuss here state assignment. The primary objective is that of reducing the power consumed by the state registers. This can be achieved by selecting the binary codes of the states so that the number of logic transitions occurring at the present-state inputs of the circuit between two consecutive clock cycles is minimized [9]. In other words, the algorithm should assign binary codes to states so as to minimize a power-oriented cost function C, that takes into account the Hamming distance (i.e., the number of bit differences) between the codes given to pairs of states among which a transition can happen. Since not all state transitions are equally probable, cost function C should consider also the probability of the state transitions. In formula, this condition can be expressed as follows:

$$C = \sum_{i,j} W_{i,j} H(S_i, S_j)$$

where $W_{i,j}$ provides the probability that a transition between states S_i and S_j (and vice versa) takes place, and $H(S_i, S_j)$ represents the Hamming distance between the binary codes of states S_i and S_j.

As an example, consider the FSM shown in Figure 14.6. The assignment of the state codes, as shown on the left-hand side of the figure, originates an evaluation of the cost function which is higher than in the case of the state assignment shown on the right-hand side of the figure. Therefore, the latter solution is more power efficient than the former one.

Other techniques are available to further attack the power problem in sequential systems; they are applicable at the FSM level (e.g., FSM decomposition) or at the technology-dependent level (e.g., retiming and guarded evaluation).

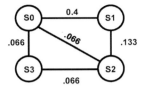

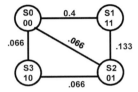

$$C = \sum_{i,j} w_{ij} \cdot H(i,j) = 1.2$$

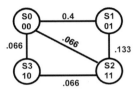

$$C = \sum_{i,j} w_{ij} \cdot H(i,j) = 0.8$$

Figure 14.6 State encoding for low power.

14.3.2
Leakage Power

A simplified model of leakage power in a CMOS gate is the following:

$$P_{Leak} = V_{dd} I_{Leak}$$

where V_{dd} is the supply voltage, and I_{Leak} is the leakage current, which consists of several contributors. The two most important ones are the subthreshold current, which is caused by low threshold voltages, and the gate current, which is caused by reduced thickness of the gate oxide. While the latter has been primarily addressed from the technological point of view, subthreshold current minimization has been faced from the design perspective; therefore, automatic synthesis solutions are available.

The most common one, which is also supported by many commercial CAD tools, exploits the availability, in modern cell libraries, of dual-threshold gates. Low-V_{th} gates can be 20%-2× faster than gates operating at the standard V_{th}. This increase in speed is not free though, and it comes at the price of a much higher leakage of the device – there could be a 10× leakage penalty between the standard and the low-V_{th} devices – and an increase in the cost of the fabrication process.

Synthesis algorithms can optimize leakage by using high threshold voltages while still meeting the timing requirements using low V_{th} devices on the more critical paths. The typical approach followed by such algorithms requires two synthesis steps; in the first, the design is synthesized and mapped onto all low-V_{th} cells, thus obtaining the fastest possible implementation. In the second pass, low-V_{th} cells not on the critical paths are replaced with high-V_{th} cells to reduce leakage without violating the timing constraints [14].

Dual-V_{th} synthesis can be fruitfully combined with gate resizing and, to some extent, dual-V_{dd} assignment, to address more aggressively subthreshold leakage reduction in nanometer-scale designs.

A second, equally popular approach to subthreshold leakage optimization, named power gating, consists of inserting sleep transistors in series to the pull-up and/or pull-down networks of CMOS gates so as to interrupt the path from V_{dd} to GND and then to reduce the sub-threshold leakage current when the circuit is in stand-by mode, while maintaining high-speed operation in active mode. In practice, only one sleep transistor is sufficient for leakage control, and the NMOS insertion scheme is preferable, since the NMOS on-resistance is smaller at the same width; therefore, the NMOS transistor can be sized smaller than the corresponding PMOS.

Normally, sleep transistor insertion is applied to clusters of gates, instead of individual gates [15], as shown in Figure 14.7.

The clustered approach helps in reducing the overhead in area and performance caused by the insertion of the sleep transistors, although it implies addressing a number of additional issues, including the granularity of the insertion (for large CMOS blocks, the size of the sleep transistors and the driving strengths of sleep signals may become large, for small CMOS blocks the number of sleep transistors and the size of the control logic may become large), the design of the sleep transistor cells (which must have different sizes and driving strengths and that must be compliant with the cells in the library), the required area and delay control (implying some constraints on the selection of gates to which sleep transistor insertion should be applied and the need of layout information), the automatic generation of the sleep signals (which implies some area, timing, and power overhead) [16].

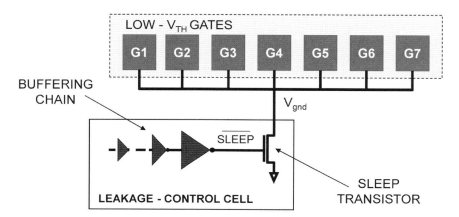

Figure 14.7 Clustered sleep transistor insertion.

14.4
Synthesis in the Nanometer Era: Variation-Aware

Working below the 100 nm mark has the main advantage of increasing the on-chip integration density and of improving speed. This allows the implementation of more functions with a lower cost per transistor. The downside of technology scaling is that the physical effects/defects that were not present, or negligible, in previous technologies now arise at the macroscopic level. This reflects into circuits with a behavior that differs from the nominal one. The worst thing is that such deviation might also change over time. We generally refer to this concept as variability. While a detailed discussion of all the sources of variability (see Figure 14.8 for a complete list) is out of the scope of this chapter, we just focus on those that can be addressed during logic synthesis.

Process variations (PVs) [17]

PVs are static variations that occur when a circuit is manufactured. Mostly due to systematic or nonsystematic impreciseness of the manufacturing process (like lithography, etching, and chemical–mechanical polishing), they induce mismatching of the electrical characteristics of transistors fabricated in the same die. PVs have a significant impact both on circuit power dissipation and performance: 20X variation in leakage power and 1.5X variation in delay between fast and slow dies have been reported in the literature. Increased variability decreases yield, with dramatic cost-per-chip implications.

Temperature fluctuation [18]

Higher transistors and current densities translate into more power consumption and power densities. The power consumed is typically converted into heat, that, if not removed at a rate greater than its rate of generation, may induce severe

		Local	Global
Statistical	Static	Intra-die (WID) random process variations	Inter-die (D2D) random process variations
	Dynamic	-	-
Deterministic	Static	Systematic process variations	Lifetime degradation (NBTI, HCI, TDDB), systematic process variations
	Dynamic	IR drop, clock jitter, coupling noise (capacitive and inductive)	Temperature and Vdd fluctuations

Figure 14.8 Types of variability in CMOS-based circuits and systems.

on-chip temperature increase. Moreover, as a consequence of power-management techniques, unbalanced switching activities across the die induce uneven spatial/temporal distribution of the power densities, which, in turn, result into hotspots and thermal gradients on the same die. High peak temperature exacerbates the problem of leakage currents (which show and exponential relationship with temperature); large temperature fluctuations induce timing mismatch and timing faults.

Aging [19]

Aging variations reflect the fact that the behavior of a circuit, especially in terms of operating frequency, degrades with time. There are two main sources of aging for transistors: bias temperature instability (BTI), and hot carrier interface (HCI). Both these physical/chemical effects result in the generation of interface traps at the silicon/oxide interface, which cause a drift of the threshold voltage over time V_{th}. This may induce delay variations during the circuit lifetime, and, in the worst case, timing faults and system failures.

Not considering variability in the earlier stages of the design flow brings to circuits that must be discarded as not compliant with their specifications because too slow or too power hungry, or, even worse, that fail before their mission time. On the other hand, mitigating variability by means of larger guard-bands might induce unacceptable overheads. Nevertheless, enabling variability-aware logic synthesis is not straightforward as it requires special efforts to push accurate models of physical-level mechanisms at a higher abstraction level.

14.4.1
Logic Synthesis for Manufacturability and PV Compensation

Due to the presence of defects and variation during the fabrication process, the final circuit may show figures that are quite far from those designed with, and predicted by the CAD tools. Compensating all these variations through a worst-case design strategy is no longer a sustainable solution; doing so would nullify the benefits of moving to a scaled node. The CAD community faced this issue by providing a suite of techniques called design for manufacturability (DFM); the majority of them are applied at a post-layout stage.

Inspired by the general rule that a higher level of abstraction implies higher possibilities of improvement of a cost function, people tried to identify viable paths to integrate manufacturability within logic synthesis. However, the work done in this field is very limited. This reflects the complexity of the problem.

Noteworthy, the contribution in [20] proposes a new paradigm that introduces yield to replace area in the cost function. As authors claim, this approach is transparent to designers, who only need to choose whether to minimize area or to minimize manufacturing cost. The basic strategy applies during the technology-mapping stage (just before physical synthesis) and focuses on choosing the best logical mapping in terms of yield; this is implemented by means of a technology-mapping heuristic that exploits libraries containing yield-optimized variants of standard cells. The key limitation is that dedicated libraries optimized for yield are not a common option in modern design kits.

14.4.2
Thermal-Aware Logic Synthesis

Most of the techniques adopted to reduce thermal effects are applicable at a higher level of abstraction, that is, at the RTL, where power-management can alter the switching activity distribution, or at a lower level, that is, floorplanning and placement, during which the power densities are defined. Reducing the peak temperature as a direct metric is not relevant at the logic synthesis stage.

A more critical aspect, instead, is to generate gate-level netlists that are timing-compliant in the full range of operating temperatures, from room temperature (or less) to 125 °C (or more, depending on design specs). Logic synthesis is done at the worst-case, which, ignoring process variation, is defined at low-voltage and high-temperature. This assumption holds when the supply voltage V_{DD} is quite far from transistors' threshold voltage V_{th} ($V_{DD} \gg V_{th}$). However, when lowering the V_{DD} (a common practice in modern ultralow power circuits), this general assumption does not necessarily hold. Delay may decrease as temperature increases [21–23], resulting in low-temperature worst-case. We refer to this effect as the inverted temperature dependence (ITD) effect . If not considered, as done in classical logic synthesis, circuits affected by ITD may fail when warming-up.

The picture gets more complicated when considering dual-V_{th} designs: low-V_{th} cells experience the worst-case delay at high temperature; high-V_{th} cells show worst delay at low temperatures. Figure 14.9 (a) pictorially describes the path distribution after performing a standard dual-V_{th} logic synthesis that considers high temperature T_{max} as the common worst-case corner. The five different groups of lines indicate the length of five paths in the circuit at different temperatures, T_{max} (125 °C), T_{min} (25 °C), T_{mid} (75 °C). Depending on how the low-V_{th} and high-V_{th} cells are distributed, timing paths may fail to match the constraint (lines above D_{nom}) at different temperatures.

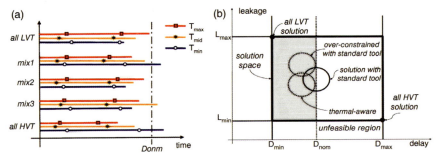

Figure 14.9 Path distribution at different temperatures (a) and solution space of the ITD-aware synthesis tool (b).

A naive solution is to design at the maximum temperature T_{max} and compensate for temperature effects by means of a larger timing constraint. In this case, even if the path delays increase for temperatures other than T_{max}, one can guarantee that the global delay constraint is still met. Obviously, guardbanded designs come at the cost of large power and area overheads.

A more efficient post-synthesis optimization strategy is proposed in Ref. [22]. It takes as a starting point an over-constrained circuit synthesized with a commercial tool and, through a dual-temperature, dual-V_{th} cell replacement optimization, it generates a leakage-optimized netlist that works for the entire temperature range.

This method is conceptually illustrated in Figure 14.9(b), which shows the delay, leakage design space of the thermal-aware dual-V_{th} assignment problem. Each possible threshold voltage configuration can be represented as a (delay, leakage) coordinate point in the design space. The all-LVT point indicates the delay/leakage coordinates when only LVT cells are used in the circuit; this point corresponds to minimum delay at the cost of highest leakage power. The all-HVT minimizes the leakage power at the cost of performance. These two points delimit the feasible solution space. Considering D_{nom} as the delay constraint specified by the designer, a standard design obtained with a classical logic synthesis might violate the timing constraints, while an overconstrained design typically shows larger leakage power.

When considering ITD by means of the proposed double-corner logic-synthesis the resulting circuits meet the delay constraint also achieving better leakage current profiles. This post-synthesis strategy can be seen as a general optimization since different dual-V_{th} policies assignment can be orthogonally applied.

Another method presented in Ref. [21] exploits the conflicting temperature dependence of dual-V_{th} cells in order to make timing paths insensitive to temperature. It reduces the temperature sensitivity of circuits by selecting a proper configuration of V_{th} such that direct- and inverse-temperature dependence compensate each other in the circuit. The obtained circuits show almost zero delay fluctuation over the whole temperature range.

14.4.3
Aging-Aware Logic Synthesis

Another side effect of the high operating temperature is the raise of wearout mechanisms, negative bias temperature instability (NBTI) being the most critical. The key to account for NBTI during logic synthesis is the availability of an abstract model that bridges the gap between the physical parameters that regulate NBTI and the metrics that can be used as knobs at the logic level.

The NBTI effect induces an increase over time of the threshold voltage V_{th} of a pMOS device, that can be approximated as $\Delta V_{th} = K(\beta \cdot t)^{1/4}$, where K is a constant that lumps all the technological parameters (e.g., oxide electric field, thermal voltage, etc.), β is the stress probability, that is, the fraction of time the gate voltage fed to a pMOS transistor is at the logic "0," and t is the lifetime. The term $\beta \cdot t$ can be seen as the stress time.

The delay of a logic gate, using the alpha-power law, is approximately given by $d = \frac{C_L \cdot V_{dd}}{(V_{gs} - V_{th})^\alpha}$ where C_L is the load capacitance, V_{gs} is the gate voltage, V_{th} is the threshold voltage, and α is a technology-related exponents that can be approximated to 1 for sub-90 nm technologies. Accounting for the NBTI-induced V_{th} shift, the new delay $d' > d$ becomes

$$d'(t) = d \cdot \left(1 + \frac{K \cdot (\beta \cdot t)^{1/4}}{V_{GT} - K \cdot (\beta \cdot t)^{1/4}}\right) \tag{14.1}$$

where $V_{th,0}$ is the (nominal) threshold voltage at time 0 and $V_{GT} = V_{gs} - V_{th,0}$ the (nominal) gate overdrive voltage. Figure 14.10 plots the normalized delay over time using Eq. (14.1) as a function of β, assuming a value of $K = 10^{-3}$ (corresponding to a delay increase of about 15% after 3 years), and $V_{GT} = 0.7$ (i.e., $V_{gs} = 1V$ and $V_{th,0} = 0.3V$).

Through the above model, design tools mitigate NBTI effects by taking into account the signal static probability β during optimization [25,26].

Aging-aware logic synthesis methods mainly differ in terms of "when" aging mitigation is applied. Three possible options have been recently explored [27,28]: after technology mapping, during technology mapping, and before technology mapping.

Concerning the first class, that is, after tech-mapping, a mapped netlist is taken as the starting point and then reconstructed by means of (i) topology manipulation and (ii) pin reordering. The former exploits the masking properties of logic gates and, similarly to what is done for low-power mapping, it tries to merge signals with higher "0" probability (the stress condition for NBTI) inside the gates. The latter performs a signal-to-pin reassignment such that the probability to propagate a "0" reduces, still maintaining the same logic behavior. Another option is to apply gate-resizing to decrease NBTI impact and achieve timing closure. Most of these approaches might experience scalability issues because the level of details after mapping is so high that applying complicated multivariable optimization algorithms becomes unpractical. For instance, the

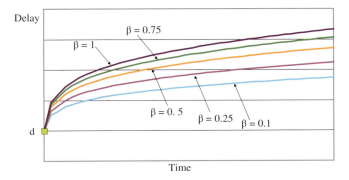

Figure 14.10 Normalized delay over time as a function of stress probability [24].

topological reconstruction problem has a complexity of $O(n^3)$ with n the number of gates. Moreover, the optimizations that are applied are local, namely, they focus on a single cell, or, in the best case, a local portion of the netlist. This prevents global optimization.

Techniques belonging to the second class, that is, during tech-mapping, try to shift the problem at a slightly higher level. The basic idea is to consider the static probability of each internal node as a direct variable of the mapping process. Since the logic cone of each node can be realized in various structurally different topologies, each of them with differently skewed zero static-probabilities, the algorithm performs a best-search mapping such that all the timing constraints are met under NBTI. This matching process is supported by precharacterized timing-libraries containing the exact NBTI-induced delay degradation as a function of the static probabilities. Internal nodes statistics are estimated by means of probabilistic propagation of the primary inputs. Though very effective, the above solution suffers from structural bias, that is, the quality of the results is strongly affected by the Boolean network used as starting point.

Inspired by the latter observation, the third class of solutions, that is, before tech-mapping, moves the problem at a even higher level. The Boolean network used during the technology mapping is preprocessed and reconstructed in order to make it more NBTI-friendly. Then, the NBTI-aware technology mapping process is run for achieving better results. The Boolean network reconstruction procedure consists of three main stages: (i) identify logic cones that are more prone to NBTI effects, (ii) extract those cones, and (iii) add back optimized NBTI cones to the original graph. Cone optimization is done using a guard-banded resynthesis procedure.

14.5
Emerging Trends in Logic Synthesis and Optimization

In order to accommodate new social needs, the electronics market is experiencing a radical shift from common standards. In particular, new requirements, like pervasive connectivity, ubiquitous computing and wearable computers, pose new extreme design challenges that EDA tools should be able to address. The most relevant are: (i) improve energy-performance efficiency of portable devices, (ii) get ready to replace silicon by means of emerging materials.

To accomplish this task, there are two strategies to be followed. First, investigate new computational paradigms that are intrinsically more efficient than conventional ones. Second, but not for importance, exploit the characteristics of new logic primitives made available by the advent of nanotechnologies. The former are exemplified by approximate computing, quantum computing, and neurocomputing, while the latter relate to device models which are different from traditional transistors, that is, ambipolar silicon-nanoWires, graphene p–n junctions, magnetic tunnel junctions. In both cases, new logic abstractions and synthesis techniques are key to achieve efficient designs, large scale integration and massive production.

For what concerns new computational paradigms, as the target of this chapter is on logic synthesis, we bypass considerations related to analog computing and other nondigital computational paradigms, that is, quantum computing and neurocomputing. Regarding post-CMOS and beyond-silicon solutions we concentrate on new digital technologies whose elementary devices have enhanced functionality as compared to standard MOS transistors and CMOS gates. For such devices, alternative implementation styles can better fit the characteristics of nanomaterials, so as to achieve better area, power, energy, and performance.

14.5.1
Logic Synthesis for Approximate Computing

The peculiarities of the "typical" inputs to be supplied to a circuit offer significant opportunities to optimize the figures of merit of the synthesis process. In a traditional synthesis context, an example of this strategy is the use of external do not cares: the knowledge of sets of particular input conditions that cannot occur can be exploited to carry out more aggressive circuit optimizations.

In a more general scenario, however, it might be possible to use some sort of "functional" properties of the circuit to be designed that are related to the application domain in which it will operate. In particular, there are many applications commonly referred to as recognition, mining, and synthesis (RMS), which are characterized by an intrinsic error resilience, that is, they do not require a perfect result and an approximate one is sufficient [29].

Image processing and multimedia are obvious examples of this class of applications, where the limited perception of the user allows tolerating approximations in the computation. However, limited user perception is only one of the possible factors affecting the resiliency of RMS applications.

Another source of tolerance to errors comes from the fact that some systems deal with inputs affected by errors or by some form of environmental noise. In this situation, errors in computation can be tolerated, as long as they are negligible with respect to the imprecision experienced by the inputs. Circuits processing data coming from sensors represent an example of this category.

A third and more subtle source of resilience is the lack, for some applications, of a unique output. This may happen because multiple outcomes are equivalently valuable for the intended purpose. This property is typical of circuits implementing some form of pattern matching; two similar outputs might be considered equally "good" for the user, and there is no explicit concept of exactness.

This general paradigm based on the possibility of implementing "approximate" versions of a specification (called approximate computing) has recently become very popular, in particular, in the area of low-power design. As a matter of fact, power is the most natural metric for which this paradigm can be exploited: approximate implementations of a circuit translate to reduced complexity, reduced number of operations, and thus reduced activity, which directly affects power consumption. However, an approximate design also implies fewer devices (i.e., smaller area) and possibly faster performance, depending on how the approximation is implemented.

Approximate computing can be addressed at any level of abstraction, from circuit to software. For a general overview of the various embodiments of approximate computing, the reader is referred to the various surveys on the topic [29–31]. In this chapter, we will focus on the implications of this paradigm on the synthesis process, and in particular on extensions of existing synthesis tools for its support.

14.5.2
Approximate Logic Synthesis (ALS)

Generally speaking, ALS can be formally expressed as the problem of synthesizing a minimum-cost Boolean network whose behavior deviates within some constraints from a specified exact Boolean function; the deviation being expressed in terms of either the magnitude or the frequency of the error [32].

A few approaches directly tackle the approximation issue in the synthesis step, using *ad hoc* algorithms either to reduce the number of literals or to simplify some nets as if they were redundant, based on a given quality constraint.

The approach of GALS [32] and its extension to multilevel logic [33] follow a traditional synthesis approach based on direct synthesis of a logic expression. The issue of error tolerance is addressed by establishing a formal equivalence between the problem of the approximate synthesis (under error magnitude constraint) and the minimization of Boolean relations (BR). This allows one to use efficient BR solvers to directly synthesize the implementation. In case error frequency is used as a constraint, a different heuristic algorithm is developed.

A different strategy is followed by the tool named SALSA (Systematic methodology for Automatic Logic Synthesis of Approximate circuits) [34]; SALSA decouples synthesis from a specific quality metric, allowing designers to set their preferred metric as input. The basic idea behind SALSA is to instantiate a so-called quality constraint circuit (QCC), as depicted in Figure 14.11.

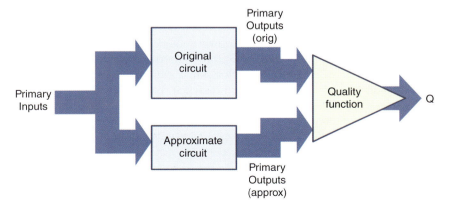

Figure 14.11 ALS template used in SALSA.

The triangular block contains a user-defined implementation of the quality function; its single-bit output Q is asserted when quality constraint is met. SALSA uses the observability do not cares (ODCs) of Q with respect to PO_{approx} to simplify the approximate circuit. The ODCs represent, in fact, the input conditions for which Q is independent of a certain bit in the approximate circuit outputs PO_{approx}, and therefore they will not violate the quality constraint. The process is repeated iteratively on all bits of PO_{approx}, and progressively updating the approximate circuit in the QCC. ODC construction and the corresponding simplifications are carried out with commercial tools for classic logic synthesis.

14.5.3
Design of Approximate IPs

The approximate paradigm can be extended to the design of IP blocks, consistently with state-of-the art synthesis frameworks, which rely on the instantiation of reusable IP blocks that are tightly integrated into synthesis environments, such as the Synopsys DesignWare library. In this case, the link to synthesis is not in the algorithms used for generating the approximate blocks, but rather in their use in a general synthesis flow; for instance, one could add a reconfiguration option for a DesignWare block to synthesize an approximate version of a function.

In this context, approximate design concerns the implementation of datapath elements, and most typically integer adders and multipliers. For these types of operators, the quantification of errors is quite immediate since they represent values in a positional system, and therefore bit positions directly map to the error magnitude.

Approximate adders leverage their modular structure, and follow two main strategies. The first one tries to implement various types of approximate full-adders (FA) and appropriately connects them according to the chosen adder architecture for the LSBs that can be approximated. For example, some FAs in the LSBs can be replaced by a simple OR gate without removing the carry logic. More in general, the truth table of the accurate FA can be modified by removing selected minterms, so as to minimize the approximation error [35].

The second strategy is more sophisticated and aims at breaking the carry propagation chain. Since the latter represents the critical path, breaking it enables delay reduction at the expense of possible errors in the sum output. An example of adder based on this principle can be found in Ref. [36].

Approximate multipliers, conversely, are less studied, in spite of their relative complexity. Some solutions build multipliers based on approximate adders, while others modify existing array multiplier architectures removing or simplifying the FAs involved in MSBs calculations. The approach proposed in Ref. [37] uses a 2×2 multiplier as basic building block, which is approximated by changing a single entry in the truth table (Figure 14.12a), resulting in about 50% reduction in the cost of the implementation by removing the carry output p_3 and using only an OR gate instead of two XORs and two ANDs (Figure 14.12b and c).

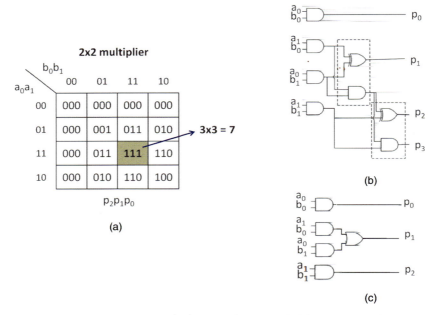

Figure 14.12 Approximate 2×2 multiplication to be used as building block: modified truth table (a), implementation of the exact circuit (b), and of the approximate one (c).

14.5.4
Post-CMOS and Beyond Silicon

In this section we review on two specific emerging devices, ambipolar silicon nanowires (SiNW) [38] and graphene p–n junctions [39]. They actually represent the two corners of a wide range of other possible emerging alternatives recently appeared. On the one side, SiNW are still silicon-based and so, they can be seen as the ultimate link between silicon and beyond-silicon. On the opposite side, graphene p–n junctions represent the cutting line, as they are built on a new 2-D flexible material. Besides their intrinsic difference, both SiNW and graphene p–n junctions share very common properties when abstracted at the logic level, and so do all the other emerging technologies. That is why the following contents are a representative sample of what logic synthesis can do in the nanotechnology era.

14.5.4.1 Emerging Devices
Figure 14.13 depicts the conceptual schemes of vertically stacked double-gate SiNW transistors [38] and graphene p–n junctions [40].

The SiNW is divided into three sections, which are in turn biased by two metal-all-around gate regions. The center gate (CG) works as in a conventional MOSFET switching conduction in the device channel by means of a potential barrier. The side regions are electrostatically controlled by the polarity gate (PG)

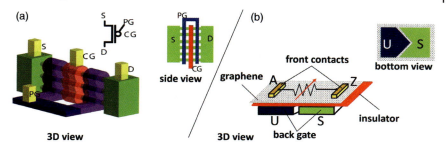

Figure 14.13 Emerging devices: silicon NWFET [38] (a) and graphene p–n junction [40] (right).

that controls Schottky barrier thicknesses at the S/D junctions and selects the majority carrier type, thus "programming" the device to be either n- or p-type. When PG = 0, it behaves as a p-MOS, hence, if CG = 0 (CG = 1) the device is ON (OFF); when PG = 1, it behaves as a n-MOS, hence, if CG = 1 (CG = 0) the device is ON (OFF). The availability of such an online polarity configuration enhances the logic functionality of the device with respect to standard MOSFET.

The graphene p–n junction consists of a graphene sheet on top of which two metal-to-graphene contacts (A and Z) serve as signal input and output, respectively, and a thick layer of oxide that isolates the two back-gates (S and U) from the graphene itself. Exploiting the electrostatic doping, voltage potentials on the back-gates (S and U) work as a control knob to tune the equivalent doping profile of the graphene sheet. A negative voltage induces p-type doping for the graphene region on top of the gate, whereas a positive voltage makes a n-type doping. From an electrical point of view, a p–n junction behaves like a voltage controlled passive resistor [41]. When opposite voltages are concurrently applied on S and U, that is, $\pm V(S) = \mp V$, the device shows a high resistance R_{OFF} between in-to-out front path A–Z (the OFF state); when same voltage is applied on S and U, that is, $V(S) = V(U) = \pm V$, the device shows a low A to Z resistance R_{ON} (the ON state)

14.5.4.2 New Logic Primitive and Possible Implementation Styles

Complementary-Like Static Logic using Standard Multilevel Synthesis
The enhanced functionality of SiNW can be exploited to enable the design of Boolean logic primitives with higher expressive power than CMOS counterparts. Following a CMOS-like implementation strategy, it is therefore possible to implement an ambipolar gate library [42]. The latter, if properly characterized, can be fed to commercial multilevel logic synthesis engines [43] to generate gate-level netlists with reduced area and better performance.

The same integration strategy could be used, at least ideally, for the synthesis of graphene digital circuits. Unfortunately, a more accurate analysis suggests that, differently from SiNW, a graphene p–n junction behaves as a voltage-controlled passive resistor rather than a transistor switch [44]. More specifically, due to the semi-metal nature of graphene, p–n junctions show a R_{ON}/R_{OFF}, which is much lower than standard technologies. Indeed, graphene is a semi-metal and it weakly implements the OFF state. This prevents the implementation of static complementary graphene standard cells since pull-up/down networks made with passive resistors would imply excessive static power consumption [44]. However, integration strategies better than CMOS-like ones are possible that solve the integration problem.

Transmission Gate Logic and One-Pass Synthesis

Reasoning in the binary domain, both SiNWs and graphene p–n junctions naturally implement the same device, namely, a switch controlled by a logic function [45]. When the control signals (PG and PC for the SiNW, S and U for the p–n junction) are fed with the same logic value (either "0" or "1"), the switch is ON (i.e., low resistance between S/D, or A/Z); with opposite logic values the device turns OFF (i.e., ideally high-impedance between S/D, or A/Z). An alternative way to describe the logic behavior of these two devices is to think of them in terms of standard transmission gates with embedded EXNOR functionality, Figure 14.14a.

This observation suggests a natural extension to transmission gate logic circuits where the information is carried out by means of signal propagation through logic paths rather than, as for the CMOS style, charges stored in parasitic capacitances across the logic gates.

Unfortunately, state-of-the-art EDA tools, which have been optimized for semi-custom standard-cell circuits, do not support the synthesis of transmission gate logic. To bridge this gap, recent academic works resort to an old research branch of the logic synthesis, the pass-transistor-logic (PTL) synthesis [46–48]. Also referred to one-pass synthesis (OPS), such a method consists of a synthesis flow where logic optimization and technology mapping are brought, simultaneously, on a common data structure that can efficiently represent both the logic behavior and the physical structure of the final circuit implementation.

Binary decision diagrams [2] (BDDs) have been particularly suited for this purpose. A BDD represents a set of binary-valued decisions, culminating in an overall decision that can be either TRUE or FALSE. A BDD can be represented by a tree where decisions are associated with vertices, and vertices are associated with

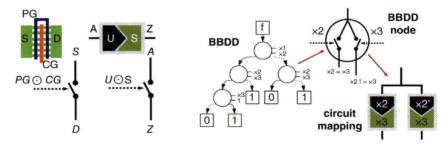

Figure 14.14 Behavioral view of SiNW and graphene p–n junction (left), their abstraction into BBDD (center), and node mapping using graphene p–n junction(right).

decision variables. Nevertheless, BDDs are not particularly efficient when used for the technology mapping of circuits described by EXNOR primitives.

Biconditional binary decision diagrams [45] (BBDDs), Figure 14.14 (center), a variant of BDDs, has been recently proposed to better fit this requirements. A BBDD is a canonical BDD extension in which the single-variable Boole's expansion is replaced with a two-variable biconditional expansion, that is, the EXNOR-based decomposition. This has two major effects. First, since the branching condition at each decision node depends on two variables per time, the logic expressive power of the decision diagram improves; this helps reducing the vertex-set cardinality of the data structure. Second, each node is a natural abstraction of the *EXOR* Boolean comparator; this allows a direct mapping of a BBDD's node into a couple of nanoswitches, either SiNWs or graphene p−n junctions, Figure 14.14 (right). Several reductions rules similar to those developed for BDDs can be seamless extended to BBDDs.

Pass-diagrams [40] (PDs) finally allow better implementation for pass-gate logic. A PD, shown in Figure 14.15, is a polarized directed acyclic graphs where series connections of two-input logic conjunctions (the nodes) form parallel root-to-sink logic branches which are mutually activated when the function is TRUE. Even better than BDDs, PDs are a native logic abstraction not just of the single circuit logic primitives, but also of the physical structure of the resulting nanoswitch network. In fact, internal vertices describe the propagation of signals through the EXNOR primitives rather than a logic decision.

This has two major positive effects on the logic synthesis process. First, more efficient EXNOR decomposition can be implemented, which results into lower vertex-set cardinality. Second, their 1-to-1 physical mapping onto nanoswitch networks enables efficient representation of a more compact logic style implementation well suited for the graphene p−n junctions, the pass-XNOR logic (PXL).

The PXL style can be seen as a middle way between pass-transistor logic and CMOS logic; like PTL, the information is carried out by means of signal propagation through root-to-sink paths, like CMOS, connections of series and parallel branches of logic devices implement AND and OR logic conjunctions. Referring

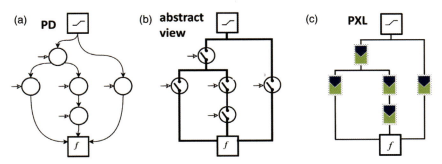

Figure 14.15 Pass-diagram (a), behavioral view (b), final PXL circuit implementation mapped on graphene p−n junctions (c).

to Figure 14.15b, the function is evaluated as TRUE when all the switches of root-to-sink paths are closed.

A PXL circuit implemented with graphene p-n junctions (the same structure can be used for SiNW), Figure 14.15c, is a dynamic logic style. It makes use of a clocked power supply rather than a static one. This well fits the resistive nature of graphene p–n junctions. Its operation encompasses two stages: the configuration phase, where the input logic signals are fed to the p–n junctions which turn ON or OFF accordingly; the evaluation phase where a clocked-power signal is injected into the root and, eventually, reaches the sink (which represents the output). It is worth noticing that, using such a dynamic style, the leakage currents are suppressed and, under reasonable slow pulse signals, the evaluation phase completes at zero-power, namely, adiabatically. These two important considerations highlight the ultralow power nature of pass-gate graphene logic circuits and the efficiency of PDs.

14.6
Summary

Traditional logic synthesis, as described in Section 14.2, represents the seed of a whole discipline, which, over the years, has evolved hands-in-hands with electronic technologies, making the design of complex systems feasible. Although very mature, the combinational and sequential synthesis algorithms for area and delay optimization conceived in the 1980s and 1990s are still widely used in commercial EDA tools; they also represent the basis for more recent design methodologies, whose objectives are much wider in scope and more ambitious, since they target a variety of design parameters ranging from power consumption to variability and temperature as described in Sections 14.3 and 14.4, respectively. The role of logic synthesis as unique tool that supports designers to face complexity issues is not limited to CMOS technologies. Recent trends have shown how standard logic synthesis methodologies and strategies can be easily adapted to support new emerging computing paradigms and beyond-CMOS technologies, as discussed in Section 14.5. The capability of logic synthesis to evolve with the technology and new design needs lies in its capability to abstract and simplify. This is what makes it a flexible and versatile tool that is unlikely to be replaced, an evergreen topic that will attract the research community interest for many years to come.

References

1 Moore, G. (1965) Moore's law. *Electronics Magazine.*
2 De Micheli, G. (1994) *Synthesis and Optimization of Digital Circuits*, McGraw-Hill Higher Education.
3 Hachtel, G.D. and Somezi, F. (1996) *Logic Synthesis and Verification Algorithms*, Kluwer Academic Publishers.
4 Brayton, R.K., Hachtel, G.D., McMullen, C., and Sangiovanni-Vincentelli, A.L.

(1984) *Logic Minimization algorithms for VLSI synthesis*, Kluwer Academic Publishers.

5 Brayton, R.K., Hachtel, G.D., and Sangiovanni-Vincentelli, A.L. (1990) Multilevel logic synthesis. *Proc. IEEE*, **78** (2), 264–300.

6 Ashar, P., Devadas, S., and Newton, A. (1992) *Sequential Logic Synthesis*, Kluwer Academic Publishers.

7 Chandrakasan, A.P., Sheng, S., and Brodersen, R.W. (1992) Low-power CMOS digital design. *IEICE T. Electron.*, **75** (4), 371–382.

8 Vrudhula, S.B. and Xie, H.Y. (1994) Techniques for CMOS power estimation and logic synthesis for low power. IWLSPD'94: International Workshop on Low Power Design, pp. 21–26.

9 Roy, K. and Prasad, S. (1992) SYCLOP: synthesis of CMOS logic for low power applications. ICCD'92: IEEE International Conference on Computer Design: VLSI in Computers and Processors, pp. 464–467.

10 Shen, A., Ghosh, A., Devadas, S., and Keutzer, K. (1992) On average power dissipation and random pattern testability of CMOS combinational logic networks. ICCAD'92: IEEE/ACM International Conference on Computer-Aided Design, pp. 402–407.

11 Tsui, C.Y., Pedram, M., and Despain, A.M. (1993) Technology decomposition and mapping targeting low power dissipation. DAC'93: International Design Automation Conference, pp. 68–73.

12 Bahar, R., Hachtel, G.D., Macii, E., and Somenzi, F. (1994) A symbolic method to reduce power consumption of circuits containing false paths. ICCAD'94: IEEE/ACM International Conference on Computer-Aided Design, pp. 368–371.

13 Bahar, R., Cho, H., Hachtel, G.D., Macii, E., and Somenzi, F. (1994) Timing analysis of combinational circuits using adds. EDAC'94: The European Conference on Design Automation., pp. 625–629.

14 Wei, L., Chen, Z., Roy, K., Johnson, M.C., Ye, Y., and De, V.K. (1999) Design and optimization of dual-threshold circuits for low-voltage low-power applications. *IEEE Trans. VLSI Syst.*, **7** (1), 16–24.

15 Long, C. and He, L. (2004) Distributed sleep transistor network for power reduction. *IEEE Trans. VLSI Syst.*, **12** (9), 937–946.

16 Babighian, P., Benini, L., Macii, A., and Macii, E. (2004) Post-layout leakage power minimization based on distributed sleep transistor insertion. ISLPED'04: IEEE International Symposium on Low Power Electronics and Design, pp. 138–143.

17 Borkar, S. (2005) Designing reliable systems from unreliable components: the challenges of transistor variability and degradation. *IEEE Micro*, **25** (6), 10–16.

18 Skadron, K., Stan, M.R., Huang, W., Velusamy, S., Sankaranarayanan, K., and Tarjan, D. (2003) Temperature-aware computer systems: opportunities and challenges. *IEEE Micro*, **23** (6), 52–61.

19 Alam, M. (2008) Reliability-and process-variation aware design of integrated circuits. *Microelectron. Reliab.*, **48** (8), 1114–1122.

20 Nardi, A. and Sangiovanni-Vincentelli, A.L. (2004) Logic synthesis for manufacturability. *IEEE Des. Test Comput.*, **21** (3), 192–199.

21 Calimera, A., Macii, E., Poncino, M., and Bahar, R. (2008) Temperature-insensitive synthesis using multi-vt libraries. GLSVLSI'08: ACM Great Lakes Symposium on VLSI, pp. 5–10.

22 Calimera, A., Bahar, R., Macii, E., and Poncino, M. (2010) Temperature-insensitive dual-synthesis for nanometer CMOS technologies under inverse temperature dependence. *IEEE Trans. VLSI Syst.*, **18** (11), 1608–1620.

23 Liu, W., Calimera, A., Nannarelli, A., Macii, E., and Poncino, M. (2010) On-chip thermal modeling based on SPICE simulation. PATMOS'10: Power and Timing Modeling, Optimization and Simulation, pp. 66–75.

24 Calimera, A., Macii, E., and Poncino, M. (2010) Nbti-aware clustered power gating. *Trans. Des. Autom. Electron. Syst.*, **16** (1), 3–1.

25 Calimera, A., Macii, E., and Poncino, M. (2012) Design techniques for NBTI-tolerant power-gating architectures. *IEEE Trans. Circuits Syst. II Express Briefs*, **59** (4), 249–253.

26 Ferri, C., Papagiannopoulou, D., Bahar, R.I., and Calimera, A. (2012) NBTI-aware data allocation strategies for scratchpad based embedded systems. *J. Electron. Test.*, **28** (3), 349–363.

27 Lin, C.H., Roy, S., Wang, C.Y., Pan, D.Z., and Chen, D. (2015) Csl: Coordinated and scalable logic synthesis techniques for effective nbti reduction. ICCD'15: IEEE International Conference on Computer Design, pp. 236–243.

28 Ebrahimi, M., Oboril, F., Kiamehr, S., and Tahoori, M.B. (2013) Aging-aware logic synthesis. ICCAD'13: IEEE International Conference on Computer-Aided Design, pp. 61–68.

29 Xu, Q., Mytkowicz, T., and Kim, N.S. (2016) Approximate computing: a survey. *IEEE Des. Test*, 8–22.

30 Han, J. and Orshansky, M. (2013) Approximate computing: an emerging paradigm for energy-efficient design. 18th IEEE European Test Symposium (ETS), pp. 1–6.

31 Jahier Pagliari, D., Macii, E., and Poncino, M. (2016) Energy-efficient digital processing via approximate computing, in *Smart Systems Integration and Simulation*, Springer, pp. 57–92.

32 Miao, J., Gerstlauer, A., and Orshansky, M. (2013) Approximate logic synthesis under general error magnitude and frequency constraints. ICCAD'13: IEEE/ACM International Conference on Computer-Aided Design, pp. 779–786.

33 Miao, J., Gerstlauer, A., and Orshansky, M. (2014) Multi-level approximate logic synthesis under general error constraints. ICCAD'14: IEEE/ACM International Conference on Computer-Aided Design, pp. 504–510.

34 Venkataramani, S., Sabne, A., Kozhikkottu, V., Roy, K., and Raghunathan, A. (2012) Salsa: systematic logic synthesis of approximate circuits. DAC'12: ACM Design Automation Conference, pp. 796–801.

35 Gupta, V., Mohapatra, D., Raghunathan, A., and Roy, K. (2013) Low-power digital signal processing using approximate adders. *IEEE Trans. Comput. Aided Des.*, **32** (1), 124–137.

36 Zhu, N., Goh, W.L., Zhang, W., Yeo, K.S., and Kong, Z.H. (2010) Design of low-power high-speed truncation-error-tolerant adder and its application in digital signal processing. *IEEE Trans. VLSI Syst.*, **18** (8), 1225–1229.

37 Kulkarni, P., Gupta, P., and Ercegovac, M. (2011) Trading accuracy for power with an underdesigned multiplier architecture. VLSI'14: IEEE International Conference on VLSI Design, pp. 346–351.

38 De Marchi, M., Sacchetto, D., Frache, S., Zhang, J., Gaillardon, P., Leblebici, Y., and De Micheli, G. (2012) Polarity control in double-gate, gate-all-around vertically stacked silicon nanowire fets. IEDM'92: IEEE International Electron Devices Meeting, pp. 8.4.1–8.4.4.

39 Chiu, H.Y., Perebeinos, V., Lin, Y.M., and Avouris, P. (2010) Controllable p–n junction formation in monolayer graphene using electrostatic substrate engineering. *Nano Lett.*, **10** (11), 4634–4639.

40 Tenace, V., Calimera, A., Macii, E., and Poncino, M. (2015) One-pass logic synthesis for graphene-based pass-xnor logic circuits. DAC'15: Design Automation Conference, pp. 1–6.

41 Miryala, S., Montazeri, M., Calimera, A., Macii, E., and Poncino, M. (2013) A verilog-a model for reconfigurable logic gates based on graphene pn-junctions. DATE'13: Conference on Design, Automation and Test in Europe, pp. 877–880.

42 Ben-Jamaa, M.H., Mohanram, K., and De Micheli, G. (2011) An efficient gate library for ambipolar cntfet logic. *IEEE Trans. Comput. Aided Des.*, **30** (2), 242–255.

43 Miryala, S., Calimera, A., Poncino, M., and Macii, E. (2013) Exploration of different implementation styles for graphene-based reconfigurable gates. ICICDT'13: IEEE International Conference on IC Design & Technology, pp. 21–24.

44 Tenace, V., Calimera, A., Macii, E., and Poncino, M. (2014) Pass-xnor logic: a new logic style for pn junction based graphene circuits. DATE'14: Conference on Design, Automation and Test in Europe, pp. 1–4.

45 Amarú, L., Gaillardon, P.E., Mitra, S., and De Micheli, G. (2015) New logic synthesis as nanotechnology enabler. *P. IEEE*, **103** (11), 2168–2195.

46 Buch, P., Narayan, A., Newton, A.R., and Sangiovanni-Vincentelli, A. (1997) Logic synthesis for large pass transistor circuits. ICCAD'97: IEEE/ACM International Conference on Computer-Aided Design, pp. 663–670.

47 Ferrandi, F., Macii, A., Macii, E., Poncino, M., Scarsi, R., and Somenzi, F. (1998) Symbolic algorithms for layout-oriented synthesis of pass transistor logic circuits. ICCAD'98: IEEE/ACM International Conference on Computer-aided Design.

48 Macchiarulo, L., Benini, L., and Macii, E. (2001) On-the-fly layout generation for ptl macrocells. DATE'01: Conference on Design, Automation and Test in Europe, DATE '01, pp. 546–551.

15

System Design in the Cyber-Physical Era

Pierluigi Nuzzo[1] and Alberto Sangiovanni-Vincentelli[2]

[1]*University of Southern California, Ming Hsieh Department of Electrical Engineering, 3740 McClintock Ave, Los Angeles, CA 90089, USA*
[2]*University of California at Berkeley, Department of Electrical Engineering and Computer Sciences, 253 Cory Hall MC#1770, Berkeley, CA 94720-1770, USA*

15.1
From Nanodevices to Cyber-Physical Systems

Recent advances in nanotechnologies are enabling a plethora of new applications, virtually in every technology area, from informatics and communications, to medicine, transportation, energy, and security [1]. The discovery of innovative materials and new material properties has made it possible to realize extremely compact and energy-efficient sensor and actuator interfaces, which can be used for a broad variety of purposes, for example, for security, early detection of serious illnesses, in autonomous cars, "smart" industrial processes, and for the detection of nanosubstances in food and in the environment. As an example, micro- and nanoelectromechanical systems (MEMS/NEMS) [2] that can sense motion, such as accelerometers, gyroscopes, and compasses, are increasingly being integrated into small, wearable devices that can provide motion-tracking capabilities at very low power consumptions [3]. Wearable sensors can then stream data to computing units, for example, smartphones, via low-power wireless interfaces, and eventually deliver real-time data for health and fitness purposes.

By relying on higher levels of integration and fully integrated system-on-chip (SoC) solutions, NEM sensors can be equipped with multicore digital processors, embedded controllers, memory, and associated software frameworks, to provide embedded sensor fusion and calibration capabilities for accurate movement identification and processing. The combination of wireless connectivity and powerful processing and storage capabilities, which are already available in a mobile device today, allows envisioning new use cases, such as remote patient monitoring, where data analysis for a patient can be directly sent to the physician for effective health care management. From wearable devices, a new generation

Nanoelectronics: Materials, Devices, Applications, First Edition. Edited by Robert Puers, Livio Baldi,
Sebastiaan E. van Nooten, and Marcel Van de Voorde.
© 2017 Wiley-VCH Verlag GmbH & Co. KGaA. Published 2017 by Wiley-VCH Verlag GmbH & Co. KGaA.

of minimally invasive neuroprosthetic devices is currently object of intense investigation [4], which can make it possible to smoothly assist, repair, or even augment human cognitive or sensory-motor functions via implantable nanodevices.

In addition to health care and fitness, "intelligent" platforms and "rich" user interfaces as the ones above are gradually revolutionizing entertainment, "smart" home control, automotive and aircraft vehicles, eventually paving the way toward complex, multiscale cyber-physical systems (CPSs) [5–7] characterized by the tight integration of computation, communication, and control with mechanical, electrical, and chemical processes. By gathering information from a multitude of sources, and processing and applying it while interacting with other applications, these systems are changing the way entire industries operate, and have the potential to radically influence how we deal with a broad range of crucial problems facing our society today, from national security and safety, to energy management, efficient transportation, and affordable health care.

However, CPS complexity and heterogeneity, originating from combining what in the past have been separate worlds, tend to substantially increase system design and verification challenges. The majority of CPSs consists of distributed sense-and-control systems destined to run on highly heterogeneous platforms, combining large "swarms" of microscopic sensors with high-performance compute clusters (the "cloud") and broad classes of mobile devices [8]. Industry observers predict that the number of smart sensing devices on the planet will reach the order of thousands per person by 2020, embedded in the environment around us and on or in our bodies [9]. A major challenge is then building systems that provide the desired functionality out of such a heterogeneous ensemble of sensing, actuation, connectivity, computation, storage, and energy. While in traditional embedded system design the physical system is regarded as a given, the emphasis of CPS design is, instead, on managing dynamics, time, and concurrency by orchestrating networked, distributed, computational resources together with the physical systems. Only by leveraging the availability and cooperation of all these elements, can we present unifying experiences and fulfill common goals, thus outperforming a system in which such elements are kept separated.

The inability to rigorously model the interactions among heterogeneous components and between the "physical" and the "cyber" sides is a serious obstacle to the efficient realization of CPSs [10]. CPS design entails the convergence of several sub-disciplines, ranging from computer science, which mostly deals with computational aspects and carefully abstracts the physical world, to automatic control, electrical, and mechanical engineering, which directly deal with the physical quantities involved in the design process. Moreover, a severe limitation in common design practice is the lack of formal specifications. Requirements are written in languages that are not suitable for mathematical analysis and verification. Assessing system correctness is then left for simulation and, later in the design process, prototyping. Thus, the traditional heuristic design process based on informal requirement capture and designers' experience can lead to implementations that are inefficient and sometimes do not even satisfy the requirements, yielding long redesign cycles, cost overruns, and unacceptable delays.

The development of CPSs in a variety of industries is indeed becoming increasingly more expensive and time-consuming. The cost of being late to market or of product malfunctioning is staggering as witnessed by the recent recalls and delivery delays that system industries had to bear. Toyota's infamous recall of approximately 9 million vehicles due to the sticky accelerator problem, Boeing's Airbus delay bringing an approximate toll of $6.6 billion are examples of devastating effects that design problems may cause. If this is the present situation, the problem of designing planetary-scale swarm systems appears insurmountable, or doomed to potentially catastrophic outcomes, unless bold steps are taken to advance significantly the science of design.

This chapter offers an overview of system design for CPSs and its possible evolution. By building on our previous work [7,10], we articulate the main design challenges caused by the complexity and heterogeneity of CPSs. We discuss how such challenges can be addressed by employing structured and formal design methodologies that provide the appropriate abstractions to represent the various dimensions of the multiscale design space and can coherently transition between them, as inspired by the success of electronic design automation (EDA) in taming the design complexity of billion-transistor VLSI chips [11]. We then illustrate a design methodology that combines the platform-based design (PBD) [12] paradigm with assume-guarantee contracts [13] to enable concurrent development of system architectures and control algorithms in a modular and scalable way [7,14]. The proposed framework has recently been applied to industrial design examples, facilitating early requirement validation, scalable system-level design exploration, and interoperability of domain-specific tools. We finally follow the structure of this methodology to review the main formalisms, techniques, and tools that can be leveraged to support it. A list of all the acronyms and abbreviations used in the text can be found in Table 15.1.

15.2
Cyber-Physical System Design Challenges

We build on previous elaborations [8,10,15] to discuss the main CPS design challenges that are relevant to system engineers and tool developers. We categorize them in terms of system modeling, specification, and integration.

15.2.1
Modeling Challenges

Model-based design (MBD) approaches [16,17] are today generally accepted as a key enabler for the design and integration of complex systems. However, because CPSs tend to stress all existing modeling languages and frameworks, a set of

Table 15.1 Abbreviations used in the chapter.

Abbreviation	Definition
CPS(s)	Cyber-physical system(s)
VLSI	Very large scale integration
MEMS	Microelectromechanical system(s)
NEMS	Nanoelectromechanical system(s)
NEM	Nanoelectromechanical
SoC(s)	System(s)-on-chip
EDA	Electronic design automation
PBD	Platform-based design
MBD	Model-based design
ISO	International Organization for Standardization
UML	Unified modeling language
HDL(s)	Hardware description language(s)
RTL	Register transfer logic
IP	Intellectual property
CBD	Contract-based design
A/G	Assume-guarantee
SMT	Satisfiability modulo theories
LTL	Linear temporal logic
STL	Signal temporal logic
DE	Discrete event
HRELTL	Hybrid linear temporal logic with regular expressions
CTL	Computation tree logic
ODE(s)	Ordinary differential equation(s)
MILP(s)	Mixed integer linear program(s)
SysML	Systems modeling language
AADL	Architecture analysis design language
HTL	Hierarchical timing language
CIF	Compositional interchange format
FMI	Functional mockup interface
FMU	Functional mockup unit
GR(1)	Generalized reactivity of rank 1

modeling challenges stems by the difficulty in accurately capturing the interactions between them. While in software design logic is emphasized rather than dynamics and processes follow a sequential semantics, physical processes are generally represented using continuous-time dynamical models, expressed as differential equations, which are acausal, concurrent models. Therefore, a first difficulty lies in accurately capturing timing and concurrency; a second difficulty lies in expressing the interactions between higher-level functional models and lower-level implementation models.

Challenge 1: Modeling Timing and Concurrency
A first set of technical challenges in analysis and design of real-time embedded software stems from the need to bridge its inherently sequential semantics with the intrinsically concurrent physical world. All the general-purpose computation and networking abstractions are built on the premise that execution time is just an issue of performance, not correctness. Therefore, timing of programs is not repeatable, except at very coarse granularity, and programmers have hard time to specify timing behaviors within the current programming abstractions. Moreover, concurrency is often poorly modeled. Concurrent software is today dominated by threads, performing sequential computations with shared memory. Incomprehensible interactions between threads can be the sources of many problems, ranging from deadlock and scheduling anomalies, to timing variability, nondeterminism, buffer overruns, and system crashes [18]. Finally, modeling distributed systems adds to the complexity of CPS modeling by introducing issues such as disparities in measurements of time, network delays, imperfect communication, consistency of views of system state, and distributed consensus [15].

Challenge 2: Modeling Interactions of Functionality and Implementation
Functional models are particularly suitable for prototyping control and data processing algorithms, since they are able to abstract unnecessary implementation details, and can be evaluated more efficiently. However, computation and communication do take time and do utilize physical resources. In this respect, pure functional models tend to be inaccurate, in that they implicitly assume that data are computed and transmitted in zero time, so that the dynamics of the software and networks have no effect on the system behaviors. Therefore, to evaluate the overall correctness of a design, it is necessary to provide implementation models capturing the system architecture and the dynamics of software and network components.

While implementation is, in general, orthogonal to functionality and should, therefore, not be an integral part of a functional model, it is essential to provide mechanisms that still capture the interactions between functionality and implementation while preserving their separation. Specifically, it should be possible to conjoin two distinct design representations with each other, namely, a functional model and an implementation model. The latter allows for design space exploration, while the former supports the design of control strategies; the conjoined models enable evaluation of interactions across these domains.

15.2.2
Specification Challenges

Depending on application domains, up to 50% of all errors result from imprecise, incomplete, or inconsistent, and thus unfeasible requirements. The overall system product specification is somewhat of an art today, since to verify its completeness and its correctness there is little that it can be used to compare with.

We categorize the specification challenges with respect to capturing system requirements (Challenge 3) and managing them (Challenge 4).

Challenge 3: Capturing System Requirements

To cope with the inherently unstructured problem of completeness of requirements, industry has set up domain- and application-class specific methodologies, including learning processes, such as the one employed by Airbus to incorporate the knowledge base of external hazards from flight incidents. Use-case analysis methods as advocated for development processes based on the unified modeling language (UML) [19] follow the same objective. A common theme of these approaches is the intent to systematically identify those aspects of the environment of the system under development whose observability is necessary and sufficient to achieve the system requirements. However, the most efficient way of assessing completeness of a set of requirements is by executing it, which is only possible if semi-formal or formal specification languages are used, where the particular shape of such formalizations may be domain dependent.

Challenge 4: Managing Requirements

Design specifications tend to move from one company (or one division) to the next in nonexecutable and often unstable and imprecise forms, thus yielding misinterpretations and consequent design errors. In addition, errors are often caught only at the final integration step as the specifications were incomplete and imprecise; further, nonfunctional specifications (e.g., timing, power consumption, size) are difficult to trace.

It is common practice to structure system level requirements into several "chapters," "aspects," or "viewpoints," quite often developed by different teams using different skills, frameworks, and tools. However, these viewpoints, for example, including function, safety, timing, energy, are not unrelated. Without a clean approach to handle multiple viewpoints, the common practice today is to discard some of the viewpoints in a first stage, for example, by considering only functions and safety. Designs are then developed based on these only viewpoints. Other viewpoints are subsequently taken into account (e.g., timing, energy), which may result in late or costly design modifications, or even redesigns.

Requirement engineering is a discipline that aims at improving the situation described above by paying close attention to the management of the requirement descriptions and their traceability (e.g., using commercial tools such as IBM Doors [20]) and by inserting, whenever possible, precise formulation and analysis methods and tools. However, the support for formal approaches in requirement structuring and analysis is still largely missing.

15.2.3

Integration Challenges

CPSs integrate diverse subsystems by often composing pieces that have been predesigned or designed independently by different groups or companies. While this is done routinely, for example, in the avionics and automotive sectors, system integration largely remains a heuristic and *ad hoc* process. In fact, integrating very complex parts or component models to develop holistic views of the system becomes very challenging, because of the difficulties in capturing the interplay of the different subsystems. On the other hand, it is essential that all the fundamental steps of system design (functional partitioning, allocation on computational resources, integration, and verification) be supported across the entire design development cycle and across different disciplines. We categorize model-based integration challenges in terms of enforcing compatibility of models in compositions (Challenge 5), preserving consistency of models across the design phases (Challenge 6), and improving scalability and accuracy of analysis techniques (Challenge 7).

Challenge 5: Enforcing Compatibility of Interconnected Models

The bigger a model becomes, the harder it is to check for correctness of connections between components. Typically, model components are highly interconnected and the possibility of errors increases. Errors may be due to different units between a transmitting and a receiving port (unit errors), reversed connections among ports (transposition errors), or different interpretation of the exchanged data (semantic errors). Since none of these errors would be detected by a type system [15], specific measures should be enabled to automatically check for them.

Challenge 6: Preserving Consistency of Model Components

Inconsistency may arise when a simpler (more abstract) model evolves into a more complex (refined) one, where a single component in the simple model becomes multiple components in the complex one. Moreover, nonfunctional aspects such as performance, timing, power, or safety analysis are typically addressed in dedicated tools using specific models, which are often evolved independently of the functional ones (capturing the component dynamics), thus also increasing the risk of inconsistency. In a modeling environment, a mechanism for maintaining model consistency is needed to allow components to be copied and reused in various parts of the model while guaranteeing that, if later a change in one instance of the component becomes necessary, the same change is applied to all other instances that were used in the design. Additionally, more sophisticated mechanisms would be needed to maintain consistency between the results of specialized analysis and synthesis tools operating on different representations of the same component.

Challenge 7: Improving Scalability and Accuracy of Analysis Techniques

CPSs may be modeled as hybrid systems integrating solvers that numerically approximate the solutions to differential equations with discrete models, such as state machines, dataflow models, synchronous-reactive models, or discrete event models [21]. While MBD tools go a long way toward facilitating several design tasks, a major set of challenges for CPS integration originates from the scarce scalability and interoperability of traditional analysis techniques. Conventional verification and validation techniques tend not to scale to highly complex or adaptable systems (i.e., those with large or infinite numbers of possible states or configurations). Moreover, simulation techniques may also be affected by modeling artifacts that impair their accuracy, such as solver-dependent, nondeterminate, or Zeno behaviors [15]. Therefore, a major bottleneck in CPS design remains the inability to accurately quantify the impact of design decisions made early in the design process, for example, during the concept design phase, when dealing with trade-offs across the overall system, on the final implementation.

15.3
A Structured Methodology to Address the Design Challenges

Several languages and tools have been proposed over the years to provide support for different design tasks and enable model-based development of CPSs, including checking system level properties or exploring alternative architectural solutions for a set of requirements. Some of these focus on behavioral modeling and simulation, while others are geared toward performance modeling, analysis, and verification. However, the largest benefits in design technologies have usually arrived by addressing the entire system design process, rather than just considering point solutions of tools and models that ease only part of the design. Specifically, to address the issues in Section 15.2, the success of EDA for VLSI systems suggests that CPS design automation efforts should be framed in structured design methodologies and in a formalization of the design process in a hierarchical and compositional way, as further discussed below.

15.3.1
Coping with Complexity in VLSI Systems: Lessons Learned

Over the past decades, a major driver for silicon microelectronics research has been Moore's law, conjecturing the continued shrinkage of critical chip dimensions. Continued scaling of electronic devices to smaller dimensions has been indeed supported by a steady stream of results in material science and material processing, up to the appearance of nanoscale devices. By taking full advantage of the availability of billion-transistor chips, increasingly higher performance SoCs are being fabricated, thus enabling new architectural approaches to information processing and communication. Arguably, handling such a steady increase in complexity over the decades would have not been possible without the continuous increase of productivity due to EDA, which pushes us to investigate how the next generation of design tools could similarly help cross the billion-device frontier of the "Internet of Things" [9].

By reflecting on the history of achievements of EDA, we observe how abstraction and decomposition have been the hallmark of successful design methods, making it possible to manage complexity and reduce the number of items to consider, by either aggregating design objects and eliminating unnecessary details with respect to the goal at hand, or by breaking objects into semi-independent parts. Hierarchy has been instrumental to scalable VLSI design, where boosts in productivity have always been associated with a rise in the level of abstraction of design capture. In 1971, the highest level of abstraction for digital integrated circuits was the schematic of a transistor; 10 years later, it became the digital gate; by 1990, the use of hardware description languages (HDLs) was pervasive, and design capture was done at the register transfer level (RTL). Dealing with blocks of much coarser granularity than in the past became essential in order to cope with the productivity increase the industry was asked to provide. The recent emphasis on SoCs, increasingly bringing system-level concerns into chip design, is a witness to this trend.

On the other hand, large and complex systems are typically assembled from smaller and simpler components, such as predesigned intellectual property (IP) blocks. These components cannot be designed in isolation; their behavior and performance should, instead, be guaranteed once they are integrated in a larger system. In this respect, compositional approaches aiming at offering such guarantees provide a "natural" perspective that should inform the whole design process, starting from its earlier stages [7].

Finally, to maximally benefit from abstractions and decompositions, it has often been essential to "artificially" constrain the design space to regular or modular styles that can ease design verification (e.g., by enforcing regular layout or synchronous design), or even make algorithmic synthesis approaches possible. In this respect, complexity can also be managed by construction, via structured methodologies that combine sets of modeling abstractions and related tools to support analysis and synthesis steps across the whole design flow. As major productivity gains are needed today in the design of complex systems, and better verification and validation is a necessity, as the safety and reliability requirements become more stringent, methodologies and tools start to also be on the critical path to CPS design, from nanoscale SoCs to large-scale, infrastructural systems [8].

Researchers from both academia and industry have chartered the field of design methodologies with increasing clarity over the past decades. However, an all-encompassing framework for CPS design that helps interconnect different tools, possibly operating on different system representations, is very difficult to assemble, and most designers still resort to patched flows [10]. Among the strategies to cope with the exponential growth in system complexity, starting from the iterative and incremental development approach proposed several decades ago [22], we recall: model-based design, the V-model, layered design, component-based design, virtual integration, and platform-based design. We refer the reader, for example, to Ref. [13] for a survey of all these methods.

The V-model was proposed several years ago by the German defense companies [23] and is often cited today, together with its variations, as the reference process in some industrial domains, such as automotive and aerospace [8]. In its basic structure, as sketched in Figure 15.1a, this methodology entails a top-down

(a)

(b)

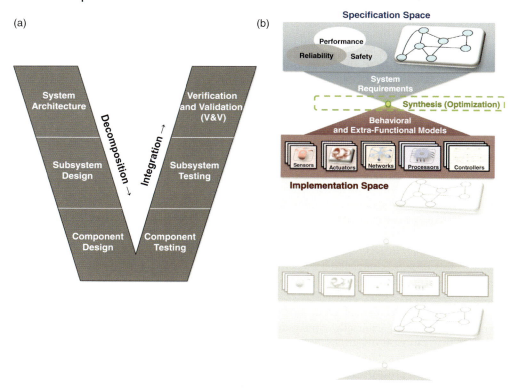

Figure 15.1 Pictorial representations for the V-Model (a) and the platform-based design methodology (b).

design phase that ends with system implementation (the left arm of the V) followed by an integration and verification phase that ends with the verification of the entire system (the right arm of the V). This "waterfall" method has produced good results when the complexity of the designs was relatively small. On the other hand, when complexity scales up, it is no longer possible to initiate the verification phase after the design is completed. Rather, early verification and continuous monitoring of the design should be favored while the refinement steps are taken. In addition, "formality" should be favored in all aspects of the design flow to allow analysis and even synthesis with guaranteed properties of the final outcome of the process.

In this respect, layered design, model-based design, and virtual integration have typically focused on the formalization of the vertical process of the design flow, that is, on abstraction and refinement between different layers of abstraction. Component-based design has, instead, principally dealt with the formalization of the horizontal process, that is, composition and decomposition at the same level of abstraction. Both the horizontal and vertical steps are considered within the PBD paradigm, which aims at combining both the aspects of the design process in a unified framework for system design [12]. More recently, a novel, cross-cutting paradigm has been advocated, contract-based design (CBD) [13], which can be applied to all methodologies proposed thus far to provide a rigorous scaffolding for verification, analysis, and abstraction/refinement.

In the following, we discuss a formal framework capable of combining PBD with contracts, as introduced in our previous work [7,14], to encompass the whole design flow, from top-level requirement formalization to a lower-level representation of the design, including synthesis and optimization-based design methods. In the remainder of this section, as well as in Section 15.4, we show how this integrated approach, which has been successfully applied to industrial design examples, can be used as a unifying framework to address the design challenges in Section 15.2, reason about the core design steps of any compositional and hierarchical methodology, and categorize the supporting formalisms, languages, and tools. To do so, we start by summarizing the basic tenets of PBD and CBD below.

15.3.2
Platform-Based Design

PBD was introduced in the late 1980s as a rigorous framework to reason about design that could be shared across industrial domain boundaries [12] and support the supply chain as well as multilayer optimization, by encompassing both horizontal and vertical decompositions, and multiple viewpoints. Indeed, PBD concepts have been applied to a variety of very different domains: from automotive [24] to system-on-chip [25], from analog circuit design [26] to building automation [27], to synthetic biology [28].

In PBD, at each step, top-down refinements of high-level specifications are mapped into bottom-up abstractions and characterizations of potential implementations. Each abstraction layer is defined by a design platform, which is the set of all architectures that can be built out of a library (collection) of components according to composition rules. A pictorial representation of a design step in PBD is shown in Figure 15.1b and can be seen as a meet-in-the-middle process consisting of two phases.

The bottom-up phase consists in building and modeling the component library, which, in a CPS, includes both the system plant and the controller. In the top-down phase, the high-level system requirements are formalized and an optimization (refinement) phase called mapping is performed, where the requirements are mapped into the available implementation library components and their composition. Mapping is cast as an optimization problem where a set of performance metrics and quality factors are optimized over a space constrained by both system requirements and component feasibility constraints. Mapping is the mechanism that allows moving from a level of abstraction to a lower one using the available components within the library. The different viewpoints of the system and of the components can be considered in the mapping phase. When some constraint cannot be satisfied using the available library components or the mapping result is not satisfactory for the designer, additional elements can be designed and inserted into the library. For example, when implementing an algorithm with code running on a processor, the functionality of the algorithm is assigned to a processor and the code is the result of mapping the "equations" describing the algorithm into the instruction set of the processor. If the processor is too slow, then real-time constraints may be violated. In this case, a new processor has to be found or designed that executes the code fast enough to satisfy the real-time constraint. After each mapping step, the current representation of the design platform serves as a specification for the next mapping step, until the physical implementation is reached. This recursive process is pictorially represented in Figure 15.1b.

In such a layered approach, it is essential to provide formal guarantees about the correctness of each refinement step from the most abstract representation of the platform (the top-level requirements) to its most concrete representation (the physical implementation). Specifically, as illustrated in Figure 15.2, mechanisms must be provided to prove or enforce that: (i) a set (or conjunction) of requirements is consistent, that is, there exists an implementation satisfying all of them; (ii) an aggregation (or composition) of components is compatible, that is, there exists a context (environment) in which they can correctly operate; (iii) an aggregation of components refines a specification, that is, it implements the

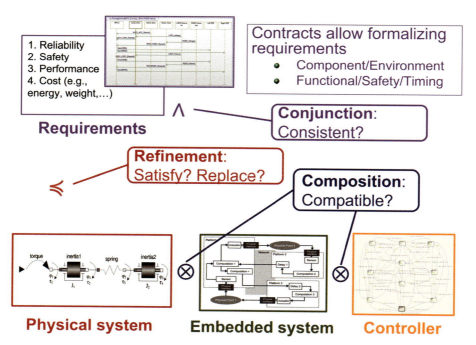

Figure 15.2 The role of contracts and their relations and operations in hierarchical and compositional methodologies.

specification and is able to operate in any environment admitted by it. Moreover, whenever possible, the properties above should be proved (or enforced) efficiently and automatically. In what follows, we show how a contract-based approach can be adopted to provide formal foundations for these and other design and verification tasks.

15.3.3
Contracts: An Overview

The notion of contracts originates in the context of compositional assume-guarantee reasoning [29], which has been known for a long time, mostly as a verification technique for the design of software and hardware. Informally, a contract can be seen as a pair $C = (A, G)$ of properties, assumptions, and guarantees, respectively representing the assumptions on the external environment of a system, and the promises of the system itself under these assumptions. In a contract framework, design and verification complexity is reduced by decomposing system-level tasks into more manageable subproblems at the component level, under a set of assumptions. System properties can then be inferred or proved based on component properties.

Contract frameworks were widely developed in the context of software engineering and object oriented programming [30,31], and further extended in the context of model-based design [32]. However, their adoption for reactive systems, that is, systems that maintain an ongoing interaction with their environment, considered in CPS design, as opposed to transformational system, considered in object oriented programming, has been advocated only recently [8,13]. Contracts (and interfaces) for reactive systems need to encompass a richer set of behaviors and models of computation. Different contract theories have been developed over the years for such systems, broadly grouped into two major families: assume-guarantee contracts [33] and interface theories [34–36]. The investigation of the relations between these families as well as their applications to industrial-strength designs are active research fields [7,37].

To be concrete, in this chapter, we follow the assume-guarantee (A/G) contract framework [13,33] to illustrate the benefits of a contract-based approach for the specification, design, and verification of complex systems. Since A/G contracts are centered around behaviors, they are expressive enough to encompass all kinds of models encountered in system design, from hardware and software models to representations of physical phenomena. Over the last few years, A/G contracts have been demonstrated in different domains, such as defense [38] and automotive [13,39,40] systems, analog integrated systems [41], and synthesis and optimization of system architectures and control algorithms in CPSs [14,42–44].

15.3.3.1 Assume-Guarantee Contracts
We start our analysis of assume-guarantee contracts with a simple, generic representation of a component and we associate to it a set of properties that the component satisfies expressed with contracts. The contracts will be used to

verify the correctness of the composition and of the refinements. A component can be seen as an abstraction representing an element of a design, characterized by a set of (input or output) variables, a set of (input or output) ports, and a set of behaviors over its variables and ports. Components can be connected together by sharing certain ports under constraints on the values of certain variables. Behaviors are generic and could be continuous functions that result from solving differential equations, or sequences of values or events recognized by an automaton. In what follows, to simplify, we use the same term variables to denote both component variables and ports. We also use $[\![M]\!]$ to denote the set of behaviors of component M. A system can then be assembled by parallel composition and interconnection of components. We denote the composition of two components M_1 and M_2, when it is defined, as $M_1 \times M_2$. Then, the behaviors of the composition can be described, in general, as the intersection of the behaviors of its components, that is, $[\![M_1 \times M_2]\!] = [\![M_1]\!] \cap [\![M_2]\!]$.

A contract C for a component M is a triple (V, A, G), where V is the set of component variables, and A and G are sets of behaviors over V [33]. A represents the assumptions that M makes on its environment, and G represents the guarantees provided by M under the environment assumptions. A component M satisfies a contract C whenever M and C are defined over the same set of variables, and all the behaviors of M are contained in the guarantees of C once they are composed (i.e., intersected) with the assumptions, that is, when $[\![M]\!] \cap A \subseteq G$. We denote this satisfaction relation by writing $M \vDash C$, and we say that M is an implementation of C. However, a component E can also be associated to a contract C as an environment. We say that E is a legal environment of C, and write $E \vDash_E C$, whenever E and C have the same variables and $[\![E]\!] \subseteq A$.

A contract $C = (V, A, G)$ is in canonical form if the union of its guarantees G and the complement of its assumptions A is coincident with G, that is, $G = G \cup \overline{A}$, where $\overline{A}$ is the complement of A. Any contract C can be turned into a contract in canonical form C' by taking $A' = A$ and $G' = G \cup \overline{A}$. We observe that C and C' have identical variables and identical assumptions. Moreover, while their guarantees are generally different, C and C' possess identical sets of environments and implementations. In fact, for any set of behaviors B we obtain: $B \cap A \subseteq G$ if and only if $B \subseteq G \cup \overline{A}$. Such two contracts C and C' are then equivalent. In what follows, we assume that all contracts are in canonical form, since every contract C can be turned into an equivalent contract C' in canonical form, without changing its semantics.

A contract is consistent when the set of implementations satisfying it is not empty, that is, it is feasible to develop implementations for it. For contracts in canonical form, this amounts to verifying that $G \neq \varnothing$, where $\varnothing$ denotes the empty set. Let M be any implementation, that is, $M \vDash C$, then C is compatible if there exists a legal environment E for M, that is, if and only if $A \neq \varnothing$. The intent is that a component satisfying contract C can only be used in the context of a compatible environment.

Contracts associated with different components can be combined according to different rules. Similar to parallel composition of components, parallel composition ($\otimes$) of contracts can be used to construct composite contracts out of simpler ones. Let M_1 and M_2 be any pair of components that are composable and satisfy, respectively, contracts C_1 and C_2. Let E be any legal environment of $C_1 \otimes C_2$, that is, $E \vDash_E C_1 \otimes C_2$. Then, the contract composition $C_1 \otimes C_2$ is defined such that

- the composition of M_1 and M_2 satisfies $C_1 \otimes C_2$, that is, $M_1 \times M_2 \vDash C_1 \otimes C_2$,
- E provides a legal environment for both C_2, when composed with M_1, and C_1, when composed with M_2, that is, $E \times M_1 \vDash_E C_2$ and $E \times M_2 \vDash_E C_1$.

Let $C_1 = (V, A_1, G_1)$ and $C_2 = (V, A_2, G_2)$ be contracts over the same set of variables V. It is possible to show [7,33] that the composite contract $C_1 \otimes C_2$ can be computed as the triple (V, A, G) where

$$A = (A_1 \cap A_2) \cup \overline{(G_1 \cap G_2)}, \tag{15.1}$$

$$G = G_1 \cap G_2. \tag{15.2}$$

To reason about component compatibility, reasoning on the composite contract, and checking whether $C_1 \otimes C_2$ is compatible, may be more convenient than directly computing the composition $M_1 \times M_2$. In fact, contracts generally offer an abstract, more compact, and usually more tractable representation of the design.

Contracts can be ordered by establishing a refinement relation, which is a pre-order on contracts. We say that C refines C', written $C \preccurlyeq C'$, if and only if $A \supseteq A'$ and $G \subseteq G'$. Refinement amounts to relaxing assumptions and reinforcing guarantees, therefore strengthening the contract. Clearly, if $M \vDash C$ and $C \preccurlyeq C'$, then $M \vDash C'$. On the other hand, if $E \vDash_E C'$, then $E \vDash_E C$. In other words, contract C refines C', if C admits less implementations than C', but more legal environments than C'. We can then replace C' with the stronger, and more concrete, contract C.

Finally, to compose multiple views of the same component that need to be satisfied simultaneously, the conjunction ($\wedge$) of contracts can also be defined so that, if a component M satisfies the conjunction of C_1 and C_2, that is, $M \vDash C_1 \wedge C_2$, then it also satisfies each of them independently, that is, $M \vDash C_1$ and $M \vDash C_2$. We can compute the conjunction of C_1 and C_2 by taking their greatest lower bound with respect to the refinement relation. That is, $C_1 \wedge C_2$ is guaranteed to refine both C_1 and C_2. Furthermore, $C_1 \wedge C_2$ is the "weakest" contract that refines C_1 and C_2, that is, for any contract C' such that $C' \preccurlyeq C_1$ and $C' \preccurlyeq C_2$, we have $C' \preccurlyeq C_1 \wedge C_2$. For contracts in canonical form and on the same variable set V, we have

$$C_1 \wedge C_2 = (V, A_1 \cup A_2, G_1 \cap G_2). \tag{15.3}$$

Further details on the mathematical expressions for computing contract composition and conjunction, as well as application examples, can be found in Refs [7,33]. For the sake of generality, we expressed the operations and relations above using operations and relations between sets (e.g., intersection, union, complement, and containment). These operations and relations can then be implemented based on the concrete formalism used to represent assumptions and guarantees in the design flow.

15.3.3.2 Horizontal and Vertical Contracts

Traditionally, contracts have been used to specify components and aggregations of components at the same level of abstraction; for this reason, they are often denoted as horizontal contracts. However, contracts can also be used to formalize and reason about refinement between two different abstraction levels, and capture mechanisms for mapping a specification over an execution platform, such as the ones adopted in the PBD process [13,41] and in the Metropolis [45], Metro II [46], and Metronomy [47] frameworks. This type of contracts is denoted as vertical contracts.

To illustrate this concept, we consider the problem of mapping a specification platform of a system at level $l + 1$ into an implementation platform at level l. In general, the specification platform architecture (interconnection of components) may be defined in an independent way, and may not directly match the implementation platform architecture. Such a heterogeneous architectural decomposition will also reflect on the contracts associated with the components and their aggregations. For instance, the contract describing the specification platform $C = \bigwedge_{k \in K} \left(\bigotimes_{i \in I_k} C_{ik} \right)$ may be defined as the conjunction of K different viewpoints, each characterized by its own architectural decomposition into I_k contracts. On the other hand, the contract describing the implementation platform $M = \bigotimes_{j \in J} \left(\bigwedge_{n \in N_j} M_{jn} \right)$ may be better represented as a composition of J contracts, each defined out of the conjunction of its different viewpoints. Because there may not be, in general, a direct matching between contracts and viewpoints of M and C, checking that $M \preccurlyeq C$ in a compositional way, by reasoning on the elements of M and C independently, as discussed in Section 15.3.3.1, may not be effective.

However, it is still possible to reason about refinement between M and C by resorting to the vertical contract $C \wedge M$, which specifies the composition of a model and its vertical refinement, even though they are not directly connected, by connecting them indirectly through a mapping, for example, by synchronizing pairs of events, as if cosimulating a model and its refinement. Informally, this kind of composition captures the fact that the actual satisfaction of all the design requirements and viewpoints (in C) by a deployment depends on the supporting execution platform and the underlying physical system (in M) as well as on the way in which system functionalities are mapped into them. Note that $C \wedge M \preccurlyeq C$ is assured to hold by construction; however, $C \wedge M$ can still be a source of inconsistencies. Therefore, to guarantee that the design can be implemented, the consistency of $C \wedge M$ must be checked or enforced. Checking consistency on the composite contract can be, however, easier than checking $M \preccurlyeq C$, since some of the assumptions made by the specification platform on the implementation platform can be discharged by the guarantees of the implementation platform, and vice versa, when considering $C \wedge M$.

Vertical contracts were previously adopted in the design of analog and mixed-signal integrated circuits [41,48] by leveraging approximations of implementation constraints to represent different viewpoints (e.g., timing, energy, noise), and then checking their compatibility or consistency during design space exploration. More recently, a similar approach has also been advocated in the context of Autosar [13]. Alternatively, when vertical assumptions and guarantees cannot be effectively expressed by compact models, compatibility and consistency of vertical contracts can be checked by cosimulation of the application and implementation platforms under a mapping mechanism, such as the one in the Metronomy framework [47], which synchronizes tuples of signals in the two platforms.

Finally, vertical contracts are particularly useful for the design of embedded controllers, in that they can formalize the agreement between control, software, and hardware engineers when specifying both system functionality and timing requirements [7,8,49]. In a typical scenario, as represented in Figure 15.3, a controller takes as assumptions several aspects that include the timing behavior of the control tasks and of the communication between tasks, for example, delay, jitter, as well as the accuracy and resolution of the computation (vertical assumptions in $\mathcal{C}$). On the other hand, the controller provides guarantees in terms of the amount of requested computation, activation times, and data dependencies (vertical assumptions in $\mathcal{M}$). As a result, several design guidelines can be derived by formulating and enforcing consistency of vertical contracts across the hardware, software, and control layers.

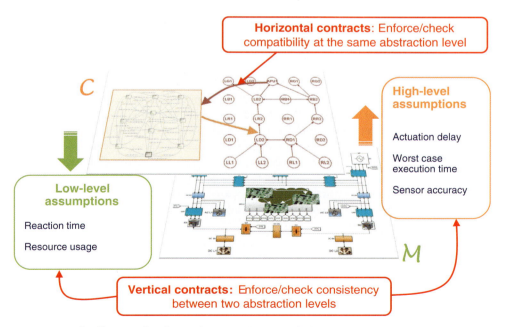

Figure 15.3 Role of horizontal and vertical contracts in system design.

15.4
Platform-Based Design with Contracts and Related Tools

We now discuss how the challenges in Section 15.2 can be addressed by placing the design process in the form of platform-based design and by using contracts to verify the design and to build refinements that are correct by construction. To do so, we consider a CPS scenario that incorporates most, if not all, of the features of general CPSs: a control system, composed of a physical plant, including sensors and actuators, and an embedded controller. The controller runs a control algorithm to restrict the behaviors of the plant so that all the remaining (closed-loop) behaviors satisfy a set of system requirements. Specifically, we consider reactive controllers, that is, controllers that maintain an ongoing relation with their environment by appropriately reacting to it. Our goal is to design the system architecture, that is, the interconnection among system components, and the control algorithm, to satisfy the set of high-level requirements.

We therefore focus on two main steps, namely, system architecture design and control design. The system architecture design step instantiates system components and interconnections among them to generate an optimal architecture while guaranteeing the desired performance, safety, and reliability. Typically, this design step includes the definition of both the embedded system (software, hardware, and communication components) and the plant architectures (e.g., mechanical, electrical, hydraulic, or thermal components). Given an architecture, the control design step includes the exploration of the control algorithm and its deployment on the embedded platform.

The above two steps are, however, connected. The correctness of the controller needs to be enforced in conjunction with the assumptions on the plant. Similarly, performance and reliability of an architecture should be assessed for the plant in closed loop with the controller. A schematic representation of the design flow is offered in Figure 15.4.

At the highest level of abstraction, the starting point is a set of requirements, predominantly written in text-based languages that are not suitable for mathematical analysis and verification. The result is a model of both the architecture and the control algorithms to be further refined in subsequent design stages. In what follows, we provide details on the three main phases of a PBD methodology that can be deployed to address the design steps above: top-down requirement formalization, bottom-up library generation, and design exploration, where requirements are mapped into implementations out of the available component libraries.

15.4.1
Requirement Formalization and Validation

Contracts are particularly effective to guide top-level requirement formalization, allocate requirements to lower-level components, and analyze them for early validation of design constraints, thus helping address the specification

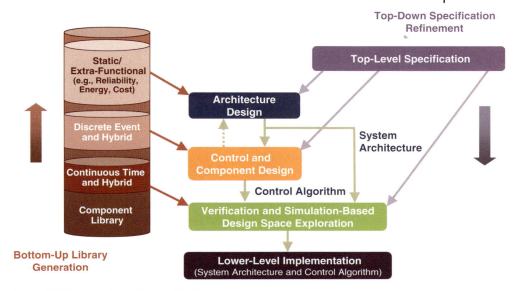

Figure 15.4 The structure of the methodology.

challenges 3 and 4 in Section 15.2. Formal requirement analysis can often be difficult, because of the lack of familiarity with formal languages among system engineers. Moreover, it is significantly different from traditional formal verification, where a system model is compared against a set of requirements. Since there is not yet a system at this stage, requirements themselves are the only entity under analysis or exploration. By formalizing requirements as contracts, it is instead possible to provide effective tests to check for requirement consistency, that is, whether a set of contracts is realizable, or whether, in contrast, facets of these are inherently conflicting, and thus no implementation is feasible. Moreover, it is possible to exclude undesired behaviors, for example, by adding more contracts, by strengthening assumptions, or by considering additional cases for guarantees. Contract-based requirement engineering can then offer substantial support to tackle the integration challenges 5 and 6 in Section 15.2. Moreover, since contracts are abstractions of components, their concrete representations are typically more compact than a fully specified design [43]. The above tests can then be performed more efficiently than traditional verification tasks, thus partially improving on the scalability issues in challenge 7 of Section 15.2.

A framework for requirement engineering has been recently developed by leveraging modal interfaces, an automata-based formalism, as the underlying specification theory [13]. However, to retain a correspondence between informal requirements and formal statements, declarative, "property-based" approaches using some temporal logic are gaining increasing interest. They contrast imperative, "model-based" approaches, which tend to be impractical for high-level requirement validation, since they often entail considering all possible combinations of system variables to construct the model, and substantial changes in the underlying models to support small changes in the requirements. A/G contracts, as introduced in Section 15.3.3, allow specifying different kinds of requirements using different formalisms, following both the declarative and imperative styles, to reflect the different viewpoints and domains in a heterogeneous system, as well as the different levels of abstraction in the design flow [7].

To facilitate reasoning at the level of abstraction of requirement engineers, a viable strategy is to drive engineers toward capturing requirements in a structured form. This can be achieved using a set of predefined high-level primitives, or patterns, from which formal specifications can be automatically generated [7,14]. Patterns can also be combined together to form higher-level domain-specific languages (DSL), as exemplified in Ref. [42]. A similar approach was also advocated in the Statemate verification environment [50] and within the European projects SPEEDS and CESAR [40].

From a set of high-level primitives, different kinds of contracts can be generated. When specifying the system architecture, steady-state (static) requirements, interconnection rules, and component dimensions can be captured by static contracts, expressed via arithmetic constraints on Boolean and real variables to model, respectively, discrete and continuous design choices. Then, compatibility, consistency, and refinement checking translate into checking feasibility of conjunctions or disjunction of constraints, which can be solved via queries to satisfiability modulo theories (SMT) solvers [51] or mathematical optimization software, such as mixed integer linear, mixed integer semidefinite positive, or mixed integer nonlinear program solvers.

When specifying the control algorithm, representing dynamic behaviors becomes the main concern; safety and real-time requirements can then be captured by contracts expressed using temporal logic constructs. For instance, linear temporal logic (LTL) [52] can be used to reason about the temporal behaviors of systems characterized by Boolean, discrete-time signals or sequences of events (discrete event abstraction in Figure 15.4). Signal temporal logic (STL) [53] can deal with continuous-time real signals and continuous dynamical models (continuous abstraction in Figure 15.4). Sometimes, discrete and continuous dynamics are so tightly connected that a discrete-event (DE) abstraction would result inaccurate, while a continuous abstraction would turn out to be inefficient, thus calling for a hybrid system abstraction, mixing discrete and continuous behaviors, such as hybrid linear temporal logic with regular expressions (HRELTL) [54] and hybrid automata [55].

Whether contract operations and relations are expressed using temporal logic or automata, they eventually translate into fundamental verification tasks. The complexity and run time of such tasks are, however, heavily dependent on the selected formalisms. While formal verification of hybrid models generates, in general, intractable problems, several approaches to cope with this issue have been reported in the literature, and can be categorized as follows:

- When discrete-time, discrete-state system abstractions are available, the wealth of results and tools in temporal logic and model checking can provide a substantial technological basis for requirement analysis [29].
- When the continuous dynamics can be abstracted using timed or rectangular automata [56], most of the verification techniques for finite-state models can still be used to obtain exact answers. Tools in this category, such as UPPAAL [57], support networks of timed automata and complex properties expressed in a subset of computation tree logic (CTL) [29]. UPPAAL can handle models with up to 100 clock variables, used to capture the continuous dynamics in timed automata.
- When the dynamics are more complex, exact computations on conservative abstractions are often not effective. Approximation techniques can, instead, be used to obtain an answer in some cases, for example, when dealing with safety requirements, which prescribe that "nothing bad shall happen" in the design. Among the tools in this category, we recall SpaceEx [58], Flow* [59], and Ariadne [60]. SpaceEx handles systems with piecewise affine, nondeterministic dynamics, while Flow* and Ariadne can also support polynomial and generic nonlinear dynamics, respectively. Scalability is, however, limited to a dozen continuous variables for Flow* and Ariadne, while SpaceEx can support up to 100 variables.
- A set of techniques leverage automatically generated abstractions and refinements of hybrid models to efficiently answer complex verification tasks. If a finite-state discrete abstraction is not accurate enough to provide a conclusive answer, it is refined until either an answer is found or the maximum number of refinement steps is reached [61,62]. In some cases, the verification problem can be solved with few refinement steps. In this category, we recall Hybridsal [63], which handles polynomial hybrid systems with up to 10 continuous variables, and Hycomp [64], which uses an SMT approach to create discrete, infinite-state abstractions, and was tested successfully on models with up to 60 continuous variables with piecewise constant dynamics.
- Automated theorem proving techniques can also be used to solve verification tasks and perform requirement validation, as in Keymaera [65], which handles specifications in the temporal logic $d\mathcal{L}$, and has been successfully used to verify collision avoidance in case studies from train control to air traffic management.
- Finally, simulation-based approaches can be used to verify black-box models (when the internal dynamics are unknown), or models of complex systems, for which only simulation is computationally feasible. Because simulation does not cover, in general, the entire state space, it cannot be used to certify the satisfaction of a contract, but rather to monitor and detect possible violations. Among the simulation-based tools, we recall Breach [66], a Matlab/C++ toolbox for the simulation and verification of signal temporal logic properties on systems defined by ordinary differential equations (ODEs) or by external tools, such as Simulink, and S-TALIRO [67], another Matlab toolbox for the analysis of

continuous and hybrid dynamical systems supporting complex properties in metric temporal logic.

While discrete-time finite-state abstractions are typically preferred, since they result into exhaustively analyzable models, the computational complexity of analyzing temporal logic contracts for these abstractions can still make industrial-scale problems difficult to solve with current tools. There are, however, a number of directions to overcome these scalability issues. A possible approach is to rely on a library of precharacterized contracts, enriched with refinement assertions, to automatically generate system abstractions during the verification process, thus breaking the main verification task into a set of smaller tasks. The application of this technique to verify controllers for aircraft electrical power systems showed up to two orders of magnitude improvement with respect to standard implementations, for example, based on solving a single, large LTL satisfiability problem [43]. Since the library characterization process must be performed only once, outside of the main design flow, the benefits of having a richer library in terms of refinement assertions largely repay the overhead of building it.

A second approach leverages encodings of the verification tasks into SMT or optimization problems to reason about Boolean combinations of linear and nonlinear constraints over the reals, albeit over a finite time horizon. A framework for refinement checking of HRELTL contracts based on SMT solving was reported by Cimatti *et al.* [54,68]. More recently, we have proposed diagnosis and repair algorithms for certain types of STL contracts based on the formulation of a set of Mixed Integer Linear Programs (MILPs), where infeasibility of a MILP indicates contract inconsistencies. The proposed algorithms provide feedback on the reasons for unrealizability of the STL contracts and suggestions for making them realizable. The resulting toolbox DiaRY has been applied to debugging controllers for various CPSs, including an autonomous driving application and an aircraft electric power system [69].

Finally, in addition to leveraging efficient, off-the-shelf SMT solvers and optimization tools, combining the SMT paradigm with optimization methods has recently served as the inspiration for the construction of novel, scalable decision procedures, and algorithms for cyber-physical system verification and design [70–75].

15.4.2
Platform Component-Library Development

To assemble the implementation platforms, a library of components, models, and related contracts must be developed for the plant and the embedded system. As shown in Figure 15.4, components and associated contracts should be hierarchically organized to represent the system at different levels of abstraction, for example, steady-state, discrete-event, and hybrid. Reflecting the taxonomy of requirements, the component library should also be viewpoint and domain dependent, following a similar approach as in the "rich component" libraries that were first proposed for automotive embedded systems [39]. At each level of abstraction, components should be capable of exposing multiple, complementary viewpoints, reflecting different design concerns, and associated, in general, with different formalisms (e.g., graphs, linear temporal logic, algebraic differential equations). Moreover, models should include extra-functional (performance) metrics, such as timing, energy, and cost, in addition to the description of their behaviors. Such a library organization is key to address the modeling challenges 1 and 2 in Section 15.2.

Components and contracts can then be expressed using the same formalisms introduced in Section 15.4.1 in the context of requirement analysis and validation. A major challenge in this multiview and hierarchical modeling scenario remains to maintain consistency among models and views, often developed using domain-specific languages and tools, as the library evolves [10] (challenges 5 and 6 in Section 15.2). In this respect, the algebra of contracts can offer an effective way to incrementally check consistency or refinement among models. This information can then be stored in the library to speed up verification tasks at design time [43]. Moreover, vertical contracts can be used to establish conditions for an abstract, approximate model, to be a sound representation of a concrete model, that is, to define when a model still retains enough precision to address specific design concerns, in spite of the vagueness required to make it manageable by analysis tools [41].

A number of languages and tools have been proposed over the years to enable capturing different system descriptions and exchanging them between different tasks of the design flow (e.g., controller design, validation, verification, testing, and code generation). An exhaustive survey is out of the scope of this chapter. Among the several languages and tools, we recall the following:

- Generic modeling and simulation frameworks, such as Matlab/Simulink [76] and Ptolemy II [77]
- Hardware description languages, such as Verilog [78], VHDL [79], or transaction-level modeling languages, such as SystemC [80] together with their respective analog-mixed-signal extensions (Verilog-AMS, VHDL-AMS, and SystemC-AMS)
- Modeling languages specifically tailored for acausal multiphysics systems, such as Modelica [81], supported by tools such as Dymola [82]
- Languages for architecture modeling, such as the systems modeling language (SysML) [83], an object oriented modeling language largely based on the unified modeling language (UML) 2.1, which also provides useful extensions for systems engineering, and the architecture analysis and design language (AADL) [84].

A number of proposals have also appeared toward modeling languages specifically tailored to CPSs. One of the first examples of these languages is Charon [85], supporting the hierarchical description of system architectures via the operations of instantiation, hiding, and parallel composition. Continuous behaviors can be specified using differential as well as algebraic constraints, all of which can be declared at various levels of the hierarchy. A few years later, Giotto [86] provided an abstract programming model for the implementation of embedded control systems with real-time constraints, while, more recently, the hierarchical timing language (HTL) [87] allows specifying critical timing constraints within the language, and directly forcing them via the compiler.

While the above languages generally lack the capability of easily interfacing with other tools, a few proposals for platform-independent interchange formats have also advanced in the literature. A few examples include the Metropolis meta-model [88], which also accounts for implementation considerations, such as equation sorting and event detection, and more recently, the compositional interchange format (CIF), which integrates compositional semantics with automata, process communication, and synchronization based on shared events, differential algebraic equations, different forms of urgency, and process definition and instantiation to support reuse and large scale system modeling. CIF can interface with a number of other languages and tools (e.g., UPPAAL, Phaver, Ariadne, Modelica, Matlab), and is currently used in both academia and industry [89].

Finally, particularly appealing for CPS modeling and simulation is the functional mockup interface (FMI), an evolving standard for composing component models, which are better realized and characterized using distinct modeling tools [90,91]. Initially developed within the Modelisar project, and currently supported by a number of industrial partners and tools, FMI shows promise for enabling the exchange and interoperation of model components. The FMI standard supports both cosimulation, where a component called FMU (functional mock-up unit) implements its own simulation algorithm, and model exchange, where an FMU exports sufficient information for an external simulation algorithm to execute simulation. However, while in principle FMI is capable of composing components representing timed behaviors, including physical dynamics and discrete events, several aspects of the standard, for example, to guarantee that a composite model does not exhibit nondeterministic and unexpected behaviors, are currently object of investigation [92].

15.4.3
Mapping Specifications to Implementations

During design space exploration, horizontal and vertical contracts can be used to define the specification and implementation platforms as well as the mechanisms needed to transition between them, that is, to map the specifications to an aggregation of library components (challenges 2 and 7 in Section 15.2). Contracts can then provide the infrastructure to reason about different aspects or representations of the design by coordinating specialized analysis, synthesis, and optimization frameworks that can operate with different formalisms.

At each abstraction layer, the mapping to a lower layer can be performed by either leveraging a synthesis tool, or by solving an optimization problem that uses constraints from both the specification and the implementation layers to evaluate global tradeoffs among components. In the following, we provide examples of mapping techniques and tools that can be useful for different design tasks.

15.4.3.1 Architecture Design

The architecture exploration problem can be, in general, intractable because of the large number of discrete alternatives, the expensive simulations often required to achieve high accuracy in performance and cost estimations, and the complex, nonlinear behaviors of the components, whose models are not always available in analytic form.

However, in several application domains, an architecture can be conveniently represented as a labeled graph, where nodes represent the (parametrized) components and edges represent their interconnections [42,71,93]. Based on this representation, a variety of system requirements, such as interconnection, safety, reliability, and energy balance requirements, expressing the high-level specification contract, can be captured by mixed integer linear constraints on the graph parameters. The same formalism can be used to represent the steady-state models for the architecture components and their aggregation. It is then possible to map system requirements to such a representation of the system architecture by enforcing the consistency of a set of mixed integer linear constraints. The ArchEx framework [42,93] implements two approaches for the efficient selection of cost-effective and reliable architectures using such a MILP formulation to minimize a cost function (e.g., component number, weight, cost, energy) while satisfying the specification and implementation contracts. Both the approaches are demonstrated on the design of aircraft power system architectures.

The first technique formulates a monolithic mixed integer linear program by using efficient approximations of the nonlinear constraints for which quantitative error bounds can be computed. The second technique, inspired by SMT solving [94], adopts, instead, an iterative scheme that breaks the complex architecture selection task into a sequence of smaller optimization tasks. At each iteration, a candidate architecture is proposed based on the subset of linear constraints. The satisfaction of the nonlinear constraints is then checked using an exact analysis method. If the constraints are satisfied, the candidate architecture is returned as the final one. Otherwise, the current solution is discarded as infeasible and new architectures are generated until a feasible one is found. By relying on efficient mechanisms to traverse the design space and prune out large portions that are inconsistent with the requirements, the latter technique has shown to outperform the former on large problem instances. An extension of this iterative scheme, combining a discrete and a continuous optimization engine, has been recently proposed to perform both topology selection and sizing, while supporting a broader category of architectures and design concerns, including the effects of system dynamics and transients [71].

The selected architecture serves as a high-level specification for the next design steps, including the definition of the control algorithm, which we detail in Section 15.4.3.2.

15.4.3.2 Control Design

Control design can be framed as the problem of mapping a specification contract and a description of the plant into a controller that implements the contract once it is composed with the plant. When contracts are expressed using a discrete-time temporal logic (e.g., LTL or CTL), the design of control algorithms for CPSs can be solved using techniques from reactive synthesis, which has been an active area of research since the late 1980s, and it is still attracting a considerable attention today [95–98]. In this case, the specification is mapped to a DE implementation of the controller, for example, in terms of a state machine. For general LTL, the synthesis problem has a doubly exponential complexity. However, a subset of LTL, namely, generalized reactivity (1) (GR(1)), generates problems that are polynomial in the number of valuations of the environment and system variables [95].

In the case of hybrid controllers, discrete abstractions, often obtained by partitioning the continuous state space into polytopes, make it still possible to reduce the synthesis problem within the realm of reactive synthesis, or other established DE control synthesis methods [99,100]. Alternatively, controller design for discrete-time CPSs can also be cast within a model predictive control framework [101,102]. For instance, some kinds of STL specifications can be encoded as mixed integer linear constraints on the system variables of an optimization problem that can be solved at each step, following a receding horizon approach [102,103]. While reactive synthesis approaches become usually impractical for systems with more than five continuous states [104], MILP-based control synthesis can leverage the empirical performance of state-of-the-art solvers and has shown to support over 30 continuous states.

In fact, because of the limited scalability of existing tools to large hybrid models, constructing effective abstractions is key to the application of controller synthesis methods. The notion of approximate bisimulation [105] has been recently introduced to obtain correct and complete abstractions of differential equations that can be used to solve controller design problems. Pessoa [106] is a software toolbox, which exploits approximate bisimulation to implement efficient synthesis algorithms operating over the equivalent finite-state machine models. The resulting controllers are also finite-state and can be readily transformed into code for any desired digital platform. This transformation assigns the finite-state controller operation to a processor, where code is the result of mapping the controller equations into the instruction set of the processor.

Another approach to mapping a controller into a processor is the control software synthesis tool QKS [107]. Given the sampling time of the controller and the precision of the analog-to-digital conversion of state measurements, QKS can compute both the controllable region and an implementation in C code of a controller driving the system into a goal region in finite time.

A library-based compositional synthesis approach that directly conforms to the PBD paradigm has recently been presented to solve high-level motion planning problems for multirobot systems [108]. The desired behavior of a group of robots is specified using a set of safe LTL properties (top-down step of the flow). The closed-loop behavior of the robots under the action of different lower-level controllers is abstracted using a library of motion primitives, each of which corresponds to a controller that ensures a particular trajectory in a given configuration (bottom-up step of the flow). By relying on these primitives, the mapping problem is then encoded as an SMT problem and solved by using an off-the-shelf SMT solver to efficiently generate control strategies for the robots.

In certain cases, real-time constraints, mostly related to the physical plant and the hardware implementation of the controller, may require the full expressiveness of continuous and hybrid models. However, solving the controller synthesis problem by directly mapping to these abstractions is a very difficult task [109]. Even in the context of timed automata, where the synthesis problem is known to be solvable in an exact way [110], efficient and practical tools are lacking. One of the few exceptions is UPPAAL-Tiga [111], an extension of UPPAAL that implements on-the-fly algorithms for solving the controller synthesis problem on timed automata with respect to reachability and safety properties expressed using timed computation tree logic.

General formulations for the controller synthesis problem in hybrid systems subject to a safety specification are mostly based on solving a differential game in which the environment is trying to drive the system into its target set at the same time as avoiding the target set of the controller [112,113]. One of the few publicly available tools implementing the game approach is Phaver+ [114], an extension of Phaver that can automatically synthesize discrete controllers for linear hybrid automata with respect to safety and reachability goals.

Finally, when algorithmic synthesis is intractable, it is still possible to cast the design exploration problem as an optimization problem, where the system specifications are checked by a formal verification engine or by monitoring simulation traces. For instance, let $\mathcal{C} = (V, \phi_e, \phi_e \rightarrow \phi_s)$ be a contract that must be checked, where ϕ_e and ϕ_s are temporal logic formulas. Then, given an array of costs J, the mapping problem can be cast as a multiobjective robust optimization problem, aiming at finding a set of configuration parameter vectors that are Pareto optimal with respect to the objectives in J, while guaranteeing that the system satisfies ϕ_s for all possible traces satisfying the environment assumptions ϕ_e. Using this formulation, it is also possible to perform joint design exploration of the controller and its execution platform, while guaranteeing that their specifications, captured by vertical contracts, are consistent. This approach is also supported by frameworks such as Metronomy to evaluate the characteristics (such as latency, throughput, power, and energy) of a particular implementation by cosimulation of both a functional model and an architectural model of the system.

15.5
Conclusions

Dealing with the complexity of cyber-physical systems requires rethinking the existing design flows as well as their formal foundations. We articulated the main design challenges and showed how they can be addressed within a structured design methodology that combines platform-based design with assume-guarantee contracts. Based on the techniques surveyed in this chapter, we expect CPS design will increasingly build on a design management feature that we call a front-end orchestrator, which will use notions from contract and interface theories to coordinate a set of back-end specialized tools and consistently process their results. Such an orchestrator will increasingly rely on design aids for requirement formalization as well as algorithms that can maximally leverage the modularity offered by contracts to perform analysis and synthesis tasks on system portions of manageable size and complexity. Finally, while system engineers routinely make use of decompositions, abstractions, and approximations to assemble their designs, the formalization of these concepts and their algorithmic implications for hybrid system design is still in its infancy. Therefore, advancing our understanding of the intricacies of compositional reasoning, and its interplay with abstraction and approximation mechanisms, will be at the heart of a rigorous discipline for system design.

Acknowledgments

This work was supported in part by the TerraSwarm Research Center, one of six centers supported by the STARnet phase of the Focus Center Research Program (FCRP), a Semiconductor Research Corporation program sponsored by MARCO and DARPA.

References

1 Waldner, J.B. (2013) *Nanocomputers and Swarm Intelligence*, John Wiley & Sons, Inc.

2 Allen, J.J. (2005) *Micro Electro Mechanical System Design*, CRC Press.

3 InvenSense, Inc. (2016). Available at www.invensense.com/ (accessed May 10, 2016.

4 Wolpaw, J. and Wolpaw, E.W. (2012) *Brain-Computer Interfaces: Principles and Practice*, Oxford University Press.

5 Sztipanovits, J. (2007) Composition of cyber-physical systems. Proceedings of the IEEE International Conference and Workshops on Engineering of Computer-Based Systems, pp. 3–6. doi: 10.1109/ECBS.2007.25

6 Lee, E.A. (2008) Cyber physical systems: design challenges. Proceedings of the IEEE International Symposium on Object Oriented Real-Time Distributed Computing pp. 363–369. doi: 10.1109/ISORC.2008.25

7 Nuzzo, P., Sangiovanni-Vincentelli, A., Bresolin, D., Geretti, L., and Villa, T. (2015) A platform-based design methodology with contracts and related tools for the design of cyber-physical systems. *Proc. IEEE*, **103** (11), 2104–2132.

8 Sangiovanni-Vincentelli, A., Damm, W., and Passerone, R. (2012) Taming Dr. Frankenstein: contract-based design for cyber-physical systems. *Eur. J. Control*, **18-3** (3), 217–238.

9 Lee, E.A., Rabaey, J., Hartmann, B., Kubiatowicz, J., Pister, K., Sangiovanni-Vincentelli, A., Seshia, S.A., Wawrzynek, J., Wessel, D., Rosing, T.S. *et al.* (2014) The swarm at the edge of the cloud. *IEEE Des. Test*, **31** (3), 8–20.

10 Nuzzo, P. and Sangiovanni-Vincentelli, A. (2014) Let's get physical: computer science meets systems, in *From Programs to Systems. The Systems perspective in Computing, Lecture Notes in Computer Science*, vol. **8415** (eds S. Bensalem, Y. Lakhneck, and A. Legay), Springer Berlin Heidelberg, pp. 193–208.

11 Sangiovanni-Vincentelli, A. (2010) Corsi e ricorsi: the EDA story. *IEEE Solid State Circuits Mag.*, **2** (3), 6–26.

12 Sangiovanni-Vincentelli, A. (2007) Quo vadis, SLD? Reasoning about the trends and challenges of system level design. *Proc. IEEE*, **95** (3), 467–506.

13 Benveniste, A., Caillaud, B., Nickovic, D., Passerone, R., Raclet, J.B., Reinkemeier, P. *et al.* (2012) Contracts for System Design. INRIA, *Rapport de recherche RR*-8147.

14 Nuzzo, P., Sangiovanni-Vincentelli, A.L., and Murray, R.M. (2015) Methodology and tools for next generation cyber-physical systems: the iCyPhy approach. *Proceedings of the INCOSE International Symposium*.

15 Derler, P., Lee, E.A., and Sangiovanni-Vincentelli, A. (2012) Modeling cyber-physical systems. *Proc. IEEE*, **100** (1), 13–28.

16 Sztipanovits, J. and Karsai, G. (1997) Model-integrated computing. *IEEE Comput.*, **30** (4), 110–112.

17 Selic, B. (2003) The pragmatics of model-driven development. *IEEE Softw.*, **20** (5), 19–25.

18 Lee, E.A. (2006) The problem with threads. *Computer*, **39** (5), 33–42.

19 Booch, G., Rumbaugh, J., and Jacobson, I. (2005) *The Unified Modeling Language User Guide*, 2nd edn, Addison-Wesley Object Technology Series, Addison-Wesley Professional, San Jose.

20 IBM Doors, (2016) Available at http://www-03.ibm.com/software/products/en/ratidoorfami (accessed May 10, 2016).

21 Lee, E.A. and Seshia, S.A. (2015) Introduction to Embedded Systems: A Cyber-Physical Systems Approach, 2nd edn, Lulu.com.

22 Larman, C. and Basili, V.R. (2003) Iterative and incremental development: a brief history. *Computer*, **36** (6), 47–56.

23 Das V-Modell (2016). Available at v-modell.iabg.de/ (accessed May 10, 2016).

24 Lin, C.W., Zhu, Q., and Sangiovanni-Vincentelli, A. (2014) Security-aware mapping for TDMA-based real-time distributed systems. *Proceedings of the 2014 IEEE/ACM International Conference on Computer-Aided Design*, IEEE Press, pp. 24–31.

25 Densmore, D., Simalatsar, A., Davare, A., Passerone, R., and Sangiovanni-Vincentelli, A. (2009) UMTS MPSoC design evaluation using a system level design framework. *IEEE Design, Automation & Test in Europe Conference & Exhibition*, pp. 478–483.

26 De Bernardinis, F., Nuzzo, P., and Sangiovanni-Vincentelli, A. (2005) Mixed signal design space exploration through analog platforms. *Proceedings of the IEEE/ACM Design Automation Conference*, pp. 1390–1393. doi http://doi.acm.org/10.1145/1403375.1403710

27 Yang, Y., Pinto, A., Sangiovanni-Vincentelli, A., and Zhu, Q. (2010) A design flow for building automation and control systems, *31st IEEE Real-Time Systems Symposium*, pp. 105–116.

28 Densmore, D., Van Devender, A., Johnson, M., and Sritanyaratana, N. (2009) A platform-based design environment for synthetic biological systems, in *The Fifth Richard Tapia Celebration of Diversity in Computing Conference: Intellect, Initiatives, Insight, and Innovations*, ACM, pp. 24–29.

29 Clarke, E.M., Grumberg, O., and Peled, D.A. (2008) *Model Checking*, The MIT Press, Cambridge, MA.

30 Meyer, B. (1992) Applying "design by contract". *Computer*, **25** (10), 40–51.

31 Abadi, M. and Cardelli, L. (1996) *A Theory of Objects*, Springer-Verlag.

32 Schmidt, D.C. (2006) Guest editor's introduction: model-driven engineering. *Computer*, **39** (2), 25–31.

33 Benveniste, A., Caillaud, B., Ferrari, A., Mangeruca, L., Passerone, R., and Sofronis, C. (2008) Chap. Multiple viewpoint contract-based specification and design, in *Formal methods for components and objects*, Springer-Verlag, Berlin, pp. 200–225.

34 de Alfaro, L. and Henzinger, T.A. (2001) Interface automata, in *Proceedings of the ACM SIGSOFT Symposium on Foundations of Software Engineering*, ACM Press, pp. 109–120.

35 Doyen, L., Henzinger, T.A., Jobstmann, B., and Petrov, T. (2008) Interface theories with component reuse. *Proceedings of the ACM IEEE International Conference on Embedded Software*, pp. 79–88.

36 Raclet, J.B., Badouel, E., Benveniste, A., Caillaud, B., Legay, A., and Passerone, R. (2009) Modal interfaces: unifying interface automata and modal specifications. *Proceedings of the ACM IEEE International Conference on Embedded Software*, ACM, New York, NY, USA, pp. 87–96. doi http://doi.acm.org/10.1145/1629335.1629348. URL http://doi.acm.org/10.1145/1629335.1629348.

37 Nuzzo, P., Iannopollo, A., Tripakis, S., and Sangiovanni-Vincentelli, A.L. (2014) Are interface theories equivalent to contract theories?. *12th ACM-IEEE International Conference on Formal Methods and Models for System Design (MEMOCODE)*.

38 Masin, M., Sangiovanni-Vincentelli, A., Ferrari, A., Mangeruca, L., Broodney, H., Greenberg, L., Sambur, M., Dotan, D., Zolotnizky, S., and Zadorozhniy, S. (2011) Meta II: Lingua Franca Design and Integration Language. *Tech. Rep.* Available at http://www.darpa.mil/uploadedFiles/Content/Our_Work/TTO/Programs/AVM/IBMMETAFinalReport.pdf.

39 Damm, W., Votintseva, A., Metzner, A., Josko, B., Peikenkamp, T., and Böde, E. (2005) Boosting reuse of embedded automotive applications through rich components. *Proceedings of Foundations of Interface Technologies*.

40 Damm, W., Hungar, H., Josko, B., Peikenkamp, T., and Stierand, I. (2011) Using contract-based component specifications for virtual integration testing and architecture design. *Proceedings of Design, Automation and Test in Europe*, pp. 1–6.

41 Nuzzo, P., Sangiovanni-Vincentelli, A., Sun, X., and Puggelli, A. (2012) Methodology for the design of analog integrated interfaces using contracts. *IEEE Sens. J.*, **12** (12), 3329–3345.

42 Nuzzo, P., Xu, H., Ozay, N., Finn, J., Sangiovanni-Vincentelli, A., Murray, R., Donze, A., and Seshia, S. (2014) A contract-based methodology for aircraft electric power system design. *Access, IEEE*, **2**, 1–25.

43 Iannopollo, A., Nuzzo, P., Tripakis, S., and Sangiovanni-Vincentelli, A.L. (2014) Library-based scalable refinement checking for contract-based design. *Proceedings of Design, Automation and Test in Europe*.

44 Maasoumy, M., Nuzzo, P., and Sangiovanni-Vincentelli, A. (2015) Smart buildings in the smart grid: Contract-based design of an integrated energy management system, in *Cyber Physical Systems Approach to Smart Electric Power Grid*, Power Systems (eds S.K. Khaitan, J.D. McCalley, and C.C. Liu), Springer, Berlin, pp. 103–132.

45 Balarin, F., Hsieh, H., Lavagno, L., Passerone, C., Sangiovanni-Vincentelli, A.L., and Watanabe, Y. (2003) Metropolis: an integrated electronic system design environment. *Computer*, **36** (4) 45–52.

46 Balarin, F., Davare, A., D'Angelo, M., Densmore, D., Meyerowitz, T., Passerone, R., Pinto, A., Sangiovanni-Vincentelli, A., Simalatsar, A., Watanabe, Y., Yang, G., and Zhu, Q. (2009) Platform-based design and frameworks: Metropolis and Metro II, in *Model-Based Design for Embedded Systems* (eds G. Nicolescu and P.J. Mosterman), CRC Press/Taylor and Francis Group, Boca Raton, p. 259.

47 Guo, L., Qi, Z., Nuzzo, P., Passerone, R., Sangiovanni-Vincentelli, A., and Lee, E.A.

(2014) Metronomy: a function-architecture co-simulation framework for timing verification of cyber-physical systems. *Proceedings of the International Conference on Hardware–Software Codesign and System Synthesis.*

48 Nuzzo, P. and Sangiovanni-Vincentelli, A. (2011) Robustness in analog systems: design techniques, methodologies and tools. *Proceedings of the IEEE Symposium on Industrial Embedded Systems.*

49 Derler, P., Lee, E.A., Tripakis, S., and Törngren, M. (2013) Cyber-physical system design contracts. *Proceedings of the International Conference on Cyber-Physical Systems*, pp. 109–118.

50 Bienmüller, T., Damm, W., and Wittke, H. (2000) The Statemate verification environment, in *Proceedings of the International Conference on Computer-Aided Verification, Lecture Notes in Computer Science*, vol. **1855** (eds E.A. Emerson and A.P. Sistla), Springer, Berlin, pp. 561–567.

51 Barrett, C., Sebastiani, R., Seshia, S.A., and Tinelli, C. (2009) Satisfiability modulo theories, in *Handbook of Satisfiability*, vol. **4** (eds A. Biere, H. van Maaren, and T. Walsh), IOS Press.

52 Pnueli, A. (1977) The temporal logic of programs. *Proceedings of the 18th IEEE Symposium on Foundations of Computer Science*, vol. **31**, pp. 46–57.

53 Maler, O. and Nickovic, D. (2004) Monitoring temporal properties of continuous signals, in *Formal Modeling and Analysis of Timed Systems*, Springer, pp. 152–166.

54 Cimatti, A., Roveri, M., and Tonetta, S. (2009). Requirements validation for hybrid systems, in *Computer Aided Verification, Lecture Notes in Computer Science*, vol. **5643** (eds A. Bouajjani and O. Maler), Springer, Berlin, pp. 188–203.

55 Alur, R., Courcoubetis, C., Henzinger, T.A., and Ho, P.H. (1993) Hybrid automata: an algorithmic approach to the specification and verification of hybrid systems, in *Hybrid Systems, LNCS*, vol. **736**, Springer, pp. 209–229.

56 Alur, R. and Dill, D.L. (1994) A theory of timed automata. *Theor. Comput. Sci.*, **126** (2), 183–235.

57 Behrmann, G., David, A., Larsen, K.G., Pettersson, P., and Yi, W. (2011) Developing UPPAAL over 15 years. *Softw., Pract. Exper.*, **41** (2), 133–142.

58 Frehse, G., Le Guernic, C., Donzé, A., Cotton, S., Ray, R., Lebeltel, O., Ripado, R., Girard, A., Dang, T., and Maler, O. (2011) SpaceEx: scalable verification of hybrid systems, in *Proceedings of the International Conference on Computer-Aided Verification, LNCS*, vol. **6806**, Springer, Berlin, pp. 379–395.

59 Chen, X., Ábrahám, E., and Sankaranarayanan, S. (2013) Flow*: An analyzer for non-linear hybrid systems, in *Proceedings of the International Conference on Computer-Aided Verification*, Lecture Notes in Computer Science, vol. **8044**, Springer, Berlin pp. 258–263.

60 Benvenuti, L., Bresolin, D., Collins, P., Ferrari, A., Geretti, L., and Villa, T. (2012) Ariadne: Dominance checking of nonlinear hybrid automata using reachability analysis, in *Reachability Problems*, Lecture Notes in Computer Science, vol. **7550** (eds A. Finkel, J. Leroux, and I. Potapov), Springer, Berlin, pp. 79–91.

61 Alur, R., Dang, T., and Ivančić, F. (2006) Counterexample-guided predicate abstraction of hybrid systems. *Theor. Comput. Sci.*, **354** (2), 250–271.

62 Clarke, E.M., Fehnker, A., Han, Z., Krogh, B.H., Ouaknine, J., Stursberg, O., and Theobald, M. (2003) Abstraction and counterexample-guided refinement in model checking of hybrid systems. *Int. J. Found. Comput. Sci.*, **14** (4), 583–604.

63 Tiwari, A. (2008) Abstractions for hybrid systems. *Form. Method Syst. Des.*, **32** (1), 57–83.

64 Cimatti, A., Mover, S., and Tonetta, S. (2013) SMT-based scenario verification for hybrid systems. *Form. Method Syst. Des.*, **42** (1), 46–66.

65 Platzer, A. (2010) *Logical Analysis of Hybrid Systems: Proving Theorems for Complex Dynamics*, Springer, Heidelberg. doi: 10.1007/978-3-642-14509-4

66 Donzé, A. (2010) Breach, a toolbox for verification and parameter synthesis of hybrid systems, in *Proceedings of the*

International Conference on Computer-Aided Verification, Springer-Verlag, Berlin, pp. 167–170.

67 Annpureddy, Y., Liu, C., Fainekos, G.E., and Sankaranarayanan, S. (2011) *S-TaLiRo: A Tool for Temporal Logic Falsification for Hybrid Systems*, Springer, pp. 254–257.

68 Cimatti, A. and Tonetta, S. (2012) A property-based proof system for contract-based design, in *EUROMICRO Conference on Software Engineering and Advanced Applications*, pp. 21–28.

69 Ghosh, S., Sadigh, D., Nuzzo, P., Raman, V., Donze, A., Sangiovanni-Vincentelli, A., Sastry, S.S., and Seshia, S.A. (2016) Diagnosis and repair for synthesis from signal temporal logic specifications. In *Proceedings of the 19th International Conference on Hybrid Systems: Computation and Control*, pp. 31–40. doi: 10.1145/2883817.2883847.

70 Nuzzo, P., Puggelli, A., Seshia, S., and Sangiovanni-Vincentelli, A. (2010) CalCS: SMT solving for non-linear convex constraints. *Proceedings of 10th International Conference on Formal Methods in Computer-Aided Design*, pp. 71–79.

71 Finn, J., Nuzzo, P., and Sangiovanni-Vincentelli, A. (2015) A mixed discrete-continuous optimization scheme for cyber-physical system architecture exploration. *Proceedings of the IEEE/ACM International Conference on Computer-Aided Design*.

72 Shoukry, Y., Puggelli, A., Nuzzo, P., Sangiovanni-Vincentelli, A.L., Seshia, S.A., and Tabuada, P. (2015) Sound and complete state estimation for linear dynamical systems under sensor attack using satisfiability modulo theory solving. *Proceedings of the IEEE American Control Conference*, pp. 3818–3823.

73 Shoukry, Y., Nuzzo, P., Puggelli, A., Sangiovanni-Vincentelli, A.L., Seshia, S.A., Srivastava, M., and Tabuada, P. (2015) Imhotep-SMT: a satisfiability modulo theory solver for secure state estimation. *Proceedings of the International Workshop on Satisfiability Modulo Theories*.

74 Shoukry, Y., Nuzzo, P., Bezzo, N., Sangiovanni-Vincentelli, A.L., Seshia, S.A., and Tabuada, P. (2015) Secure state reconstruction in differentially flat systems under sensor attacks using satisfiability modulo theory solving. *Proceedings of the International Conference on Decision and Control*, pp. 3804–3809.

75 Shoukry, Y., Chong, M., Wakaiki, M., Nuzzo, P., Sangiovanni-Vincentelli, A.L., Seshia, S.A., Hespanha, J.P., and Tabuada, P. (2016) SMT-based observer design for cyber physical systems under sensor attacks. *Proceedings of the International Conference on Cyber-Physical Systems*.

76 Simulink (2016). Available at http://www.mathworks.com/products/simulink (accessed May 10, 2016).

77 Ptolemy II (2016). Available at ptolemy.eecs.berkeley.edu (accessed May 10, 2016).

78 Verilog (2016). Available at www.verilog.com/ (accessed May 10, 2016).

79 Ashenden, P.J. (2010) *The Designer's Guide to VHDL*, vol. **3**, Morgan Kaufmann.

80 SystemC (2016). Available at http://www.accellera.org/downloads/standards/systemc (accessed May 10, 2016).

81 Modelica (2016). Available at www.modelica.org/ (accessed May 10, 2016).

82 Dymola (2016). Available at http://www.dynasim.se/ (accessed May 10, 2016).

83 SysML (2016). Available at http://www.omg.org/spec/SysML (accessed May 10, 2016).

84 AADL (2016). Available at http://www.aadl.info/aadl/currentsite (accessed May 10, 2016).

85 Alur, R., Grosu, R., Hur, Y., Kumar, V., and Lee, I. (2000) Modular specification of hybrid systems in Charon, in *Hybrid Systems: Computation and Control, LNCS*, vol. **1790**, Springer, pp. 6–19.

86 Henzinger, T., Horowitz, B., and Kirsch, C. (2003) Giotto: a time-triggered language for embedded programming. *Proc. IEEE*, **91** (1), 84–99.

87 Ghosal, A., Sangiovanni-Vincentelli, A., Kirsch, C.M., Henzinger, T.A., and Iercan, D. (2006) A hierarchical

coordination language for interacting real-time tasks, in *Proceedings of the ACM IEEE International Conference on Embedded Software*, ACM, New York, pp. 132–141.

88 Pinto, A., Carloni, L.P., Passerone, R., and Sangiovanni-Vincentelli, A.L. (2006) Interchange format for hybrid systems: abstract semantics, in *International Workshop on Hybrid Systems: Computation and Control*, Springer, pp. 491–506.

89 Agut, D.E.N., van Beek, D.A., and Rooda, J.E. (2013) Syntax and semantics of the compositional interchange format for hybrid systems. *J. Log. Algebr. Program.*, **82** (1), 1–52.

90 Blochwitz, T., Otter, M., Akesson, J., Arnold, M., Clauss, C., Elmqvist, H., Friedrich, M., Junghanns, A., Mauss, J., Neumerkel, D., Olsson, H., and Viel, A. (2012) Functional mockup interface 2.0: The standard for tool independent exchange of simulation models.

91 MODELISAR Consortium and Modelica Association (2010) *Functional Mock-up Interface for Co-Simulation. Version* 1.0, Retrieved from www.fmi-standard.org.

92 Broman, D., Brooks, C., Greenberg, L., Lee, E.A., Masin, M., Tripakis, S., and Wetter, M. (2013) Determinate composition of FMUs for co-simulation, in *Proceedings f the ACM IEEE International Conference on Embedded Software*, IEEE Press, Piscataway, pp. 2:1–2:12.

93 Bajaj, N., Nuzzo, P., Masin, M., and Sangiovanni-Vincentelli, A.L. (2015) Optimized selection of reliable and cost-effective cyber-physical system architectures. *Proceedings of the Design, Automation and Test in Europe.*

94 Hang, C., Manolios, P., and Papavasileiou, V. (2011) Synthesizing cyber-physical architectural models with real-time constraints. *Proceedings of the International Conference on Computer-Aided Verification.*

95 Piterman, N., Pnueli, A., and Sa'ar, Y. (2006) Synthesis of reactive(1) designs, in *Proceedings of the Verification, Model Checking, and Abstract Interpretation*, Springer, Berlin, pp. 364–380.

96 Kloetzer, M. and Belta, C. (2008) A fully automated framework for control of linear systems from temporal logic specifications. *IEEE Trans. Automat. Contr.*, **53** (1), 287–297.

97 Kress-Gazit, H., Fainekos, G., and Pappas, G. (2009) Temporal-logic-based reactive mission and motion planning. *IEEE Trans. Robot.*, **25** (6), 1370–1381.

98 Wongpiromsarn, T., Topcu, U., Ozay, N., Xu, H., and Murray, R.M. (2011) TuLiP: a software toolbox for receding horizon temporal logic planning, in *Proceedings of the International Conference on Hybrid Systems: Computation and Control*, ACM, New York, pp. 313–314.

99 Ramadge, P. and Wonham, W. (1989) The control of discrete event systems. *Proc. IEEE*, **77** (1), 81–98.

100 Cassandras, C. and Lafortune, S. (2008) *Introduction to Discrete Event Systems*, SpringerLink Engineering, Springer.

101 Maasoumy, M., Nuzzo, P., Iandola, F., Kamgarpour, M., Sangiovanni-Vincentelli, A., and Tomlin, C. (2013) Optimal load management system for aircraft electric power distribution. *International Conference on Decision and Control*, pp. 2939–2945.

102 Raman, V., Donze, A., Maasoumy, M., Murray, R.M., Sangiovanni-Vincentelli, A., and Seshia, S.A. (2014) Model predictive control with signal temporal logic specifications. *International Conference on Decision and Control.*

103 Raman, V., Donzé, A., Sadigh, D., Murray, R.M., and Seshia, S.A. (2015) Reactive synthesis from signal temporal logic specifications, ACM, New York, *HSCC '15*, pp. 239–248. doi: 10.1145/2728606.2728628, URL http://doi.acm.org/10.1145/2728606.2728628.

104 Rungger, M., M., MazoJr, and Tabuada, P. (2013) Specification-guided controller synthesis for linear systems and safe linear-time temporal logic, in *Proceedings of the International Conference on Hybrid Systems: Computation and Control*, ACM, pp. 333–342.

105 Girard, A., Pola, G., and Tabuada, P. (2010) Approximately bisimilar symbolic

models for incrementally stable switched systems. *IEEE T. Automat. Contr.*, **55** (1), 116–126.

106 Mazo, M. Jr., Davitian, A., and Tabuada, P. (2010) PESSOA: A tool for embedded controller synthesis, in *Computer Aided Verification, LNCS*, vol. **6174**, pp. 566–569.

107 Mari, F., Melatti, I., Salvo, I., and Tronci, E. (2014) Model based synthesis of control software from system level formal specifications. *ACM T Softw. Eng. Methodol.*, **23** (1), Article 6. doi: 10.1145/ 2559934

108 Saha, I., Ramaithitima, R., Kumar, V., Pappas, G.J., and Seshia, S.A. (2014) Automated composition of motion primitives for multi-robot systems from safe LTL specifications. *Proceedings of the IEEE/RSJ International Conference on Intelligent Robots and Systems (IROS)*.

109 Bresolin, D., Di Guglielmo, L., Geretti, L., Muradore, R., Fiorini, P., and Villa, T. (2012) Open problems in verification and refinement of autonomous robotic systems. *Euromicro Conference on Digital System Design*, pp. 469–476.

110 Maler, O., Pnueli, A., and Sifakis, J. (1995) On the synthesis of discrete controllers for timed systems (an extended abstract). *STACS*, pp. 229–242.

111 Behrmann, G., Cougnard, A., David, A., Fleury, E., Larsen, K.G., and Lime, D. (2007) UPPAAL-Tiga: time for playing games! in *Proceedings of the International Conference on Computer-Aided Verification*, Springer, pp. 121–125.

112 Tomlin, C., Lygeros, J., and Sastry, S. (2000) A game theoretic approach to controller design for hybrid systems. *Proc. IEEE*, **88** (7), 949–970.

113 Balluchi, A., Benvenuti, L., Villa, T., Wong-Toi, H., and Sangiovanni-Vincentelli, A.L. (2003) Controller synthesis for hybrid systems with a lower bound on event separation. *Int. J. Control*, **76** (12), 1171–1200.

114 Benerecetti, M., Faella, M., and Minopoli, S. (2013) Automatic synthesis of switching controllers for linear hybrid systems: safety control. *Theor. Comput. Sci.*, **493**, 116–138.

16
Heterogeneous Systems

Daniel Lapadatu

Alfa Rom Consulting SRL, Str. Casin 8/9E, RO-012266 Bucharest, Romania

16.1
Introduction

In this chapter, heterogeneous system will be discussed to understand any microsystem composed of devices that cannot be integrated monolithically. Examples are all the microsystems that contain a number of technologically different parts: MEMS and NEMS for sensing, CMOS circuitry for processing, RF/wireless components for communication, photonics, PCB-s, mounting, encapsulation and packaging units, and so on.

In the past decades, the IC packaging technology has not been able to match the advances in IC design and technology, which continued to follow Moore's law. The continuous miniaturization of microelectronic components is a serious challenge to the traditional packaging and materials. Consequently, the mismatch in performance of the IC and the package is considered to be the leading limitation to system performance [1,2].

The heterogeneous systems offer the opportunity to integrate dissimilar devices and materials into 3D systems. By combining various semiconductor packaging technologies and silicon fabrication processes with optimized interconnects and packaging hierarchy, the heterogeneous systems benefit of smaller size, reduced footprint, lower cost, and overall improved electrical performance. Furthermore, the integration within the heterogeneous systems of fundamentally different materials, such as inorganic, organic, and biological systems, is expected to provide new opportunities for the development of new products based on new technologies and new system-level architectures.

This chapter will present specific issues and challenges of heterogeneous systems design, integration, and testing. As examples and case studies, three heterogeneous systems will be discussed in more detail. The first two are highly specialized products, manufactured in very low volumes (in the order of

Nanoelectronics: Materials, Devices, Applications, First Edition. Edited by Robert Puers, Livio Baldi, Marcel Van de Voorde, and Sebastiaan E. van Nooten.

Figure 16.1 Sensonor's STIM300.

thousand of units per year), while the latter is intended for high-volume production (millions of units per year).

The first example, illustrated in Figure 16.1, is STIM300, Sensonor's inertial measurement unit [3]. It is a small, tactical grade, low weight, high performance Inertial Measurement Unit (IMU). STIM300 contains, among others, three MEMS angular rate sensors, integrated as bare dies, their analog and digital ASICs, integrated also as bare dies, three MEMS accelerometers, encapsulated with their ASICs, two 2-axis non-MEMS inclinometers, also encapsulated with their ASICs, various microelectronic ICs and several discrete components, mounted on flex-rigid PCBs and housed in a rigid, custom-made aluminum case. The IMU is factory-calibrated and compensated over its temperature operating range. STIM300 can be further integrated in even higher level, complex system, such as iMAR's iATTHEMO-C inertial reference [4].

The second example, illustrated in Figure 16.2, is Sensonor's SAR500 angular rate sensor [5]. It is a high-precision, low-noise, high-stability, calibrated, and compensated digital oscillatory gyroscope. SAR500 contains a MEMS angular rate sensor die and an analog ASIC, each housed in rigid custom-made ceramic packages soldered on top of each other, several discrete components, metal lids and a protective plastic cap. A digital ASIC contains the needed control and functional algorithms to achieve the superior performance. The device is factory-calibrated and compensated for temperature effects to provide high-accuracy digital output over a broad temperature range.

Referring to Figure 16.2, the main components of the system are as follows:

1) MEMS angular rate sensor (bare die)
2) 16 pin side brazed ceramic package for housing the angular rate sensor
3) Metal lids for hermetical encapsulation of the angular rate sensor die
4) Analog ASIC (bare die)
5) 36 pin grid array ceramic package for housing the analog ASIC die
6) Metal lid for hermetical encapsulation of the analog ASIC die
7) Discrete components (encapsulated capacitors)
8) Protective plastic cap

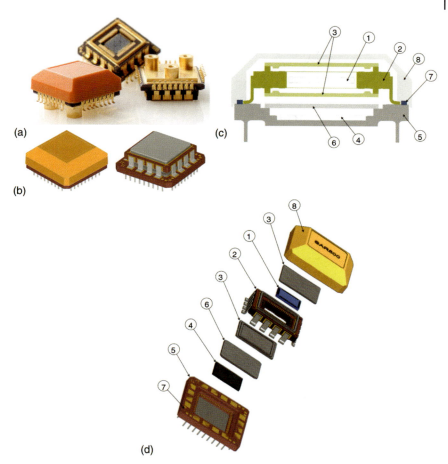

Figure 16.2 Sensonor's SAR500. (a) Product view of an earlier version; (b) schematic view, with and without the protective plastic cap; (c) schematic cross-sectional view; (d) exploded view.

The third example, illustrated in Figure 16.3, is poLight's Packaged TLens® [6]. The Packaged TLens is a tunable MEMS lens, whose active optical components are based on deformable polymers integrated onto a silicon chip and housed in a customized plastic package. It has no moving parts and is ideally suited for designing small form autofocus camera modules. Some of the unique capabilities of the Packaged TLens include: high optical axis stability, constant field of view, extremely quick autofocus – enabling to take several images and combine all into one, touch and refocus, and "all-in-focus" images in full resolution.

Referring to Figure 16.3, the main components of the system are as follows:

1) Glass back window (bare glass die)
2) Polymer optic (elastomer)

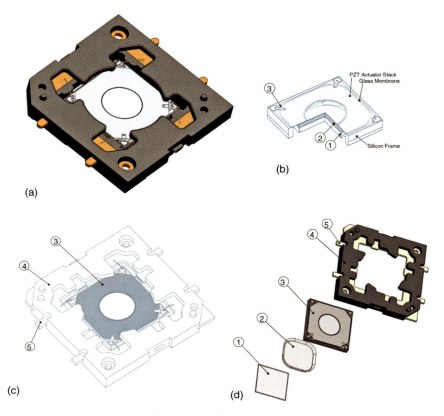

Figure 16.3 poLight's Packaged TLens®. (a) Product view (mounted on a camera module); (b) schematic cross-sectional view of TLens® actuator; (c) schematic view of Packaged TLens®; (d) exploded view.

3) MEMS actuator (a silicon frame with a glass membrane and a PZT actuator stack)
4) Plastic package
5) Metal inserts

16.2
Heterogeneous Systems Design

As in the case of microsystems, the design of heterogeneous systems requires several different levels of description and detail [7,8]. In general, it involves three mutually coupled tasks, as illustrated in Figure 16.4.

*   *Design considerations*, in which the designer must evaluate the design constraints, the available devices, technologies and materials, the different fabrication and assembly methods and, eventually, the expected manufacturing costs.

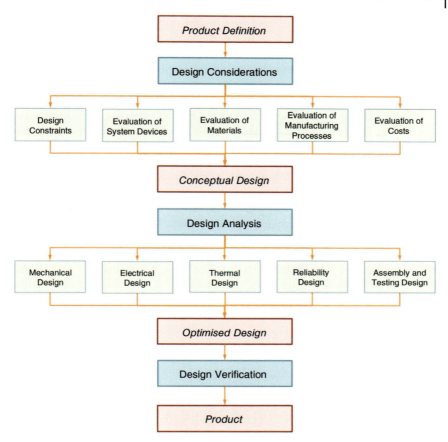

Figure 16.4 Overview of heterogeneous system design.

- *Design analysis*, in which the designer partitions the system into components, analyses each proposed approach and selects the signal mapping and transduction method, materials, technologies, fabrication sequence for each component, methods for packaging and assembly.
- *Design verification*, consisting of system assembly, packaging, testing, calibration, and compensation.

16.2.1
Design Considerations

Regardless of its category, a technology demonstrator, a research tool, or a commercial product, once the product is properly defined by a list of specifications, the designer must first document and evaluate the design constraints.

Although they tend to vary from case to case, most of these constraints arise from marketing considerations: customer demands, market size, product impact on the market, time to market, strength of competition, environmental conditions, size and weight limitations, type of intended applications, availability of manufacturing technologies and facilities, costs. After a proper evaluation of all the above-mentioned elements, a *conceptual design* should emerge.

In the considered examples, STIM300 and SAR500 are cost and size effective alternatives for systems that otherwise would use fiber optic gyroscopes. With their small size and weight, and a robust design for tough environments, achieving the targeted performance has been the most challenging constraint.

In contrast, the Packaged TLens, with its simple and high performance autofocus technology, is produced on a wafer-scale level in very large volumes, high yield and high quality. It is a suitable alternative to the current voice coil actuator used in the camera modules of smartphones. The solution offered by the Packaged TLens improves the easiness of integration in camera module, increases the speed, lowers the power consumption, and reduces the footprint. By targeting the smartphone market, the size, performance, robustness, extremely low power consumption, and cost are the most relevant constraints, with the latter being by far the most challenging one.

16.2.2
Design Analysis

During this phase, the designer must first partition the system into components and then consider the modeling of various, complex multiphysics phenomena, then explore the design-space and optimize the synthesis process, as illustrated in Figure 16.5.

In contrast to microsystems, the design of heterogeneous systems presents specific challenges arising from the following:

- Physical heterogeneity: more than one physical domain is present
- Technology heterogeneity: more than one technology is present
- Abstraction heterogeneity: more than one level of description is needed

Consequently, the partition of the system and mapping the system functions and specifications to the underlying blocks and/or components are themselves extremely difficult tasks. In fact, one fundamental issue of the heterogeneous system design is how to represent the various physical phenomena at abstraction levels that match the complexity of the problem to be solved [9].

The design-space exploration and optimization processes, driven by the system specifications and requirements, involve the following individual, interdependent but potentially parallelizable design tasks:

- *Mechanical design*, including structural design
- *Electrical design*, including signal distribution, power distribution, electromagnetic interference (EMI), skin effect, noise, time delay, crosstalk, and so on

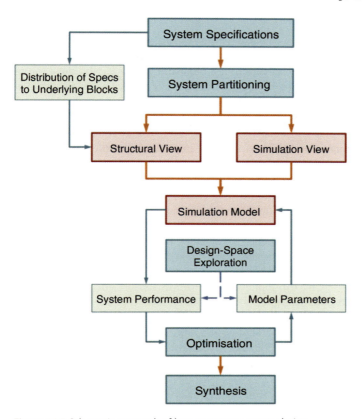

Figure 16.5 Schematic approach of heterogeneous system design.

- *Thermal design*, including thermal management and cooling methods
- *Reliability design*, including subsystem failure analysis and failure mechanisms
- *Assembly and testing design*, in which provisions for cost-effective assembly and testing are made.

Nowadays, the design-space exploration and optimization processes of any complex system are carried out almost exclusively by means of finite element method (FEM). The theories and formulation of the FEM are available in a large number of reference publications [8,10,11] and have been commercialized in many popular software products, such as ANSYS, COMSOL, or CoventorWare.

Due to the physical, technology, and abstraction heterogeneities, some blocks/ partitions may not be simulated together, they may require different simulation tools and design flows. This situation is usually solved by introduction of ports to propagate the I/O properties after they have been converted into compatible formats. The same approach allows also for running coupled electromechanical analysis, thermomechanical analysis, and dynamic analysis at system level.

16.2.2.1 **Mechanical Design**

The main objective of the mechanical design is to ensure the structural integrity and reliability of the system when it is subjected to specified loading at both normal operating and overload conditions. Exposure to shock, vibrations, variable and/or nonuniform temperature, and humidity are the typical environmental loads that affect the performance of the system.

The performance of many existing devices is degraded by the presence of nonuniform characteristics and built-in stress that can cause unintended sensitivity to external mechanical and thermal loads or unpredictable drift of characteristics. Although this problem is usually addressed directly at device level by use of stress-release structures and pedestals [12–14] and/or new fabrication methods [15], the system designer will need to ensure that the mechanical loads are kept as low, as constant, and as uniform as possible.

To prevent the negative effects of exposure to shock and vibrations and the development of undesired resonance frequencies, the system must be fitted with proper dampers and shock absorbers.

Since the variation of temperature or humidity is known to cause considerable deformation of the PCB [16], mounting the MEMS sensors and ASICs directly on PCBs, although much cheaper, is not recommended. Instead, at system level, the stress decoupling between of the sensitive MEMS parts is achieved by employing at least one stress-decoupling substrate, properly configured, as illustrated in Figure 16.6 [17–19]. The stress-decoupling substrate can be anything suitable for the considered application, from structured silicon beams to metal brackets, fixtures, and lids.

Further, stress decoupling is achieved by making use of small-sized pedestals or posts implemented either on the device itself, as illustrated in Figure 16.7a, or on the substrate, as illustrated in Figure 16.7b. In addition to stress release, these pedestals and posts provide die attach area control and stable reference alignment plane. This is particularly of high relevance in products using accelerometers, inclinometers, and angular rate sensors that need to be accurately aligned to the predefined reference axes and planes and hold this alignment for their lifetime.

In the case of SAR500, as shown in Figure 16.8, the decoupling between the package and the angular rate sensor die has been achieved by means of a double-side, fully symmetrical suspension of 75-μm thick gold wires, which in this particular product serves electrical, mechanical, and thermal purposes [21]. The wires had to be dimensioned as to maintain the sensing die in its reference position even after it has been exposed to the maximum acceptable shock and vibration levels. A compromise had to be negotiated between the thickness of the wires on one hand, and their cost and the risk of cracking the underlying substrate during the wire bonding on the other hand.

In the case of the Packaged TLens, which is driven by the high optical performance and very low cost constraints, a simpler solution has been chosen to achieve the stress-decoupling between the package and the device: an elastic glue, specifically selected and further optimized for this purpose, has been used

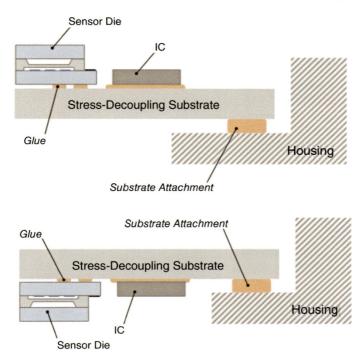

Figure 16.6 Generic configuration for stress-decoupling at system level.

to attach the actuator component to four, off-center positioned, small protrusions of the package side walls, as shown in Figure 16.9. In this configuration, any packaging-related stress will result in a harmless rotation of the actuator die around its optical axis, rather then bending and tilting out of plane and ruining the optical alignment of the product.

In addition, four cantilevers have been employed to limit the vertical movement of the actuator when subjected to extreme shock levels, thus considerably improving the shock resistance of the product.

16.2.2.2 Electrical Design

The role of the electrical design is to ensure that the system includes suitable communication paths for signals and suitable channels for power distribution. In addition to mapping the signals, the system designer needs to consider the following at system level [22,23]:

- The parasitic capacitances associated with balls, vias, wires, ground, and power planes
- The parasitic inductances associated with wires, coupled lines, ground, and power planes
- Parasitic charging of dielectrics

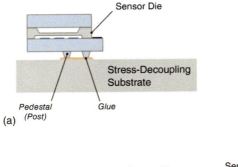

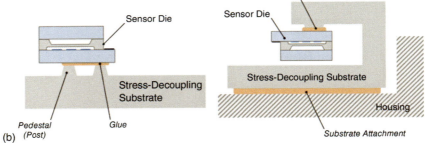

Figure 16.7 Use of pedestals or posts for stress-decoupling and die attach area control. (a) Pedestals implemented on the die. (Adapted from Ref. [20]) (b) Pedestals or posts implemented on the substrate. (Adapted from Ref. [19].)

Figure 16.8 The stress-decoupling solution implemented in SAR500: double-side suspension by means of thick Au wires.

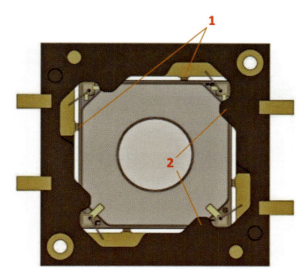

Figure 16.9 Stress-decoupling and shock absorbing solutions implemented in the Packaged TLens®: (1) glue dispensed on the side walls of the package and actuator for elastic, low stress, mechanical coupling; (2) shock absorbing, plastic cantilevers for limitation of vertical movement.

- The noise associated with parasitic capacitances and inductances, crosstalk, and the various types of contacts;
- The transmission lines, in those cases when the devices operate at very high frequencies
- The risk of electrostatic discharge
- The risk of electromagnetic interference, both from external and internal sources

Figure 16.10 shows a cross-sectional view that illustrates some of the parasitic elements associated with the ceramic package of SAR500. Beside the obvious request of minimizing the parasitic capacitances and – if the case – inductances, for devices using differential capacitive sensing and actuating, additional care has to be taken to *match* the parasitics along the paired signal ways.

For the systems containing MEMS devices, some specific electrical issues have to be carefully considered and addressed during the design phase.

Parasitic Charging of Dielectrics

The parasitic charging of dielectric surfaces, as well as the migration of bulk charges in some insulating materials, such as glass, are major factors that limit the short- and long-term stability and the reliability of capacitive systems [24,25].

To avoid the electrostatic charging of the dielectric surfaces caused by external or internal sources, custom-made packages with metal lids, guard rings and/or

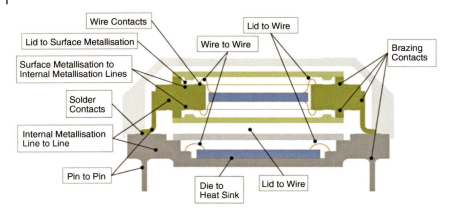

Figure 16.10 Example of package parasitic elements (capacitances, inductances, and resistances) inside SAR500.

screening electrodes have to be used. In addition, periodic switching the polarity of the applied DC signals, if implemented, substantially improves the long-term stability of the electrical characteristics of the system [5].

In the particular case of MEMS devices, adding an outer surface metallization to the sensor die and/or using electrically conductive glues to attach the die on the substrate, as shown in Figure 16.11, considerably improves the performance of the system.

Electrostatic Discharges (ESD)

At system level, air breakdown can easily occur between the sharp corners of electrodes or pads and nearby conductors, fixtures or even the chipped edge of the bare silicon dies. ESD can also cause the melting and erosion of the involved metal parts.

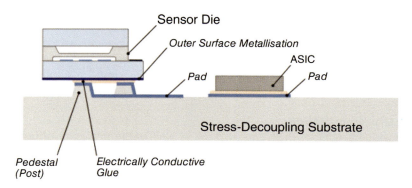

Figure 16.11 Generic arrangement for improving the bias stability of MEMS-based systems.

Strict adherence to layout and design rules, use of rounded corners and sufficient space between conductors, the insertion of guarding rings or screens safe guards against ESD.

Electromagnetic Interference (EMI)

In particular for IMUs, for their best product performance, the three orthogonal accelerometers need to be placed as close as possible to each other. However, if too close, there will almost certainly be some level of electromagnetic interference between the accelerometers' ASICs. Insertion of electromagnetic screens between ICs and components with embedded ICs is good practice.

16.2.2.3 **Thermal Design**

All microelectronic devices require power for their use. Following the tremendous progress in miniaturization, the power density in todays electronic components, if not properly managed, may result in component failure.

As will be described in the next section, higher operating temperatures is one of the largest contributor to the loss of reliability in components. Failure rate increases nearly exponentially with temperature. As such, an increase of 10 °C is expected to double the component failure rate. Therefore, the thermal design and the thermal management are critical to the reliability of the system.

The thermal design must consider the fundamental heat transfer mechanisms – conduction, convection, and radiation, all playing major roles in the cooling of the electronic components and the reduction of thermally induced noises. To avoid the development of undesired stress and strain in the system, the designer will also need to ensure that temperature and humidity are hold as constant and uniform as possible.

In order to cool down effectively the microcomponents, the designer must ensure that the generated heat is rapidly transported within the system and evacuated from the system. Including *thermal vias*, *heat sinks* is the most common and effective approach toward achieving the heat removal.

In particular for the MEMS-based systems, the thermal design must also minimize the temperature difference between the MEMS structure and the ASICs, since it may cause wrongful reading of the sensor temperature within the compensation schemes.

In the case of SAR500, the high thermal efficiency of the product is achieved by using thick gold wires to connect the sensor element to its package, then by using a large number of dedicated thermal vias and metal layers within the packages, metal lids, and metal seal rings for encapsulations and, finally, a thick CuW heat sink in direct contact to the product's ASIC, as illustrated in Figure 16.12. The thermal design of SAR500 also achieves excellent temperature uniformity within the angular rate sensor element.

In the case of the Packaged TLens, the thermal efficiency of the product is improved by the insertion of two thick metal plates, serving also as electrodes, located within the plastic package.

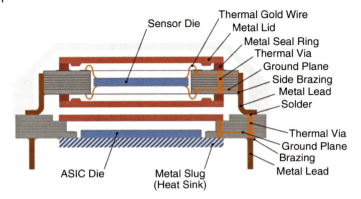

Figure 16.12 The thermal elements of SAR500. For simplicity, only the elements located in the right hand side half of the system are shown.

16.2.2.4 Reliability Design

Reliability is the ability of a system or component to perform, under stated conditions and for a specified period of time, the functions for which it was designed. To ensure that the system is reliable one needs to

- design up-front for reliability, stage known as *reliability design* and
- conduct accelerated tests, stage known as *reliability testing.*

Although the effects are felt at the system level, failure mechanisms occur at the lowest component level. The design for reliability aims to identify, understand, and prevent the underlying failures even before the system is built.

The cause of the failures may be thermal, mechanical, electrical, chemical, or a combination of these. Figure 16.13 shows the typical failure mechanisms in microsystems. Here, the temperature is considered to only accelerate and amplify the mentioned failure mechanism, without being itself a primary failure mechanism. As illustrated, the failure mechanisms fall into the following two categories:

- *Overstress mechanisms*, in which the stress within a single event exceeds the strength or capacity of a single component and causes its failure.
- *Wear-out mechanisms*, in which a gradual exposure to lower stress levels, within the strength and capacity of the component, but over an extended period of time, results in cumulative damage that eventually leads to component and, subsequently, system failure.

The unwanted effects at system level are not only the results of some primary causes but also of circumstances. While the primary causes are extremely difficult to control, the circumstances can be avoided by design. As an example, while the formation of cracks cannot be avoided, one can take care by design

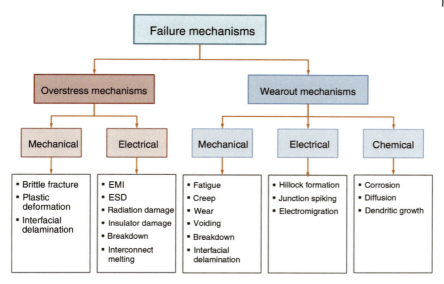

Figure 16.13 Typical failure mechanisms in heterogeneous systems.

that the affected surface is always in compression, thus preventing the further development of the crack and the subsequent structural failure.

As an example of failure after system assembly, Figure 16.14 shows a crack caused by excessive mounting mechanical stress within the 36-pin grid array ceramic package of SAR500.

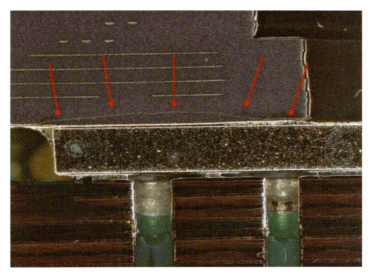

Figure 16.14 A crack, indicated by the red arrows, within the 36-pin grid array ceramic package of SAR500.

In general, the design for reliability may be achieved by either

- reducing the exposure to stresses that cause the failure or
- increasing the strength of the component.

The particular actions toward improving the reliability of the systems are usually taken as part of the mechanical, electrical, or thermal designs.

Specific to reliability design for heterogeneous systems is the fact that different failure mechanisms are dominant in different components, which may or may not interact at system level. Therefore, it is not always possible to design against all failure mechanisms. More than that, since the design against a specific failure mechanism may amplify another failure mechanism, careful analysis is always required to achieve the desired system-level reliability.

16.2.3
Assembly and Testing Design

Apart from the functional aspects of the product, the assembly of the heterogeneous system must also be considered during the design phase. Provisions for cost-effective assembly and testing have to be made as earlier as possible.

The assembly of the heterogeneous systems contains the following generic steps, which may in principle be fully automated, whose impact on the system design need to be considered:

- Part feeding
- Part grasping
- Part mating and alignment
- Part bonding and fastening
- Sensing and verification

Special care has to be taken to provide alignment features on all relevant components, without which the assembly of the system would not be possible.

Considering the example of SAR500, areas needed for grasping and proper alignment of the sensor die and the ceramic packages have been reserved on all surfaces, including the lateral ones, as illustrated in Figure 16.15. The reference areas those are located on the top surface of the sensor die ceramic package or those located between the reference points A1, A2, and B on the lateral surfaces of the ASIC die ceramic package, must be free of any contamination and chipping and must have a superior surface roughness, for precise optical orientation measurements.

Considering the example of the Packaged TLens, illustrated in Figure 16.16, features that are used for automated part feeding, grasping, and alignment had to be incorporated in the actual product.

The assembly of heterogeneous microsystems presents major challenges to engineers in terms of reliability and cost. Some of the reasons are as follows:

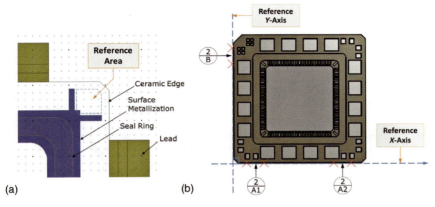

(a)   (b)

Figure 16.15 Features required for assembly of SAR500. (a) Reference area on the sensor die ceramic package, reserved for the alignment of the sensor die and the packages. (b) The reference X- and Y-axis of the ASIC package, seen from the top side.

- Lack of standards. The multiple levels of heterogeneity imply that the products are assembled according to very specific, highly customized procedures.
- Lack of full automatization, though robots can be developed in particular cases.
- Requires reliable alignment equipment.
- Lack of established methodology in setting proper tolerances of parts and assembly.

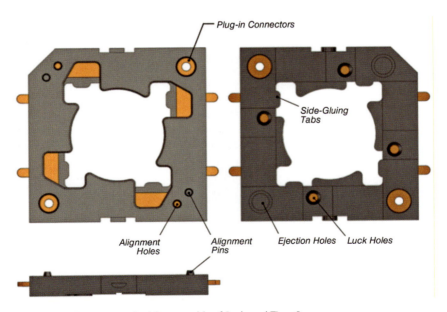

Figure 16.16 Features required for assembly of Packaged TLens®.

Figure 16.17 Example of a customized jig and test station used for the simultaneous testing of 32 STIM modules (From Ref. [26].)

Considering the above-listed reasons, it is understandable that, in the case of heterogeneous systems, in parallel to the actual system design, specific assembly and testing tools and accessories also need to be designed and developed. They may include among others feeders, grippers, aligners, probers, trays, carriers, and jigs.

Figures 16.17–16.19 show some of the STIM300-, SAR500-, and TLens-specific jigs, used for cost-effective, semi-automatic assembly, testing, and calibration of the products. The design and development of such jigs may be at times as challenging as the product itself.

16.3
Heterogeneous Systems Integration

As in the case of any system, the heterogeneous system integration deals with the process of bringing together the various components into one system and ensuring that the subsystems function together as a coordinated whole. The integration process consists of succession of packaging and assembly steps.

Not counting the bare chips/dies, which is usually considered as Level 0 (L0), there are 3 levels of heterogeneous system packaging hierarchy as followed:

- (L1) *Die-level* packaging, also known as *single-chip packaging*
- (L2) *Device-level* packaging, in which several chips/dies, and/or components are packed together, also known as *multichip packaging*
- (L3) *System-level* packaging

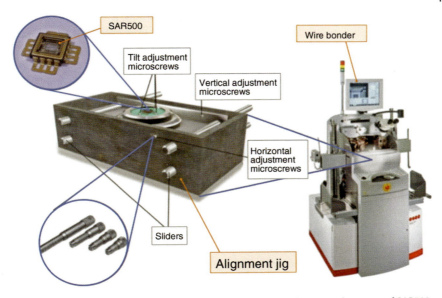

Figure 16.18 Example of a customized jig used for assembly and precise alignment of SAR500 component parts prior to and during double-sided wire bonding.

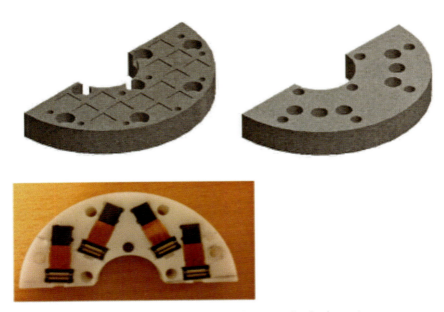

Figure 16.19 Examples of TLens®-related jigs used for testing the shock margins.

What is probably specific to all heterogeneous systems is their 3D aspect of the integration process. Therefore, it is not unusual that, due to the variety of technologies involved, the integration of heterogeneous systems includes simultaneously several of the following typical 3D-integration approaches [27]:

- Stacking of packages
- Package on package (PoP)
- Package in package (PiP)
- Wire bonded chip stack
- 3D chip-stack with TSV stacking
- Stacking of embedded dies by wafer-level packaging without TSVs
- Die-to-wafer assembly, including die-to-wafer with die cavity
- Wafer-to-wafer assembly, including wafer-level bonding technologies as anodic bonding, silicon fusion bonding, low-temperature surface bonding, SLID, and eutectic bonding [28].

Within all electronic systems, heterogeneous or not, the components have to be mechanically and electrically connected together. This could be achieved not only by conventional semiconductor packaging technologies [29], such as wire bonding, soldering and gluing, but also by more advanced system-level interconnect technologies that make use of interposers [27,30,31] or even of nanowire-polymer combinations and flip chip assembly on flexible and rigid substrates [32].

Following are three concept categories of 3D-integration [27], illustrated in Figure 16.20:

- Stacking of packages and/or substrates
- Stacking of dies, bare, or embedded, without use of TSV
- 3D-TSV technology

Regardless of the chosen method and category, the 3D-integration increases the packaging density, reduces the delays caused by long wires, and improves the interconnect bandwidth. It is worth noting that, in spite of their lack of maturity, the recent trend in the microsystem integration, has been to achieve and migrate toward 3D-vertical stacking by using TSV interconnections [30].

Although they are expected to be overcome in the near future, when intended for use in more demanding products, the current 3D-TSV technologies have the following drawbacks that need special consideration:

- Stress-induced and contamination-induced degradation of mounted components directly caused by the presence of TSV.
- Reduced electrical reliability caused by electromigration and delamination within the TSV interposers.
- Challenging testability of subsystems.

Since the heterogeneous systems contain components fabricated in different technologies, their integration is heavily limited by a string of constraints, such

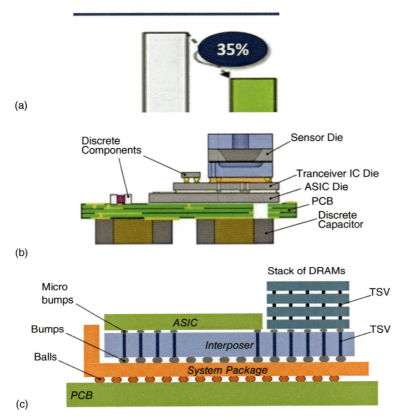

Figure 16.20 The concept categories of 3D-integration. (a) Stacking of packages [33]; (b) stacking of dies and discrete components (Adapted from Infineon Technologies and Ref. [34,35]) (c) Stacking of dies by means of 3D-TSV (Adapted from Amkor Corp. and Ref. [36].)

as low assembly temperature, low post-assembly stress (induced by TSVs, adhesives, or soldered joints), low shock resistance of fragile MEMS/NEMS devices.

Considering the integration of Packaged TLens, schematically shown in Figure 16.21, the main limitations arise from the fact that any operation shall minimize the stress applied to the optical MEMS that may result in the degradation of the optical performance. For example, the assembly processes must at no stage exceed 90 °C, which puts serious constraints on the choice of wire bonding material and process. The choice of the adhesive that mechanically attaches the package to the actuator is even more restricted. Apart from the temperature limitation, the resulting post-assembly stress had to be kept within very narrow limits over the entire operating temperature range in order to obtain, as intended, a nearly flat glass membrane. A completely new adhesive, and its dispensation and curing, had to be specifically developed for this product.

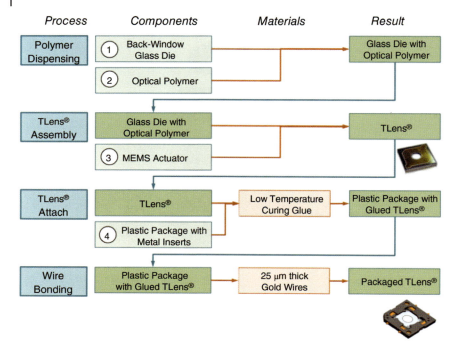

Figure 16.21 The integration sequence of the Packaged TLens®. The numbers refer to the component ID in Figure 16.3.

In the case of SAR500 integration, schematically shown in Figure 16.22, the bonding of the 75 μm thick Au wires, whose extreme thickness was required for mechanical and thermal reasons, on pads located on glass substrates has been extremely challenging from technological point of view. The subsequent welding of the lids proved also to be extremely difficult. Welding requires very high temperatures concentrated locally, but the system has been designed to evacuate fast and efficiently the heat, as described in the previous sections.

16.4
Testing the Performance and Reliability of Heterogeneous Systems

Due to their nature, the heterogeneous systems contain ready-fabricated components that are subsequently assembled together into a stand-alone product. These component parts have been already selected, tested for functionality and quality and delivered by their manufacturers. However, it is not unusual that after their assembly into the final product some of the component parts underperform or, in extreme situation, even fail. Briefly put, testing at chip- or component-level does not guaranty a functional heterogeneous system, a situation that results in extremely low yields and prohibitive costs.

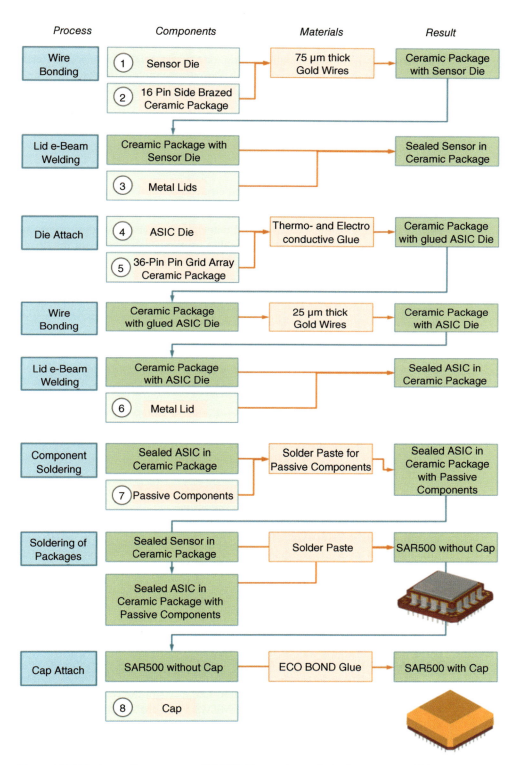

Process	Components	Materials	Result
Wire Bonding	① Sensor Die ② 16 Pin Side Brazed Ceramic Package	75 µm thick Gold Wires	Ceramic Package with Sensor Die
Lid e-Beam Welding	Creamic Package with Sensor Die ③ Metal Lids		Sealed Sensor in Ceramic Package
Die Attach	④ ASIC Die ⑤ 36-Pin Pin Grid Array Ceramic Package	Thermo- and Electro conductive Glue	Ceramic Package with glued ASIC Die
Wire Bonding	Ceramic Package with glued ASIC Die	25 µm thick Gold Wires	Ceramic Package with ASIC Die
Lid e-Beam Welding	Ceramic Package with ASIC Die ⑥ Metal Lid		Sealed ASIC in Ceramic Package
Component Soldering	Sealed ASIC in Ceramic Package ⑦ Passive Components	Solder Paste for Passive Components	Sealed ASIC in Ceramic Package with Passive Components
Soldering of Packages	Sealed Sensor in Ceramic Package Sealed ASIC in Ceramic Package with Passive Components	Solder Paste	SAR500 without Cap
Cap Attach	SAR500 without Cap ⑧ Cap	ECO BOND Glue	SAR500 with Cap

Figure 16.22 The integration sequence of SAR500. The numbers refer to the component ID in Figure 16.2.

When it comes to the reliability testing for heterogeneous systems, it is worth noting that multiple components with various technology backgrounds and materials will exhibit various short- and long-time behavior, resulting in a complicated picture at the system-level.

The degradation of performance at system-level is in the vast majority of cases caused by the following:

- Assembly-related stress, mechanically or thermally generated
- Humidity
- Electrostatic charging of insulators and electromigration of charges

Considering STIM300, some of its high-performance parameters rely on the perfect matching of the excitation and detection frequencies of the resonating structure within the angular rate sensor dies. Any mismatch between the drive-mode and detection-mode oscillators rapidly degrades the element sensitivity to angular rates [37]. Since the mechanical stress affects differently the two resonance frequencies, the system will not meet its specifications after mounting the angular rate sensor dies in STIM300 if not for the careful design of the stress-decoupling substrate and the glue dispensing process.

Considering the Packaged TLens, if no precautions were taken by design, the stress transferred from the package to the actuator during the assembly process or during regular operation as a result of temperature variations and heterogeneity of materials would seriously degrade the optical performance of the device.

In both examples it is very easy and cost efficient to test the components that go into the system, but this does not guaranty that the product will meet its specifications, hence the conundrum of the heterogeneous systems: expensive testing at system-level is still needed and low yields are highly probable. Although high-volume testing may be carried out on expensive, customized tools, it is usually slow and requires complex compensation routines for various effects arising from the heterogeneity of the system.

The high-volume calibration and compensation of the heterogeneous systems are notoriously difficult tasks. To start with, the heterogeneity of materials and functionalities among the components results in nonuniform temperature distribution inside the system, even in the case of stable environmental conditions. Referring in particular to MEMS gyroscopes or MEMS-based IMUs, the temperature in the MEMS sensing elements and the signal controlling and processing ASICs is different, even with excellent thermal design. Therefore, the schemes employed to compensate the effects of the environment temperature are very complex and extremely difficult to implement in practice, resulting in long, slow, and expensive calibration routines.

Another specific aspect related to testing and calibration of the heterogeneous systems is rather their large size compared to the size of the constituent components. This means that, in the case of MEMS-based IMUs, the position of the subcomponents directly affects the system performance; the calibration routines must take this into account.

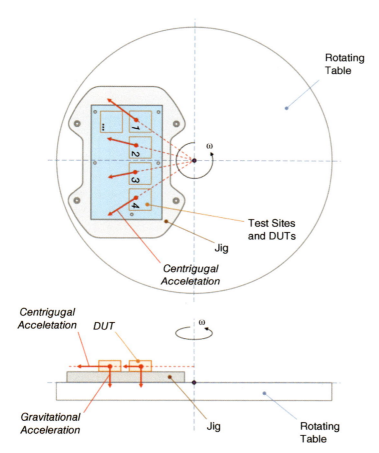

Figure 16.23 Generic arrangement for alignment of axes of several IMUs simultaneously.

As an example, let's have a look at the undesired effects arising from the existence of centrifugal and gravitational accelerations during the calibration of the angular sensors. Figure 16.23 shows a generic arrangement used to perform the alignment of axes of several IMUs, simultaneously. Due to the different gyration radii and the nonzero sensitivity to linear acceleration of all inertial devices, the angular rate sensors, three for each IMU, will be exposed to different levels of centrifugal and gravitational accelerations and, consequently, their output will be affected in a different manner, as shown in Figure 16.24. Note that the error for the Z-channel depends also on the test site in which the device has been loaded.

To remove these undesired effects, which if left uncompensated will result in a product that does not meet its accuracy specifications, a compensation scheme has to be implemented. The schemes must take into account the geometry of

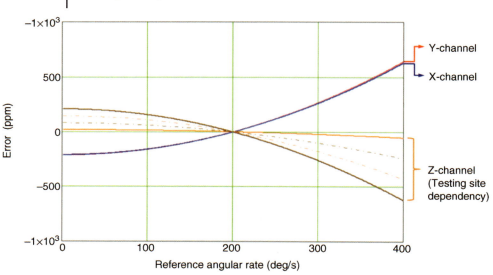

Figure 16.24 Computed systematic angular rate errors, as a fraction of full-scale output, caused by the linear accelerations present during the calibration routines for the generic arrangement shown in Figure 16.23.

the product, jig and rotation table, the location of the components within the system, as well as the log files containing the information of applied settings for the angular rates and temperature.

For instance, to remove the effects of the exposure to centrifugal and gravitational acceleration, first the values and orientations of these accelerations are computed from log files using basic physics and geometry, as schematically shown in Figure 16.25. Then the result is used in automated routines that subtract these effects from the actual measurement data, as schematically shown in Figure 16.26.

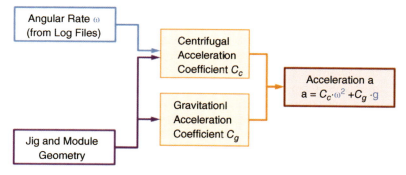

Figure 16.25 Generic method of computing the linear acceleration levels during the calibration of angular rate sensors.

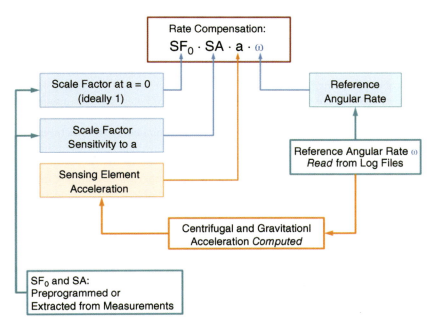

Figure 16.26 Generic scheme for compensating the effects of linear accelerations during the calibration of the angular rate sensors in IMUs.

Apart from the above-mentioned compensation of the linear accelerations effect on scale factor calibration of the angular rate sensors, there are similar complex schemes required to compensate the bias sensitivity to linear accelerations, the die-to-die bias and scale factor sensitivity to linear accelerations, and temperature and the cross-axis sensitivity to angular rates. All these results were put together in a very long and slow calibration procedure, performed on highly customized, expensive tools and equipment.

16.5
Conclusions

The emerging industry capabilities in advanced assembly and 3D chip stacking, including the TSV as the most promising one, allow the integration of components with different technological background and made out of dissimilar materials to manufacture new products based on new system-level architectures. For example, heterogeneous integration of III–V semiconductor materials with silicon-based technologies will enable the development of systems that incorporate optical sources and detectors and extremely high-speed electronics.

Additionally, the emergence new materials, such as graphene or flexible plastics and glasses that may be directly used as new sensing and/or circuit elements will be extremely important in the next decade.

As already mentioned, the migration from 2D architectures and assembly to 3D-integration opens up a window for new functionalities and applications. This is expected to be especially true in soft materials, flexible or stretchable electronics, smart sensors, distributed sensors, and nanoparticles for chemical and biological applications [38].

Another trend is the one represented by the field of heterogeneous integration for biological systems, involving the development of interfaces between biological media and electronic devices. New materials, combined in 3D structures, and integrated with measurement circuitry will provide new opportunities in the field of medical sciences.

Apart from understanding and controlling the properties of complex materials, the nanotechnology researchers will need to be able to interconnect these nanostructures and nanoparticles at scales where quantum effects can no longer be ignored. New fabrication methods and manufacturing equipment required for system integration need to be invented and developed into a reliable fabrication infrastructure.

Acknowledgments

The author would like to thank Sensonor AS, poLight AS, and Fraunhofer Research Institution for Modular Solid State Technologies EMFT for their kind support and permission to reprint the proprietary information and materials within this chapter.

References

1 Duan, C. and LaMeres, B.J. (2010) *On and Off-Chip Crosstalk Avoidance in VLSI Design*, Springer.

2 Davis, J.A. and Meindl, J.D. (2012) *Interconnect Technology and Design for Gigascale Integration*, Springer.

3 Sensonor (2016 STIM300, product brief. Available at www.sensonor.com.

4 iMAR (2016) iATTHEMO-C true north finding high precision heading, attitude, position & velocity reference based on dual antenna GNSS & MEMS IMU data fusion.Available at www.imar-navigation .de.

5 Lapadatu, D., Blixhavn, B., Holm, R., and Kvisterøy, T. (2010) SAR500 - A high-precision high-stability butterfly gyroscope with north seeking capability, Position Location and Navigation Symposium (PLANS), p. 6.

6 Polight (2016) Packaged TLens, product brief, Piezoelectrically actuated optical lens . WO/2016/009079. Available at www.poLight.com

7 Senturia, S.D. (2002) *Microsystem Design*, Kluwer Academic Publishers.

8 Hsu, T.R. (2002) *MEMS & Microsystems*, McGraw-Hill.

9 Labrak, L. (2009) Heterogeneous system design platform and perspectives for 3D integration, International Conference on Microelectronics - ICM, p. 161.

10 Hsu, T.R. (1986) *The Finite Element Method in Thermomechanics*, Allen & Unwin.

11 Pepper, D.W. and Heinrich, J.C. (2006) *The Finite Element Method: Basic Concepts and Applications*, Taylor & Francis.

12 Blomqvist, A. (2008) Oscillating micromechanical sensor of angular velocity, US
   7325451 B2. VTI Technologies Oy.

13 Blomqvist, A. (2008) Oscillating micromechanical sensor of angular velocity, US
   7454971 B2. VTI Technologies Oy.

14 Lapadatu, D., Kvisterøy, T., and Jakobsen,
   H. (2004) Micromechanical device, US
   6684699 B1. Sensonor AS.

15 Lapadatu, D., Kittilsland, G., and Jacobsen,
   S. (2013) Method for manufacturing a
   hermetically sealed structure, US
   20130146994 A1. Sensonor Technologies
   AS.

16 Ashby, D. (2012) *Electrical Engineering*,
   vol. **101**, Elsevier.

17 Foster, M., Jafri, I., Eskridge, M., and
   Zhou, S. (2009) Mechanical isolation for
   MEMS devices, EP 2075221 A2.
   Honeywell Int. Inc.

18 Eskridge, M.H. (2013) Die mounting stress
   isolator, EP 2006248 B1. Honeywell Int.
   Inc.

19 Childress, M.A., Dinh, N.T., and Golden,
   J.C. (2014) Die attach stress isolation, US
   8803262 B2. Rosemount Aerospace
   Inc.

20 Magendanz, G., Eskridge, M., and Loesch,
   M. (2012) MEMS devices and methods
   with controlled die bonding areas, US
   8240803 B2. Honeywell Int. Inc.

21 Schröder, S., Nafari, A., Persson, K.,
   Westby, E., Fischer, A., Stemme, G.,
   Niklaus, F., and Haasl, S. (2013) Stress-
   minimized packaging of inertial sensors
   using wire bonding, Transducers &
   Eurosensors XXVII, pp. 1962–1965.

22 Bakoglu, H.B. (1985) *Circuits,
   Interconnection and Packaging of VLSI*,
   Addison Wesley.

23 Tummala, R., Rymaszewski, E., and
   Klopfenstein, A. (eds) (1997)
   Microelectronics Packaging Handbook,
   Chapman and Hall.

24 Wibbeler, J., Pfeifer, G., and Hietschold,
   M. (1998) Parasitic charging of dielectric
   surfaces in capacitive
   microelectromechanical systems (MEMS).
   Sens. Actuators A, **71**, 74–80.

25 Bahl, G. *et al.* (2008) Observations of
   fixed and mobile charge in composite
   MEMS resonators, Solid-State Sensors,
   Actuators and Microsystems Workshop,
   pp. 102–105.

26 TU (2011) Supergyro i Lommeformat.
   Teknisk Ukeblad (Oct. 31).

27 Garrou, P., Bower, C., and Ramm, P.
   (2011) *Handbook of 3D-Integration:
   Volume 1-Technology and Applications of
   3D Integrated Circuits*, John Wiley & Sons.

28 Alexe, M. and Gösele, U. (eds.) (2004)
   Wafer Bonding, Springer.

29 Tummala, R.R. (2001) *Fundamentals of
   Microsystems Packaging*, McGraw-Hill.

30 Huemoeller, R. (2012) *Industry TSV
   Product Roadmap Drivers and Timing*,
   IMAPS North.

31 Amkor Technologies (2016) Amkor
   product sheet. Available at www.amkor.
   com, 2016.

32 Russell, T.P. *et al.* (2000) Ultra-high
   density nanowire array grown in self-
   assembled Di-block copolymer template.
   Science, **290**, 2126.

33 Samsung Corp (2012) 3D TSV technology
   and wide IO memory solutions, Design
   Automation Conference.

34 Ramm, P. (2009) Fabrication of 3D
   integrated fabrication of 3D integrated
   heterogeneous systems, IMAPS Nordic,
   Norway.

35 Schjølberg-Henriksen, K., Visser Taklo,
   M.M., Lietaer, N., Prainsack, J., Dielacher,
   M., Klein, M., Wolf, J., Weber, J., Ramm,
   P., and Seppänen, Timo. (2009)
   *Miniaturised sensor node for tire pressure
   monitoring (e-CUBES)*, in *Advanced
   Microsystems for Automotive Applications*,
   Springer.

36 Huemoeller, R. (2012) *Through Silicon Via
   (TSV) Product Technology*, IMAPS, NC,
   USA.

37 Akar, C. and Shkell, A. (2009) *MEMS
   Vibratory Gyroscopes*, Springer.

38 Sperling, R.A., Gil, P.R., Zhang, F., Zanella,
   M., and Parak, W.J. (2008) Biological
   applications of gold nanoparticles. *Chem.
   Soc. Rev.*, **37**, 1896.

17
Nanotechnologies Testing

Ernesto Sanchez and Matteo Sonza Reorda

Politecnico di Torino, Dip. di Automatica e Informatica, Corso Duca degli Abruzzi 24, 10129 Torino, Italy

17.1
Introduction

In an ideal world, the life of any product starts with the definition of the specifications it should match. Then, designers create a model for its implementation, and using this model the product is manufactured and delivered to customers, who will finally use it in the field. In all the above steps (specification, design, manufacturing, usage) *faults* may arise, that is, the result may differ from what was expected. As a result, the final product may behave differently than expected (*misbehavior*, or *failure*).

The role of testing is to detect faults as soon as possible, so that they can be removed, hence improving the quality of the final product. A well-known rule of thumb exists, stating that the cost for fixing a fault increases by 10 each time we move to the following step without detecting it. On the other side, cost is always a crucial parameter, and engineers do their best to identify the most convenient way to test a product with the minimum cost/effort.

Depending on the specific stage in the product lifetime we are considering, faults correspond to rather different phenomena. In the specification phase, they mainly are incongruent or missed specifications. In the design phase, they may correspond to bugs introduced by designers or by electronic design automation (EDA) tools. When addressing the above faults, test takes the specific name of *verification*.

In this chapter, we focus on the test addressing faults introduced by the manufacturing process, which are typically called *defects*. This kind of test is mainly performed at the end of the production process of an integrated circuit (IC), although different test steps can be performed along the manufacturing process (e.g., on the wafer, before it is cut in dies). The goal of all these test steps is to identify the existence of any defect inside the product, thus preventing the situation in which a faulty product is delivered to the customer. In order to achieve

Nanoelectronics: Materials, Devices, Applications, First Edition. Edited by Robert Puers, Livio Baldi, Sebastiaan E. van Nooten, and Marcel Van de Voorde.
© 2017 Wiley-VCH Verlag GmbH & Co. KGaA. Published 2017 by Wiley-VCH Verlag GmbH & Co. KGaA.

this goal, the straightforward approach is to apply proper stimuli to the device inputs, checking whether it produces the expected output behavior. The feasibility of this approach is limited by the poor controllability and observability we may have on the inner parts of a circuit if we just act on its inputs and just observe its outputs. To face this problem, starting from the 1980s designers started to take test issues into account already during the design phase, introducing hardware structures intended to support the test operations (*Design for Testability*). This approach is now very widely used in practice.

Despite the great efforts and the significant achievements of the research activities in the IC test area, test is still a major problem when producing integrated circuits and its cost often represents a significant percentage of the overall product cost. Moreover, the impact of possible flaws in the test procedure, resulting in defective products to be delivered in the field, may have very serious consequences both from a technical and an economical point of view. Not surprisingly, the first area identified by ITRS in 2014 among the five where rapid technology changes are needed in order to support the market needs is test cost. Finally, the trend for an increasing importance of test issues is expected to continue, fed by the Moore's law.

The goal of this chapter is to highlight the major challenges and solutions related to test when considering the latest VLSI technologies, with special emphasis on the new technologies that are being (or are expected to be) introduced.

This chapter starts with a section devoted to introducing the key vocabulary about testing, as well as the basic concepts and techniques in the IC test area. The interested reader can find further information about this part in Refs [1,2]. In the following section we present some key challenges existing in this area today. We focus on some major manufacturing alternatives to the current CMOS technology, and highlight the solutions proposed to support their test. Finally, we draw some conclusions.

17.2
Background

Test started to be considered right at the same time when the first integrated circuits were produced, since it has always been impossible to create a perfect manufacturing line, producing flawless devices.

Hence, a test step distinguishing fault-free products from good ones is always necessary. This step is also subject to limitations. In particular, it is possible that a fault-free circuit is marked as faulty (*overtesting*), or that a faulty circuit is marked as good (see Figure 17.1). While the former situation affects the profitability of the process (reducing the number of products to be sold), the latter impacts on the quality of the delivered products, which is often one of the terms of the contract signed between the manufacturing company and the customer. A quantitative measurement of the probability that the latter situation arises is

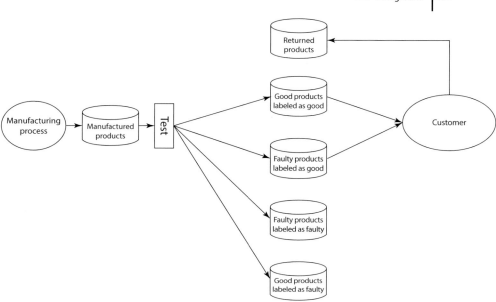

Figure 17.1 Test process.

typically required in practice. The *defect level* (*DL*) (often measured in terms of defects per million, or DPM) metric is often used, which can be estimated in different ways. One approach is based on tracking how many products return from the field, and computing their percentage with respect to the total number of delivered products. This approach has the following limitations:

- It requires some time before reaching statistical meaningfulness
- Not all faulty device may really be returned
- Some returned faulty devices may have broken in the field.

Hence, it is important to also be able to theoretically estimate the defect level. The first formula for this purpose was proposed in 1981 by Williams and Brown and requires the knowledge of the *Yield Y* (i.e., the percentage of good products coming out of the production line) and the *fault coverage FC* (i.e., the percentage of faults that the test can detect)

$$DL = 1 - Y^{1-FC}$$

Based on the above formula (for which several more effective versions have been developed in the most recent decades), the test engineer needs to estimate the percentage FC of faults detected by the adopted test procedure. This task is typically hard to be accomplished, if the list of all possible defects that may arise must be preliminarily computed. Hence, it is common to rely on *fault models*, that is, an engineering model of something that could go wrong in the IC manufacturing. Fault models are typically devised so that they can be enumerated, they are relatively easy to manage, and represent well the real defects. Several fault models have been proposed and adopted in practice.

By far the most common one is the *stuck-at*. This fault model works at the Gate-level netlist and only applies to digital blocks. It suggests considering the list of faults, each forcing a net of the circuit to a fixed 0 or 1 value, independently on the applied values. Apart from its simplicity and relatively low cardinality (there are two stuck-at faults for every net), the key advantage of this fault model lies in the fact that the input test vectors that are generated to address it have been experimentally proven to be able to detect a wide range of real defects.

Due to the growing criticality of defects affecting the circuit speed, that is, their ability to correctly work at the maximum frequency taking into account the delay of their components, the stuck-at fault model is increasingly complemented by *delay fault models*, such as the transition delay and the path delay models. Both work on the combinational blocks inside a circuit. The former models faults affecting the delay of a single gate: its detection requires activating a given transition (slow to rise or slow to fall) on a gate output, and then propagating it up to the block outputs, which are sampled at the maximum operating frequency. The latter considers the paths connecting any block input to any block output and assumes that the delay for propagating a transition on the input up to the output becomes larger than expected. Once again, detecting the fault requires triggering a transition on the path input and propagating it up to the output, where the transition is observed. These fault models are also largely used and well supported by the commercial EDA tools. The former benefits from a low cardinality (two faults for each gate) but is only able to detect defects corresponding to a local increase in the delay. The latter is much mightier, but suffers from the fact that the number of possible paths is clearly very large, and is typically unfeasible to deal with all of them. The common approach is based on considering only a subset of the possible paths, including the longest ones. Unfortunately, choosing the best percentage of paths to be considered is not easy, and current timing analysis tools are often so imprecise that there is often the risk not to include in the subset the paths that are really critical.

Other fault models are sometimes used, also working at other abstraction levels, such as the stuck-on and stuck-close fault model, working at the transistor level, as well as the bridge and open fault model, considering the possible shortcut between two lines or open on a line in the circuit layout. Efforts are also being done to define new fault models working at higher abstraction levels (e.g., the register transfer one), thus matching the trend in electronic design to work at this level (or even higher) and then use automatic synthesis tool to generate the lower level descriptions; unfortunately, the proposed fault models typically lack in effectiveness in terms of defect detection capabilities.

Finally, it is worth mentioning that all the above fault models deal with the digital parts in the circuit. Currently, and despite several efforts, there is no consensus on any metric able to provide a measure for the quality of a test set devoted to analog parts. Hence, in this area the most commonly adopted approach remains the functional one, based on listing the functionalities provided by a given circuit (or circuit block), and then checking whether they are correctly implemented by the circuit. This approach has several limitations, including the fact that it can hardly provide an objective measure, and it is hard to automate.

Once a suitable fault model is identified, the issue exists of how to generate proper stimuli for the circuit, so that a given fault coverage can be achieved. During the 1970s and 1980s, big efforts were devoted to devise the most complete and effective algorithms to automatically generate test stimuli, mainly addressing the stuck-at fault model. This finally led to very good and practically feasible theoretical solutions, which were shortly followed by commercially available *automatic test pattern generation* (ATPG) tools, able to work on combinational blocks, given that their gate-level netlist is available. On the other side, it was soon clear that the computational effort for automating the generation of test sequences for sequential circuits was out of any feasible limit, and the fast growth in circuit complexity was making the task always harder.

For this reason, starting from the 1980s, several companies begun systematically using design for testability approaches intended to introduce a test mode acting on an external signal, such that during test the generic sequential circuit turns into a combinational one. This result is achieved by transforming the circuit flip-flops into a shift register (also called *scan chain*) that can also be uploaded/downloaded from the outside. In this way the flip flop outputs (i.e., the inputs of the combinational blocks) can be controlled to any specified value, and the values on the flip-flop inputs (i.e., the outputs of the combinational blocks) can be observed. This approach, which was given the name of *scan test*, quickly became very popular, and is still the basis for the test of today circuits. Its major advantage lies in its ability to provide an easy, low cost, and fully automated way to test a circuit, at least when the stuck-at fault model is the target. Figure 17.2 outlines the basic transformation of a circuit (upper part) into a scan circuit (lower part), with the addition of Scan In and Scan Out signals. The additional signal for moving from normal to test mode is not shown in the figure. To reduce the test time, which is dominated by the time for uploading/downloading the scan chain (which are not optimized for running at full speed), it is common to connect the flip-flops in a circuit into multiple chains, often taking into account the existing partitioning in clock domains.

The design for testability approach entered into wide industrial acceptance in the 1980s, thanks to scan test. In about the same period, another issue started to become relevant: the test of embedded memories. Memories are the most highly integrated modules in the semiconductor industry. Hence, they are often those that are most prone to defects. Researchers devised effective algorithms for their test, based on a sequence of steps, each repeating the same group of read/write operations on all memory words (*March algorithms*). However, applying the sequence of input stimuli to a target memory, and observing its output behavior may simply be unfeasible, when the memory is deeply embedded in a circuit. The solution which started to be systematically applied in the 1980s lies in introducing some suitable circuitry surrounding the memory, so that from the

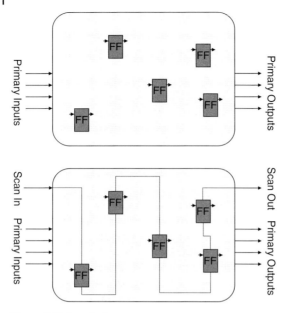

Figure 17.2 Scan test.

outside one can force it to disconnect from the rest of the circuit, trigger suitable hardware to generate the required sequence of stimuli for the memory, command the observation of its behavior (typically computing a signature of the output data), and then looking at the value of a resulting good/faulty signal. This approach, denoted by *built-in self-test* (BIST) is very suitable for the test of embedded memories, but can also be applied to other situations, following the general architecture of Figure 17.3.

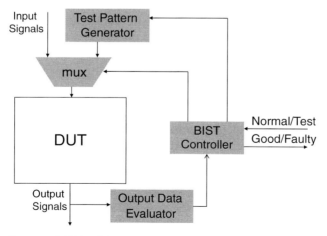

Figure 17.3 BIST architecture.

During the first decade of this century, all major EDA companies introduced commercial tools to support *Logic BIST*, that is, the possibility to adopt BIST for testing a chunk of sequential logic. The approach uses a circuit where scan chains have been already introduced to connect flip-flops when in test mode, as explained before. The chains are uploaded with values generated by a suitable circuitry on-board the chip, and the values they capture are downloaded and compressed by another circuitry. At the end, the result of the test can be checked by simply comparing the computed signature with the expected one. The test vector generation circuitry can be either a random generator, or a decompressor, receiving inputs from an outside tester, and decompressing them into a sequence of vectors to be applied to the chains that has been generated by an ATPG. In this way a very high and deterministic fault coverage can be achieved at the expense of a limited hardware cost.

It is worth mentioning that a major advantage of all the BIST approaches lies in their ability to perform a test where stimuli can be applied (and output values can be observed) at the maximum operating frequency of the circuit, thus creating the conditions for detecting delay faults. This important result can be achieved with a limited cost in terms of additional circuitry on-board the device under test (DUT). As a comparison, performing the same test resorting to an automatic test equipment (ATE) in charge of both applying the test stimuli and observing the DUT responses at-speed would require a fast and expensive test system; hence the wide success of BIST solutions.

17.3
Current Challenges

17.3.1
SoCs and Embedded Instruments

The complexity of current IC designs continues to grow, and designers are forced to reuse modules (also called intellectual property (IP), cores) coming from previous designs or from third parties. This led to a new design paradigm, where large systems are integrated into a single device, called *System on Chip* (SoC). According to IBS, the typical number of IP cores existing in each design grew from 16 for 90 nm designs to more than 200 for 20 nm ones. Turning the design effort into an assembly effort also requires some change in the approach to test. Some of the test steps (e.g., ATPG and fault simulation) cannot be performed on the design netlist as a whole, which is too large. Rather, the test solutions for the different IP cores should come with the cores themselves, and the test engineer should be able to assemble all of them into the final test set.

This requires defining standard interfaces for the test of the IP cores, so that the test assembly process can be made easier and hopefully automated. The IEEE 1500 standard describes such an interface, as well as the format for documenting it, so that EDA tools can both generate and read these

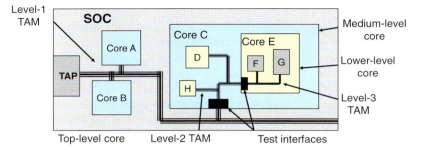

Figure 17.4 SoC test architecture.

descriptions. The interface takes the form of a *wrapper*, which isolates the core during test and allows proper stimulation and observation of its input and output signals. IEEE 1500 does not mandate any test solution, and it supports both BIST solutions and solutions where input and output test values are conveyed to the core through a test access mechanism (TAM). The designer is in charge of deciding how to connect the IEEE 1500 wrappers of the cores in the design and to design the related TAM. All these test signals are often connected to the standard IEEE 1149 test access port (TAP) that allows their interface with the outside tester.

A major advantage of this approach is that it easily supports hierarchy in a SoC (e.g., see Figure 17.4). Moreover, combining BIST with hierarchical core-based solutions supports a test that requires a much simpler ATE. In fact, the mix of the above techniques may allow a test mechanism that only requires the external tester to send/receive messages (rather than vectors) to/from the device. In this way, at-speed test can be supported even with low-cost (i.e., low-frequency) ATEs.

In recent years the number of instruments existing within a device for test, calibration, debug, and other purposes is also growing. These instruments may correspond to both sensors and actuators, as well as registers existing in the IEEE 1500 wrappers. Their common characteristic is that they do not play any role from a functional point of view, but are crucial to support ancillary activities that are crucial during different parts of the product life. From a test point of view, it is worth mentioning as an example the sensors devoted to detect the health of the device, that is, to understand when it is close to its end of life (due to the wear-out phenomena). These instruments can be used to support the prognostic activities, that is, to early detect a device which entered a critical phase: different actions can be taken when this situation has been detected, ranging from substituting the device (when possible) to simply decrease its operating frequency, thus preventing the output of wrong values.

From a designer point of view, a major challenge is how to provide an effective way for accessing these devices, either during the test phase, or during the operational phase of the device. In order to avoid custom solutions, that can hardly be supported by commercial EDA tools and prevent easy reuse of instruments, the IEEE 1687 standard has recently been approved. This standard describes a flexible way for connecting these instruments to any external interface (e.g., the IEEE 1149 TAP port), using an internal network corresponding to a set of configurable scan chains including registers to access the instruments (*test data registers*, or TDRs) and programmable switches (*segment insertion bits*, or SIBs). The typical way of accessing any instrument is to first access the internal network to configure it, specifying the status of any SIB, and then accessing the network for reading or writing the TDRs in the selected part of the network. The key advantage stemming from the new standard is that the access to the instruments can be done using a network, which can flexibly be configured according to the needs, instead of using a static network, which would typically correspond to a single scan chain. The standard also mandates the format for describing the network and its access protocol, thus supporting the automated generation of the stimuli to be applied to a device to access any instrument.

17.3.2
Process Variations

The new semiconductor technologies show an increasing impact of process variations on the manufactured devices. Practically, it is more and more difficult to tune the manufacturing process so that the fabricated transistor parameters are those mandated by the design specifications. In reality, the parameters are different from die-to-die, wafer-to-wafer, and lot-to-lot, and also between transistors on the same die. Parameter variations may be caused by a wide range of phenomena, including differences in impurity concentration densities, oxide thicknesses, diffusion depths, and similar factors. These nonuniform conditions result in deviations in transistor parameters, such as threshold voltage, W/L ratio, as well as variation in the widths of interconnect wires. In particular, the process variations are expected to impact design performance (modifying the gate and interconnect delays) and power consumption to an increasingly large extent in newer technologies.

Traditionally, the effects of process variations were faced by introducing margins and guard-bands able to guarantee the correct behavior even in the worst case. Unfortunately, this approach can hardly be followed when process parameter values become smaller, and possible variations represent a large percentage of their value.

A first alternative solution to guard-bands is represented by the adoption of fault tolerance: the systematic introduction of redundant structures in the new circuits can provide a viable solution, although its cost (mainly in terms of silicon area, and sometimes even in terms of performance) may be relevant. Examples of techniques following this approach are *razor* (based on duplicating flip-flops and checking the outputs of the combinational blocks at slightly different time instants for each clock cycle) [3], *error coding* (allowing to easily detect possible errors resulting from processing, storage, and transmission operations), and *time redundancy* (e.g., RESO, implying re-execution of an instruction with shifted operands).

A second solution lies in developing suitable test stimuli able to effectively detect faults produced by process variations. Unfortunately, this often requires repeating the test at different values of the operating parameters (e.g., voltage and frequency). Techniques for identifying the best set of operating conditions at which the test must be performed become crucial for achieving the desired product quality with an acceptable test cost. Finally, gathering extensive information about the manufacturing and test processes, and extracting statistical forecast about the outliers may allow to apply the most suitable test to each die (adaptive testing), thus producing significant benefits in terms of test accuracy and cost.

17.3.3
Combining End-of-Manufacturing and In-Field Test

Another clear trend in the test domain is the growing importance of in-field test. Two major drivers support this trend: from one side, the increased adoption of electronic systems and devices in safety-critical applications (e.g., in automotive and biomedical applications) asks for technical solutions able to guarantee the system integrity, that is, any fault possibly arising in a device can be quickly detected. In this way, the negative effects of the fault can hopefully be minimized. On the other side, the reduction in the average lifetime of ICs due to the technological evolution makes the occurrence of such faults more likely than in the past, and thus increases even more the importance of the above solutions.

As a final result, many products used in safety-critical applications do require suitable mechanisms for supporting their test during the operational phase (*in-field test*). Standards and regulations (e.g., IEC 61508 for generic safety-related systems, ISO 26262 for automotive applications, DO-254 for avionics) explicitly require the existence of such mechanisms, and specify the procedures to assess their effectiveness.

From a technical point of view, the most suitable solution to the above requirements would be to simply adapt the solutions adopted for end-of-manufacturing test, so that they can be also used during in-field test. Following this idea, some of the BIST structures existing in most microprocessors to support the end-of-manufacturing test of some internal memories can also be accessed and used during the power-on self-test of the computers using these devices, thus allowing an effective test of the same memories even during the operational life.

On the other side, a common characteristic of most of the new domains where electronic systems are adopted under strict safety-critical constraints is that several different companies are involved in the design and manufacturing process, each having its own interests and constraints. For example, the electronic control units (ECUs) composing the complex distributed electronic system existing on todays' cars are integrated by the car maker but produced by different OEM (original equipment manufacturer) companies by resorting to both commercial and custom devices. In the former case, the same device may be used for products intended to guarantee different levels of safety. Hence, it is not given that suitable mechanisms (e.g., based on design for testability) for performing in-field test are provided or disclosed by the device manufacturer. On the other side, the OEM company is often not aware of the internal structure of each device, and thus the test development may turn into a real challenge. Moreover, when considering SoC devices it may often happen that its internal structure is partially unknown even to the manufacturing company. In general, the adopted solutions are selected combining technical and commercial constraints, and must be able to guarantee that the intellectual property of each product is preserved. Additionally, since the test is often activated during the operational phase (e.g., during idle time slots), the adopted solution must match the constraints of the applications (e.g., in terms of duration and used resources). Finally, it is worth mentioning that the requirements for some of the considered domains increasingly often include security, which typically mandates blocking as much as possible the access to the inner part of each device from the outside. Unfortunately, this requirement goes in the opposite direction of testing, which benefits from a high level of controllability and observability of the inner circuitry.

The common practice in industry is to face the above problems with a proper mix of different solutions, sometimes heavily relying on design for testability (e.g., BIST), sometimes resorting to a functional approach. When processor-based systems are considered, the latter approach represents a viable solution, based on adopting a suitable test program in charge of detecting possible faults affecting either the CPU core or the other components of the system (*software-based self-test*, or SBST [4]). Hybrid solutions are also frequently adopted [5].

17.4
Testing Advanced Technologies

Several solutions to overcome the limitations of the CMOS technology are being evaluated. They include promising techniques that may allow the manufacturing of electronic devices with a higher integration level, lower power consumption, higher operational frequency than current CMOS ones. On the other side, these new technologies are still in a preliminary phase of development.

Hence, we still do not have (among the others) a deep enough knowledge about the possible defects that may occur during the corresponding manufacturing process (which is also still partly unknown and unoptimized). The practical adoption of the new technologies will only be possible when suitable solutions will be available, able to guarantee a sufficient yield of the manufacturing process, which means that the percentage of faulty products is sufficiently low, and the techniques to identify faulty products (test) are effective enough. Clearly, the test techniques have to be focused on the most likely defects that may arise in practice. The above motivations triggered several works (e.g., [6], dealing with silicon nanowires, or [7], dealing with quantum-dot cellular automata) aimed first at the identification of the most likely defects that may affect the manufacturing process when new technologies are applied, and then to generate suitable test stimuli for them (e.g., [8], for threshold logic gates).

In some cases, it has been speculated that the manufacturing process adopting the new techniques will intrinsically not be able to achieve an acceptable yield [9]. In these cases, usage of some sort of *repair* may represent an interesting approach, typically based on the availability of spare components on the chip, and on the possibility of reconfiguring it so that the faulty components are substituted with the spare ones. This solution is turn requires the ability not only to identify faulty products but also to identify the faulty components (*diagnosis*).

Finally, some of the new semiconductor technologies are likely to be more prone to defects arising during the operational life than the current CMOS technology. These defects may belong both to the category of *transient* defects, which affect the behavior of the electronic product without affecting its structure (and whose effects can disappear with time), and to the one of *permanent* defects, that modify the structure of the product, thus permanently impacting on its ability to correctly provide the expected behavior (from a functional and electrical point of view). If the probability of the resulting misbehavior becomes higher than the acceptable level, fault tolerant solutions must be adopted, typically based on introducing some sort of redundancy. An example where fault tolerance is applied to nanotechnologies is described in Ref. [10].

There is a strong connection between the test solutions aiming at identifying faulty products at the end of the manufacturing process and the fault tolerance solutions used to reduce the probability of misbehaviors during the operational life. In some cases, fault tolerance techniques may substantially reduce the need for test solutions, since they can lead to devices working despite the presence of faults. On the other side, active fault tolerance is often based on test techniques to detect the occurrence of defects during the operational phase that are very similar (and in some cases are the same) to those used for end-of-manufacturing test.

Finally, process variation effects affect new technologies, too. Being able to predict and tame these effects is crucial to guarantee acceptable quality and yield, and some preliminary works have been published in this direction (e.g., [11]).

In this section we summarize the state-of-the-art in the area of testing of electronic devices implementing some of the new technologies.

17.4.1
Resonant Tunneling Diodes and Quantum-Dot Cellular Automata

As stated by many authors, for example [12], when dealing with the synthesis of complex circuits, CMOS technologies are prone to use Boolean gates; on the contrary, new nanoscale technologies are also suitable for building circuits that implement threshold functions.

A threshold function is a logic function $f(x_1, x_2, x_3, \ldots, x_n)$ that assumes the value 1 when the sum of all of its n weighted inputs is higher or equal than a given threshold T, and the value 0 otherwise:

$$f(x_1, x_2, x_3, \ldots, x_n) = \begin{cases} 1, & \text{if } \sum_{i=1}^{n} w_i \cdot x_i \geq T, \\ 0, & \text{if } \sum_{i=1}^{n} w_i \cdot x_i < T. \end{cases}$$

Then, it is straightforward to define a threshold logic gates (TLG) as a multiple input gate that computes the weighted sum of its inputs, and compares it against a given threshold determining in this way the output value of the gate. TLGs can be designed using different techniques belonging to the nanoscale technologies [13], for example, through a monostable–bistable transition element (MOBILE) made using resonant tunneling diodes and heterostructured field-effect transistors (RTD-HFET), and by using majority gates, implemented using quantum-dot cellular automata (QCA).

Considering threshold network devices, in Ref. [12] the authors proposed a synthesis methodology oriented to synthesize digital circuits through different nanotechnologies. In particular, the authors claimed that to synthesize a circuit using threshold logic gates, it is necessary to

a) collapse the Boolean functions of a threshold node;
b) determine if a Boolean function is threshold or not;
c) split the function into smaller functions if it is not a threshold one;
d) share the shared Boolean nodes in the final threshold circuit.

The proposed tool is called *threshold logic synthesizer* (TELS), and it is possible to target different nanoscale devices once the threshold gates have been developed.

In Ref. [8], the authors proposed an automatic test patter generation technique and simulation tool for TLGs devices implemented using RTD-HFET technology. The proposed approach is based in the following steps:

Fault Model Definition
Analyzing the most likely defects (i.e., cuts and shorts) in the tackled technology [14], the authors performed a set of simulations, in order to better understand the circuit behavior in presence of a fault and concluded that every one of considered defects in the RTD-HFET technology can be modeled as a stuck-at-0 (SA0) or stuck-at-1 (SA1). Once the fault model is identified, it is possible to derive the device fault list.

Fault Simulation
The goal of the fault simulation is to determine if a considered test pattern t is able to cover one or more faults F. This process is exploited during the generation time to determine whether the fault list is empty or not. In the case of the

threshold logic circuits, the authors stated that almost any Boolean logic fault simulation technique is suitable.

Irredundant Threshold Networks and Redundant Faults Removal

Redundant parts of the circuit are removed according to Definitions 1 and 2:

Definition 1: A combinational threshold network G, is irredundant if the removal of any node or edge in G does not represent the same set of Boolean functions.

Definition 2: If no test vector exists for fault f in G, then f is redundant.

Thus, Theorem 1 states as follows:

Theorem 1: Given G, if no test vector exists to detect a fault f, then, the corresponding node or edge in the network can be removed without affecting the network functionality.

Test Pattern Generation

For any fault in the circuit, a set of constraints is derived, and then a SAT solver is used to obtain the test pattern that satisfies all the fault constrains.

Fault Collapsing

Considering the following rules, and analyzing the TLG properties, some theorems can be stated and exploited to collapse the circuit faults.

Rule 1: Two faults f_1 and f_2 are said to be equivalent if the corresponding faulty versions of the network G_{f1}, and G_{f2}, respectively, have identical input–output logic behavior. Consequently, one of the two faults is dropped from the fault list.

Rule 2: Fault f_1 is said to dominate fault f_2 if any test vector that detects f_2 also detects f_1. Consequently, f_1 is dropped from the fault list.

The authors then concluded that the cuts and shorts in a generic mobile device can be modeled as single stuck-at faults. Experimental results were performed in the MCNC benchmark, and show the suitability of the proposed method.

Later, Goparaju and Tragoudas [15] proposed a new ATPG method for threshold circuits able to detect not only short and cuts, previously modeled as stuck-at faults, but also weight-related defects. The authors extend the previous methodologies, by considering a new fault model that is able to consider weighted related circuit faults. The experimental results were run on the ISCAS'85 benchmarks and experimentally validated the method.

On the other hand, a quantum-dot cellular automata (QCA) is a nanoscale device that represents logic states not as a voltage but by the configuration or position of an electron pair confined within a quantum-dot cell. QCA promises small features sizes and ultra-low power consumption; and in addition, some research works claim that these cells may be fabricated through a self-assembly molecular implementation, allowing new devices to achieve densities of 10^{12} devices/cm^2 operating at terahertz frequencies [7,16].

Considering different research works on QCAs [17,18], the most likely manufacturing defects that may occur during the QCA manufacturing process are related to the actual cells position [7], and are defined as follows:

1) Cell displacement defect: the faulty cell is misplaced from its original direction.
2) Cell misalignment defect: the direction of the faulty cell is not properly aligned.
3) Cell omission defect: a cell is actually missing.

As a matter of fact, test pattern generation for QCA based on modeling single stuck-at faults is not enough to obtain high fault coverage on position related faults. Thus, in order to achieve high defect coverage, the specific QCA implementation of each circuit function must be considered during the test generation [7].

17.4.2
Crossbar Array Architectures

Circuit integration at nanoscale levels raises new issues regarding system reliability; in fact, even though new nanotechnologies promise to overcome the main lithography problems modifying the manufacturing paradigm form top-down to bottom-up self-assembly fabrication processes, the circuits developed using these technologies will likely suffer defect rates in the range $10^1 - 10^3$, that is, considerably higher than the actual CMOS levels ($10^9 - 10^{12}$) [19]. As proposed in Ref. [20], following some chemical synthesis processes it is possible to successfully fabricate new nanoscale devices as highly regular structures, such as those used by two-dimensional crossbar arrays.

Briefly speaking, a crossbar is made with two sets of orthogonal nanowires that propagate the input values to the device outputs according to the state of programmable switches placed on every intersection of the nanowires [21]. The crossbar architectures are very similar to programmable logic arrays (PLAs). In fact, it is possible to implement almost any combinational circuit using these structures [22]. However, the resulting device is more prone to be defective than CMOS-based ones and then, in order to make the device work despite the fabrication defects, it will be required to effectively map the design taking in consideration these new manufacturing issues.

Some testing strategies have been proposed for testing crossbar circuits. For example, in Ref. [23] the authors studied the impact of switch and nanowire faults in 2D crossbar circuits. In particular, a fault model for stuck-at open and close faults is used tackling the crossbar switches, whereas open and bridging faults are analyzed in the case of the crossbar nanowires. The authors concluded that biasing the chemical synthesis process to decrease the probability of having switch stuck-closed faults will positively impact the production yield.

On the other side, in Ref. [24], the authors proposed a testing methodology for crossbar-based nonvolatile memories. In particular, the authors defined a new set of nine fault models based on the classical definition of *self-* and *coupling-faults*, and proposed a corresponding set of patterns able to test them. Additionally, the proposed work is extended to test not only the undesirable electric paths in the crossbar (called *sneak paths*) but also multiple memory elements at a time.

In a more comprehensive approach [25], the authors propose an integrated algorithm framework that performs mapping, morphing, and hardening of the synthesized crossbar simultaneously. In particular, the authors present a set of efficient algorithmic schemes to perform logic equivalence checking (LEC) to maximize the effectiveness of logic morphing; then, they use a logic hardening technique for tuning the amount of required redundancy for a given logic maximizing the defect tolerance capability of the crossbar; and finally, an integrated algorithmic framework efficiently exploring both logic morphing and logic hardening is used in the mapping process.

17.4.3
Carbon Nanotubes

Since they have been first proposed in 1991 [26], carbon nanotubes (CNT) appeared as one of the most promising technologies to extend CMOS beyond its limitations. In principle, using the CNT technology one can build field effect transistors (called CNTFETs) [27,28] with excellent electrical properties.

Manufacturing CNTFET devices is challenging, because several defects are quite likely to arise, and it appears quite difficult to achieve good yield. Hence, test solutions are crucial to guarantee an acceptable level of quality.

More in details, the following two basic types of defects make the manufacturing of CNFET-based logic designs hard [29]:

1) Misaligned carbon nanotubes; this type of defect may cause an incorrect logic function implementation.
2) Metallic CNTs; in this case the CNFETs conductivity cannot be controlled by acting on the gate, and the CNFET simply acts as a resistor.

The work in Ref. [29] addresses the former type of defects; the authors propose a solution for designing circuits that correctly work even in the presence of misaligned CNTFETs, and show that the cost from the adoption of this solution (leading to the so-called *misaligned-carbon-nanotube-immune* circuits) is still acceptable (i.e., in the order of a few percent) in terms of area, power, and delay.

The work in Ref. [12] addresses the latter type of defects, which can be modeled as stuck-at faults on the CNTFET gate. The authors of Ref. [12] first analyze the manufacturing defects that may produce them, and then show that even when complex gates are implemented, these defects can be effectively tested resorting to the test stimuli generated by targeting the above stuck-at fault. The method described in Ref. [29] uses a different approach (named *VLSI-compatible metallic nanotube removal* or VMR) to deal with this kind of defects. To eliminate the metallic nanotubes, the Stanford team switched off all the good CNTs. Then they forced into the circuit a large amount of current, which concentrated in the metallic nanotubes, up to the point in which they burned up and disappeared from the circuit structure.

In 2013, researchers from Stanford University announced [30] that they were able to manufacture the first computing system entirely based on CNTFETs, thus proving that they were able to deal with the possible defects, with special emphasis on the two types listed above, using the solutions described in Ref. [29]. Although quite simple (the computer is composed of 178 transistors), this result shows that CNTFET devices can be manufactured.

17.4.4
Silicon Nanowires FETs

Silicon nanowires FETs (SiNWFETs) showed to have some nice electrical properties and to be manufacturable with a high level of integration, since they can use a CMOS compatible fabrication process. In some cases (e.g., Ref. [31]), they are implemented in the so-called *double-gate* version, where each transistor owns an additional gate input, which allows controlling its polarity. The resulting device is denoted as controllable-polarity (CP) silicon nanowire FET (CP-SiNWFET). Some examples of controllable polarity devices have been manufactured, belonging to the double-gate [32] and three-independent-gate [31] types. These devices have been successfully used for the fabrication of CP logic gates that provide compact hardware realization with remarkable circuit design flexibility.

For a new technology such as CP-SiNWFET, which has different geometrical structure and physics of operation than CMOS technology, it is not *a priori* known that how the manufacturing defects will impact the device and logic circuits. In Ref. [6], the authors report the results of an inductive fault analysis to investigate the specific malfunctions of CP-SiNWFETs. They used three-independent-gate silicon nanowire FETs (TIG-SiNWFETs), which have been fabricated and used as a potential candidate for manufacturing CP devices. Considering the technology process, the possible defects that can change the functionality of the CP-SiNWFETs and occur during fabrication process are modeled. According to this analysis, they can be grouped into the following categories:

- Nanowire break
- Gate oxide short
- Bridge between two or more terminals
- Bridge among interconnects
- Floating gates

Using this list of possible defects, the authors also investigate the functionality and the performance of various logic gates in the presence of defects via electrical simulation. Based on the results of this analysis, they extended the current CMOS fault models to a new hybrid model, including stuck-at n-type and stuck-at p-type, which can be efficiently used for the detection of defects in CP logic gates. The experimental results revealed that the GOS and floats on the PGs are detectable by analyzing the performance parameters like delay and leakage. Moreover, they show that the current CMOS test methods are not able to capture all faults in CP logic gates: as an example, stuck-open faults can hardly be detected. The results confirm the inefficiency of the traditional test methods and fault models for covering the defects occurring in the CP-SiNWFET technology.

Using the new fault model including the stuck-at n-type and stuck-at p-type faults, a further analysis of the fault effects is reported in [10], where it is shown that most of them can be masked by the intrinsic robustness of the gates implemented using CP-SiNWFET devices. This robustness can be fruitfully exploited to implement fault tolerant structures, which are shown to be significantly smaller than the corresponding structures using the traditional CMOS technology.

17.5
Conclusions

In this chapter, we deal with the issue of testing an integrated circuit with respect to possible faults affecting its hardware structure, both at the end of the manufacturing process, and when it is already deployed in the field. Despite the big efforts devoted to tame the complexity and cost of the test step, its relative cost with respect to the total manufacturing one tends to increase. There are already several cases (e.g., with several MEMS devices) where the test cost is greater than the manufacturing one.

We identified some major challenges in the current scenario, and summarized a few solutions. We gave special emphasis to the test of ICs manufactured resorting to new technologies that are supposed to substitute or integrate with the traditional CMOS one in the close future.

Clearly, our overview is not complete: other major challenges concern, for example, the test of MEMS, as well as the solutions to take into account power and thermal issues. Space issues prevented us from covering all the aspects of the wide test arena.

As a general conclusion, we can state that any effective solution in this area stems from a suitable mix of correct choices in terms of considered fault models, adopted design for testability solutions, and effective EDA tools. Moreover, the key choices leading to an effective test plan are more and more depending on the company ability to own a global view of the parameters characterizing both the manufacturing and the test procedures, thus allowing the real-time extraction of useful information able to quickly identify the weak points where to focus the efforts.

References

1 Bushnell, M.L. and Agrawal, V.D. (2005) *Essentials of Electronic Testing for Digital,* *Memory and Mixed-Signal VLSI Circuits,* Springer.

2 Wang, Laung-Terng, Wu, Cheng-Wen, and Wen, Xiaoqing (2006) *VLSI Test Principles and Architectures*, Elsevier Science.

3 Ernst, D., Das, S., Lee, Seokwoo, Blaauw, D., Austin, T., Mudge, T., Kim, Nam Sung, and Flautner, K. (2004) Razor: circuit-level correction of timing errors for low-power operation. *IEEE Micro*, **24** (6), 10–20.

4 Psarakis, M., Gizopoulos, D., Sanchez, E., and Sonza Reorda, M. (2010) Microprocessor software-based self-testing. *IEEE Des. Test Comput.*, **27** (3), 4–19.

5 Bernardi, Paolo, Ciganda, Lyl M., Sanchez, Ernesto, and Reorda, Matteo Sonza (2014) MIHST: a hardware technique for embedded microprocessor functional on-line self-test. *IEEE Trans. Comput.*, **63** (11), 2760–2771.

6 Mohammadi, H.G., Gaillardon, P., and De Micheli, G. (2015) From defect analysis to gate-level fault modeling of controllable-polarity silicon nanowires. *IEEE Trans. Nanotechnol.*, **14** (6), 1117–1126.

7 Tahoori, M.B., Huang, Jing, Momenzadeh, M., and Lombardi, F. (2004) Testing of quantum cellular automata. *IEEE Trans. Nanotechnol.*, **3** (4), 432–442.

8 Gupta, P., Zhang, R., and Jha, N.K. (2008) Automatic test generation for combinational threshold logic networks. *IEEE Trans. VLSI Syst.*, **16** (8), 1035–1045.

9 Tahoori, M.B., Jha, N.K., and Bahar, R.I. (2010) Testing aspects of nanothechnology trends, in *System-on-Chip Test Architectures: Nanometer Design for Testability* C (eds Laung-Terng Wang, Charles E. Stroud, and Nur A. Touba), Morgan Kaufmann.

10 Ghasemzadeh, H., Gaillardon, P.E., Zhang, J., De Micheli, G., Sanchez, E., and Sonza Reorda, M. (2016) A Fault-Tolerant Ripple-Carry Adder with Controllable-Polarity Transistors. ACM Journal on Emerging Technologies in Computing Systems, 13 (2), 1–13.

11 Mohammadi, H.G., Gaillardon, P.-E., Yazdani, M., and De Micheli, G. (2013) A fast TCAD-based methodology for variation analysis of emerging nano-devices. IEEE International Symposium on Defect and Fault Tolerance in VLSI and Nanotechnology Systems (DFT), pp. 83–88.

12 Zhang, R., Gupta, P., Zhong, L., and Jha, N.K. (2005) Threshold network synthesis and optimization and its application to nanotechnologies. *IEEE T. Comput. Aided Des.*, **24** (1), 107–118.

13 Beiu, V., Quintana, J.M., and Avedillo, M.J. (2003) VLSI implementations of threshold logic: a comprehensive survey. *IEEE Trans. Neural Netw.*, **14** (5), 1217–1243.

14 Prost, W., Auer, U., Tegude, F.-J., Pacha, C., Goser, K.F., Janssen, G., and Van der Roer, T. (2000) Manufacturability and robust design of nanoelectronic logic circuits based on resonant tunnelling diodes. *Int. J. Circ. Theor. App.*, **28** (6), 537–552.

15 Goparaju, M.K. and Tragoudas, S. (2008) A novel ATPG framework to detect weight related defects in threshold logic gates. Proceedings of the 26th IEEE VLSI Test Symposium, pp. 323–328.

16 Liu, Mo and Lent, C.S. (2003) High-speed metallic quantum-dot cellular automata. Proceedings of the Third IEEE Conference on Nanotechnology (IEEE-NANO 2003), vol. 2, pp. 465–468.

17 Armstrong, C.D., Humphreys, W.M., and Fijany, A. (2003) The design of fault tolerant quantum dot cellular automata based logic. Proceedings of the 11th NASA VLSI Design Symposium.

18 Fijany, A. and Toomarian, B. (2001) Bouncing threads: merging a new execution model into a nanotechnology memory. Proceedings of the IEEE Computer Society Annual Symposium on VLSI, vol. 3, pp. 27–37.

19 ITRS (2013) International Technology Roadmap for Semiconductors, Emerging Research Devices.

20 Butts, M., DeHon, A., and Goldstein, S.C. (2002) Molecular eletronics: devices, systems and tools for gigagate, gigabit chips. Proceedings of the IEEE International Conference on Computer-Aided Design, pp. 443–440.

21 DeHon, A. and Wilson, M.J. (2004) Nanowire-based sublithographic programmable logic arrays. Proceedings of the International Symposium on

Field-Programmable Gate Arrays (FPGA2004), pp. 123–132.

22 DeHon, A. (2003) Array-based architecture for FET-based, nanoscale electronics. *IEEE Trans. Nanotechnol.*, **2** (1), 109–162.

23 Huang, Jing, Tahoori, M.B., and Lombardi, F. (2004) On the defect tolerance of nano-scale two-dimensional crossbars. Proceedings of the 19th IEEE International Symposium on Defect and Fault Tolerance in VLSI Systems, October 10–13, pp. 96–104.

24 Kannan, S., Rajendran, J., Karri, R., and Sinanoglu, O. (2013) Sneak-path testing of crossbar-based nonvolatile random access memories. *IEEE Trans. Nanotechnol.*, **12** (3), 413–426.

25 Su, Yehua and Rao, Wenjing (2014) An integrated framework toward defect-tolerant logic implementation onto nanocrossbars. *IEEE Trans. Comput. Aided Des.*, **33** (1), 64–75.

26 Ijima, S. (1991) Helical microtubules of graphitic carbon. *Nature*, **354**, 56–58.

27 Avouris, P., Appenzeller, J., Martel, R., and Wind, S.J. (2003) Carbon nanotube electronics. *Proc. IEEE*, **91** (11), 1772–1784.

28 McEuen, P.L., Fuhrer, M.S., and Park, H. (2002) Single-walled carbon nanotube electronics. *IEEE Trans. Nanotechnol.*, **1** (1), 78–85.

29 Patil, N., Deng, Jie, Wong, H.-S.P., and Mitra, S. (2007) Automated design of misaligned-carbon-nanotube-immune circuits. Proceedings of the 44th ACM/IEEE Design Automation Conference, pp. 958–961.

30 Shulaker, MaxM., Hills, Gage, Patil, Nishant, Wei, Hai, Chen, Hong-Yu, Philip Wong, H.-S., and Mitra, Subhasish (2016) Carbon nanotube computer. *Nature*, **501**, 526–530.

31 Zhang, J., Gaillardon, P.-E., and De Micheli, G. (2013) Dual-threshold-voltage configurable circuits with three-independent-gate silicon nanowire FETs. IEEE International Symposium on Circuits and Systems, pp. 2111–2114.

32 De Marchi, M., Sacchetto, D., Frache, S., Zhang, J., Gaillardon, P.-E., Leblebici, Y., and De Micheli, G. (2012) Polarity control in double-gate, gate-all-around vertically stacked silicon nanowire FETs. IEEE International Electronic Devices Meeting, pp. 8–4.

Part Six
Nanoelectronics-Enabled Sectors and Societal Challenges

18

Industrial Applications

L. Baldi[1] and M. Van de Voorde[2]

[1]*Formerly technology development manager at STMicroelectronics, Milano – Independent consultant on Nanoelectronics*
[2]*Rue du Rhodania, 5, BRISTOL A, Appartement 31, 3963 Crans-Montana, Switzerland*

18.1
Introduction

Nanoelectronics has undergone enormous development worldwide, and its potential for future growth has been illustrated in previous chapters.

However, nanoelectronics has also completely changed the way we live, impacting all sectors from medicine to energy, security, communications, and social relations.

The main focus of this part of the book is to provide evidence of the role that nanoelectronics has already played, and can play in the future in even larger way, in a large variety of fields in industry and society.

The role of nanoelectronics is in general hidden. When we use Internet to book an hotel, to check on Wikipedia or to upload our latest pictures on Facebook, we are not aware that nanoelectronics is enabling not only the tool, smartphone, tablet, or PC that we are using but also all the data base and computing systems, where information is stored and processed, and the communication network connecting them to us.

In 2009, the European Commission realized that enabling technologies, even if poorly visible to general public, play an essential role in industrial and innovation capacity to address the societal challenges, and started an activity to identify Key Enabling Technologies (KET) and to put them at the basis of the future European Research, Development and Innovation Programme that became Horizon 2020. Nanoelectronics was one of the KETs. It should be mentioned that similar research programmes have been launched in the United States, for example, National Science Foundation (NSF) and in the Department of Energy (US-DOE) as well as in Japan, for example, NIMS (National Institute for Materials Science in Tsukuba) and so on.

In Horizon 2020, Europe identified seven Societal Challenges that should drive the applications in the Programme. Similar potential fields for Research and

Nanoelectronics: Materials, Devices, Applications, First Edition. Edited by Robert Puers, Livio Baldi,
Marcel Van de Voorde, and Sebastiaan E. van Nooten.
© 2017 Wiley-VCH Verlag GmbH & Co. KGaA. Published 2017 by Wiley-VCH Verlag GmbH & Co. KGaA.

Innovation for societal welfare were defined in all industrial countries worldwide. Nanoelectronics underpins almost every single industrial sector and plays an important role in each challenge, even if with different degrees of visibility.

The progress of microelectronics in the last 50 years has been tremendous, but its evolution into nanoelectronics, and its potential for further development, as indicated in previous parts of the book, can introduce revolutionary changes in all industrial sectors and in society. Therefore, industry and society will have to know and learn more and more about the subject of nanoelectronics and its enabling power in all sectors.

In the following chapters, the readers will be provided an understanding of the as yet unexploited potential of nanoelectronics in the fields that Horizon 2020 indicated as main societal challenges for Europe:

- Health, demographic change, and well-being
- Food security, sustainable agriculture and forestry, marine and maritime and inland water research, and the bioeconomy
- Secure, clean, and efficient energy
- Smart, green, and integrated transport
- Climate action, environment, resource efficiency, and raw materials
- Europe in a changing world – inclusive, innovative, and reflective societies;
- Secure societies – protecting freedom and security of Europe and its citizens.

In this part of the book, we plan to give an overview of how nanoelectronics is going to play an important role in solving the main challenges of our society.

18.2
Health, Demographic Change, and Well-being

Meeting the health and well-being needs of Europe's aging society not only poses immense challenges but also offers valuable to exploit Europe's strengths in technology innovation, particularly in the area of nanoelectronics.

Opportunities in all parts of the care cycle – from screening and early diagnosis to treatment, therapy monitoring, and aftercare – are analyzed in chapter 19.

Perhaps less evident is the indirect impact that nanoelectronics can have on prevention and the definition of standards for healthy living, based on the possibility to analyze large banks of data. Moreover, Internet of Things (IoT) is going to play an important role in health and assisted living, with the possibility to deploy networks of specialized autonomous sensors in home or on the body.

18.3
Food Security, Sustainable Agriculture and Forestry, Marine and Maritime and Inland Water Research, and the Bioeconomy

Insuring a proper and healthy supply of food, while saving the environment, is a challenge of which Europe is becoming increasingly aware. Events like deforestation, depletion of maritime environment and the mad cow disease have demonstrated the need to keep a close control on the alimentary chain.

Nanoelectronics is just starting to enter the field: networks of sensors, monitoring systems, and robotics can increase the efficiency of agriculture while reducing the demand on natural resources. Smart tags can help to trace the food chain, and chemical sensors could in the future replace the expiration date for packaged food, thus reducing waste. Also in this field, IoT will be the approach through which nanoelectronics will extend its presence.

Some of the benefits of nanoscience have already been seen in the agrifood sector, others are still at the research and concept stage. They include applications as follows:

- Sensory improvements (flavor/color enhancement, texture modification)
- Monitoring pesticides and dosage of fertilizers for plants
- Sensors for monitoring the health of crops
- Biosensors and pathogen detectors for use in real time of production
- Nanosensors/diagnostics for food production control
- Monitoring food safety with chemical sensors in intelligent packaging materials
- Trace and tracking in food transport and distribution

18.4
Secure, Clean, and Efficient Energy

The industrial world is well aware of the strategic impact of energy supply, and on its impact on environment. Europe has set the following three aggressive targets for 2020:

- Greenhouse gas emissions 20% lower than 1990
- 20% of energy from renewable sources
- 20% increase in energy efficiency

Micro- and nanoelectronics are going to play an important role to this purpose as illustrated in chapter 20, both on the supply side, by facilitating the introduction of renewable energy sources with power electronics, and on the consumption side, by allowing a smarter use of energy.

Ways in which energy can be saved in our everyday life are covered also in the chapter 23 on Smart Cities.

18.5
Smart, Green, and Integrated Transport

Mobility is a major issue in all industrial countries worldwide, especially in urban areas, not only because it is still one of the major sources of pollution and of fossil fuel consumption but also because car accidents are one important cause of casualties. The impact of nanoelectronics on safe transport, with the introduction of forms of assisted driving is the topic of chapter 21. The role that nanoelectronics will play in reducing the pollution, with the move to hybrid and electrical cars, is covered in chapter 23 on Smart Cities, since it is in this area that it will find more easily applications.

18.6
Climate Action, Environment, Resource Efficiency, and Raw Materials

Realizing that the era of plentiful and cheap resources is coming to an end, Europe together with other nations has set the target of build a green economy – a circular economy in which reuse has the priority over exploitation, and sustainability is a key concept. There is a strong interaction with the challenge on energy, and it is in energy efficiency where nanoelectronics can play a major role. However, the deployment of autonomous interconnected sensors will help to a continuous monitoring of the environment. On the other side, the extensive use of information society tools in manufacturing will improve efficiency and reduce the use of resources, as described in chapter 22 on Industry.

18.7
Europe in a Changing World – Inclusive, Innovative, and Reflective Societies

Europe is living in an interconnected and changing world, as dramatically demonstrated by the recent economic crisis and by the mass migration of people. Reforms are needed in social welfare system, efficiency of public services, effectiveness of the judicial system, quality of education system, financial system, economic competitiveness, and so on. The most important role that nanoelectronics has played and is going to play in the future is in ensuring a comprehensive communication system, coupled to easy access to information. Probably no single action has contributed so much to the creation of an inclusive society as the introduction of smart phones and Internet. The potential for growth in the field of communications and social life, which is far from being exploited, is covered in chapter 22 on Consumer Electronics and chapter 23 on Smart Cities. In addition, the tools that nanoelectronics is making available for the collection and analysis of Big Data will surely have an important role in analyzing and forecasting societal trends.

18.8
Secure Societies – Protecting Freedom and Security of Europe and Its Citizens

Citizens of industrial countries, businesses, and administrations are increasingly dependent on Information and Communication Technologies (ICTs) for their daily activities. Recent events have shown that the risk of disruption, accidental, or most probably intentional, is always present and it has detrimental impact with high associated economic or societal costs. Also recent terrorist attacks have increased the concern about personal security. Nanoelectronics is called to cover a key role in this field, by providing tools for personal identification, and by hardening communication and control systems against any form of cyber attack, and making them more resilient against natural disasters.

Deployment of sensors and sophisticated tools for face and motion recognition will also support the fight against crime and terrorism, as described in the chapter on Smart Cities.

In addition to these application-focussed topics, two new themes have emerged in the Horizon 2020 programme, getting increasing attention: Smart Cities and IoT.

Smart Cities is an attempt to focus the impact of nanoelectronics in different areas to the specific task of solving the increasing problems raised by the growing urbanization. It aims to include nanoelectronics for mobility, energy saving, health, security, and inclusive society in an effort to make our megalopolis safer, more compatible with environment, and a more pleasant place to live in. With respect to the coverage given to societal challenges, there is an additional aspect of integration of different sectors and planning. A dedicated chapter (chapter 23) has been dedicated to this theme.

Internet of Things is a more elusive concept, especially since several different descriptions have been given to it. In summary, it is the possibility of combining sensors, low power electronics, and wireless communications to create a network of interconnected sensors and actuators, capable to provide services without requiring a direct human intervention. Its field of application spans from mobility to health, from domotics to energy saving and environment protection. Differently from the Smart Cities theme, which can be considered as a problem looking for solutions coming from applications of nanoelectronics, Internet of Things can be rather considered as a solution, or better, an approach to solutions, based on the still only partially exploited potential of nanoelectronics. For this reason, no specific chapter has been dedicated to it, but IoT appears in almost all chapters on Societal Challenges.

This part of the book treats the importance of nanoelectronics in these fields more in-depth and gives an Outlook of what is expected in the future.

19
Health

Walter De Raedt and Chris Van Hoof

imec, Kapeldreef 75, 3001 Leuven, Belgium

19.1
Introduction

The worldwide demographic evolution and our increasing life expectancy also imply that many of us will be suffering from one or more chronic illnesses during a larger part of our lives. Medical-grade wearables show great promise, having huge untapped potential to become a cornerstone technology in the field of health care. For chronic patients, tools are needed that improve the risk stratification, follow-up, and management and are also able to monitor disease progression and prevent relapse. A massive opportunity for wearable sensing has to do with behavior change in the illness prevention and healthy living areas. Frictionless technology, personalized algorithms/feedback, and power autonomy are key requirements for a widespread user adoption in this context.

19.2
The Worldwide Context

In the Western world, the health care system is facing severe challenges. Besides, the cost of health care is increasing tremendously (see Figure 19.1). This is especially the case in the United States, where the health care cost in the last 40 years has doubled to a value of $7.290/capita (2007) [1], more than double the value in countries such as Germany, the United Kingdom, and Japan. At the same time, millions of people in, for example, the United States do not have access to health care at all.

Demographic changes are often blamed for the rise in costs. As can be seen in Figure 19.2, the percentage of 65+ year olds has steadily increased over the last 25 years and is even to accelerate in the future, reaching values between 20 and 30%. However, data in Japan suggest that there is no direct relationship, as is

Nanoelectronics: Materials, Devices, Applications, First Edition. Edited by Robert Puers, Livio Baldi, Marcel Van de Voorde, and Sebastiaan E. van Nooten.
© 2017 Wiley-VCH Verlag GmbH & Co. KGaA. Published 2017 by Wiley-VCH Verlag GmbH & Co. KGaA.

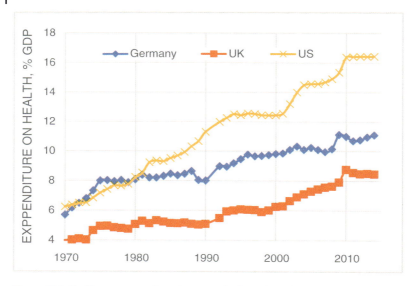

Figure 19.1 Health care expenditure increase of a few selected countries. (*Source:* OECD data [1].)

pointed out in Ref. [2], since Japan has the longest life expectancy in the world – one that European countries are not expected to reach for over two decades– and Japan spends only 8.5% of its GDP on health care, less than the OECD average. A frequently cited 2007 study of health care spending across OECD countries between 1970 and 2002 found that outside the United States, on average aging drove a rise of 0.5% per year in health expenditure, and other factors

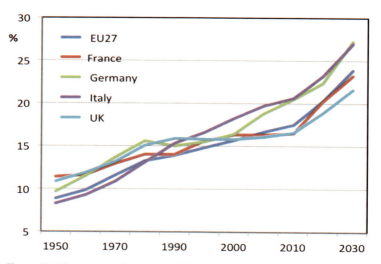

Figure 19.2 Percentage of 65+ years old in a few European countries. (*Source:* UN.)

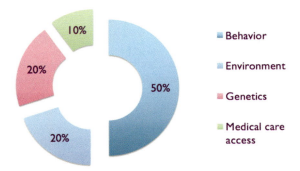

Figure 19.3 Relative importance of the factors determining personal health.

accounted for 3.2% [3]. Also, the steady progress in biomedicine and science has increased the survival from acute trauma, but this is associated with an increase in the number of individuals with severe disabilities [4].

The majority of the diseases is linked to chronic diseases. According to the OECD, between 10 and 20% of West Europeans over 65 years of age require some form of long-term care. Integrating this provision with health systems will become more important as the age increases beyond 65. Originally designed for acute, episodic care rather than ongoing treatment of chronic conditions, the current health systems are not good at providing the ongoing, coordinated care required by the latter [2].

A second important element is the need for prevention. Looking at the relative importance of the factors that determine personal health, it is striking that "behavior" scores very high (see Figure 19.3). Candace Imison, Deputy Director of Policy at the King's Fund, a British health care think tank, cites as a major challenge "behaviours within the population that are driving health needs: obesity, alcohol abuse, smoking, lack of exercise" [2]. These needs reflect four modifiable behaviors that are the root cause of much illness and premature death. Existing and emerging medical-grade wearable devices focus on diagnosis and chronic disease management. Consumer wearables are aiming to address our behavior (most notably fitness). By combining the convenience and frictionless use of consumer wearables, the signal quality of medical-grade physiological sensors and the personalized algorithms to adhere to coaching, powerful solutions will emerge that would assist our health at any age.

It is now recognized that the monitoring and prevention of diseases and follow-up of patients are two key components for improving the *quality of life*. The improvement in quality of life and the reduction of hospitalization bring associated cost savings at both the individual and society levels. On the other hand, similar wearable technologies can be used as personal trainers for the well-being of active people. Smart systems for health and well-being are specially designed to simultaneously measure many physical and physiological parameters. Captured data can be analyzed in real time to provide the physiological status of the person wearing the device(s). Such body parameters are useful for recording the physiological status of both healthy people and people with a chronic disease or frailty.

In smart systems dedicated to health care and wellness applications, the sensors collect physical, chemical, and biochemical data to enable interpretation and monitoring of a person's physiological status, in relation to the actual environmental and social context. To afford a complex multiparametric real-time sensing based on several types of sensors in portable smart systems, the power consumption per sensor element becomes a figure of merit as important as the other requirements (sensitivity, selectivity, robustness, reliability, integration on advanced silicon platform, and/or flexible substrates). The power figure should include the power required by readout electronics, signal conditioning and AD conversion, and wireless communication, but usually excludes digital signal processing such as linearization, pattern recognition, and sensor fusion.

Micro- and nanotechnology will play a key role here. Over the last 60 years, technology has progressed tremendously. Following Moore's law, huge computation power has been employed in handheld devices, sensors have shrunk in size, and wireless communication has penetrated into the consumer market, to name but a few developments. A schematic is shown in (Figure 19.4), with sensors all over and sometimes inside the body, enabling to monitor people (e.g., 24/7) or to assist them in conducting activities that help in the prevention of diseases (e.g., fitness). These devices enable monitoring an increasing amount of physiological parameters, such as ECG, EEG, blood pressure, posture, and so on.

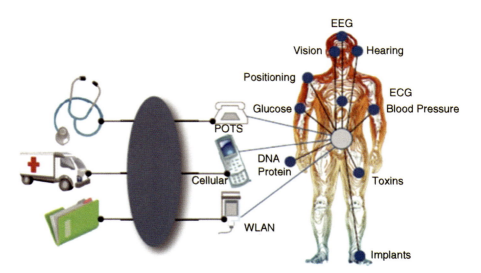

Figure 19.4 Schematic of a body sensor network. On-body sensors measure physiological parameters and send these to the cloud.

19.3
Requirements and Use Cases for Emerging Wearables

The existing and emerging medical-grade wearable devices focus on diagnosis and chronic disease management (e.g., cardiac rhythm patches) [5]. Consumer wearables are aiming to address our behavior (so far mostly fitness assessment and follow-up). By combining the convenience and frictionless use of consumer wearables, the signal quality of medical-grade physiological sensors, and the personalized algorithms to adhere to coaching, powerful solutions will emerge that assist our health at any age.

In many cases, the embodiments of smart systems for the health domain will be quasi-invisible, self-powered body area networks or, if appropriate, implantable devices, monitoring vital health signals and offering the required information for taking appropriate actions to preserve human health. They should acquire a well-defined view of the state of a person's health by using a real-time, ultralow-power, multiparametric combination of nonintrusive, biosignal sensors (ECG, MEMS-based accelerometers and gyroscopes, optical pulse oximetry, etc.) to allow early warning and subsequent enhancement of the quality of life. They can employ future technologies such as electronic skin or wearable self-powered networks of sensors with wireless interfaces communicating at a few meters. To benefit from already existing mobile communication technologies, such systems should be compatible, from the communication viewpoint, with existing gateways (such as mobile phones) to serve as smart parts in the *Internet of Things* [6]. Possible implementations of smart systems such as patchable and/or wearable sensors with communication interfaces and their own sources of energy, energy storage, and power management (integrated in smart garments, smart watches, etc.) are schematically shown in Figure 19.4.

The key characteristics of such devices and systems are as follows:

- The devices are to monitor specific body signals, intermittently or continuously and linked to an individual. Therefore, they are on or near the body, embedded in clothing (or implanted for a certain period of time).
- They should be comfortable and unobtrusive. This sets upper limits to their dimensions and weight. Typically, their size is on the order of magnitude of millimeters to centimeters.
- The devices either store their data internally and are read out regularly, or they are wirelessly linked to a central hub (e.g., smartphone), which enables to share the data externally.
- The devices should be autonomous, which means that there should be sufficient energy available to operate the device during the required lifetime of the application. This can range anywhere from a few minutes to a few months.

As an important consequence of the requirements above, the power budget is limited. There are two solutions. Either the devices are battery operated, powered by energy harvesting, or a combination of both. In case of battery-operated devices, the maximum size of the device determines the available capacity of a battery. In case of energy harvesting, power is also very limited despite many claims in literature. It is important to note that cost is an important factor here. MEMS technology enables to reduce the cost of the harvesters considerably, but this is at the expense of power delivered. As a consequence, the power budget for these devices is directly limited from several microwatts to a few milliwatts at most.

Here, we will put our focus on three types of applications domains:

- *Stay fit:* Applications for the healthy people to enable them to become or stay in a fit condition.
- *Get well:* Applications for people who suffer from a disease and where the applications can either treat or monitor the status and rehabilitation progress.
- *A better live:* Applications for the chronically ill, where technology can help to increase their quality of life.

Wearable devices are now being used (and will increasingly do) for prevention (including the promotion of a healthy lifestyle through fitness and stress monitoring), diagnostic purposes, therapy, and therapy monitoring. They can also be used at home by the user or can enable the patient to be monitored or treated at home under supervision of a professional. In the latter case, they act as a kind of an extension of the hospital care.

Another important development is related to mHealth (m-health or mobile health), which is a term used for the practice of medicine and public health, supported by mobile devices. Currently, the most important device is the smart-phone or tablet, but devices such as wrist watches are emerging due to their apparent convenience. As a consequence, all the use cases illustrated below need to interface with these devices, connecting wirelessly using protocols such as Bluetooth, WiFi, and so on. Once connected to the hub (smartphone or gateway at home), they take care of transferring the data to the user or the cloud.

Let us now have a look to the ways technology is evolving and how it will enable us to have a longer healthy life.

19.3.1
Assisted Living

Let us imagine a person called Lars, aged 85. He still is in a good condition, and fortunate enough to be able to live independently in his convenient flat in the village where his daughter lives. He enjoys to meet his close friends regularly and to organize his life. However, this is not so obvious after his fall a few months back: he slipped on a wet spot in his entrance hall and only after 2 h his daughter passed by and found him, since he was unable to stand up again due to heavy pain in his left leg: fortunately, nothing was broken but he – and his anxious daughter – got a serious alert. Next time could be worse or even. . . . Luckily, there was this new start-up in the village that was looking for candidates to evaluate a new integrated alarm and care system: He only needs to wear a wristband – a pearl of advanced technologies integrating a GPS receiver, a mobile connection, an intelligent activity tracker, and a pulse sensor – and a few wireless sensors are placed in his apartment. All information is streamed to a cloud server and processed with the latest algorithms in order to interpret his condition. He now can also give an alarm wherever he is, so that his worried daughter receives a message (the GPS tracker sends his coordinates) and can help or call emergency services. This type of solutions allows him to keep both his freedom and his privacy. Long-term follow-up of his behavior patterns will allow identifying potential threads for his health in the future.

19.3.2
Congestive Heart Failure (CHF)

A few years before, at the age of 82, Lars developed a congestive heart failure (CHF), a complex chronic condition where the heart is failing to provide adequate perfusion to the body. This is a typical case of a heart failure patient with a reduced ejection fraction and this causes a rise in pulmonary pressures forcing fluid to leave the vascular system and build up inside the lung (congestion). This congestion is one of the main drivers of disease worsening and hospital readmission. This fluid buildup is currently assessed using invasive catheters, echocardiography, and thoracic X-ray imaging. An experimental wearable bioimpedance sensor patch system (see Figure 19.5) consisting of a reusable sensor module and a disposable printed patch can be used to reflect this fluid buildup as well as the fluid drainage during treatment (see Figure 19.6) of CHF patients. Larger patient studies are currently ongoing. With such a wireless bioimpedance patch, Lars' condition might be traceable at home with the aim to prevent

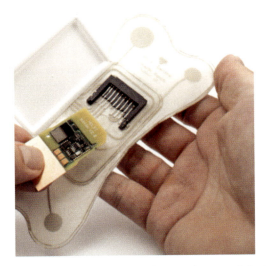

Figure 19.5 Wearable wireless bioimpedance patch for body fluid assessment (imec).

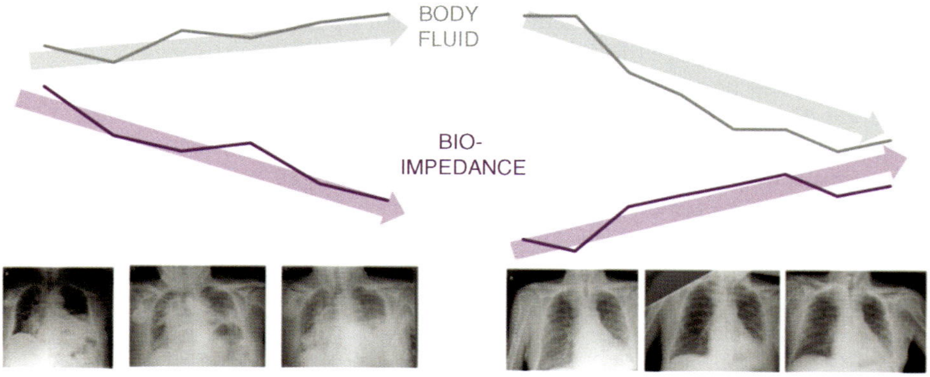

Figure 19.6 The bioimpedance signal reflects fluid buildup in the lungs (left) and during fluid drain (right). Side-by-side thorax X-ray for confirmation.

hospital readmission while it may at the same time become possible to monitor treatment efficiency.

19.3.3
Cancer and Point of Care

At the age of 70, Lars encountered strong indicators of possible problems with his prostate. After examination, his doctor performed a PSA test; after a few days Lars had to come back to discuss the results and since there were worrying indications on the possible presence of a tumor, he was directed to the hospital for taking a biopsy of his prostate in order to check the severity of the tumor. A week later this biopsy was taken and the evaluation was positive: It was a benign tumor in an early stage, and only regular follow-up was needed. But let us imagine the potential of the ongoing developments of a new generation of POC (point of care) devices in this or comparable scenarios.

Currently, the first POC devices are appearing (e.g., desktop-like diagnostic tools of Biocartis). These devices allow, for example, faster blood molecular analysis at the doctor's cabinet rather than sending blood samples to the centralized clinical laboratory, resulting in a faster, cheaper, and more personalized analysis with faster treatment iterations.

But the use of nanotechnologies will bring even more exciting solutions: At several research laboratories (e.g., Johns Hopkins [7], and imec), lab-on-chip concepts are turning into reality: The idea is that a disposable chip can be loaded with saliva, a blood sample, or other body fluid and then be analyzed fast and efficiently with a standard laptop or smartphone even at home. Application for disease monitoring and surveillance, rural health care, and clinical trials is obviously very attractive.

Such devices are enabled by the integration of several expert domains: nanotechnology, microfluidics, optics, microbiology, and so on. An example of such approach is shown in Figure 19.7.

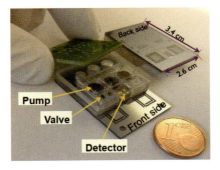 

Figure 19.7 On-chip individual cell imaging and sorting for disease detection (imec).

After applying a drop of blood on the sensor, through microfluidic channels and by means of micropumps, this blood is led to a planar integrated lens-free CMOS imaging sensor, which is able to identify different types of cells in the bloodstream such as single cancer cells. After identification, each cell type is routed toward a specific sensor for further analysis. This technology is able to deal with DNA analysis, protein identification, blood cell analysis, and so on [8].

Imagine how this technology could simplify the long-term analysis and follow-up of Lars' prostate problem.

The first smartphone-connected personal devices for single function detection are appearing on the market (e.g., the smartphone glucose sensor Glucodoc from Medisana), and more complexity toward multiple complex molecules will appear soon.

19.3.4
Sleep Monitoring – Sleep Apnea

Let's now go back further in time. Lars is 52 years old and is developing curable risk factors such as obstructive sleep apnea (OSA) and hypertension. Sleep apnea is a respiratory condition with an intermittent collapsing and opening of the upper airway that causes Lars to stop breathing. There can be hundreds of these Apnea–Hypopnea Index events (AHI) during the night that can last up to a minute or longer. The opening of the airway causes a typical snoring sound. While this sound is innocent, the episodes of nonbreathing cause chronic stress on the body that can drastically increase other risk factors (i.e., hypertension, heart disease, stroke, drowsiness, etc.). Today, polysomnography (PSG), a multichannel vital sign monitoring system, is used in a sleep clinic to diagnose for sleep apnea. A wearable system based on an integrated multisensor system-on-chip (see Figure 19.8a) [9] is able to track relevant vital signs such as heart rate, respiration rate, and respiration volume that can reflect episodes of apnea during sleep.

A small patient study (see Figure 19.8b) showed a high correlation in AHI events compared with gold-standard PSG testing. When such a system is

(a)

(b)

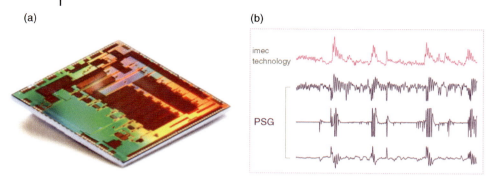

Figure 19.8 (a) The multi-sensor system-on-chip (MUSEIC) for low-power medical signal acquisition and processing. (b) Respiration rate and volume during sleep apnea episodes.

successfully validated, Lars could avoid costly clinical diagnostics and long waiting times by having his sleep apnea diagnosed at home with a convenient health patch. In order to check if Lars's blood pressure is normalizing after the treatment of his sleep apnea, he would benefit from blood pressure tracking. Cuffless wearable wristband devices that create a blood pressure proxy by measuring pulse-arrival time (PAT) are in development and are undergoing rigorous clinical validation: Such system would be a valuable tool for Lars and his peers.

19.3.5
Presbyopia

Let's go further back: Lars is now 45 years old and is developing presbyopia, a disorder affecting the intraocular lens in his eye that refocuses the light for near-vision tasks such as reading. Instead of using reading glasses, a smart contact lens platform containing smart electro-optic lenses that can tune their focal length would be a tremendous increase of comfort for him. Such a device could mimic the original functionality of the defective intraocular lens while intrinsically maintaining a full field of view. Getting the entire system in the form factor of a contact lens while ensuring sufficient oxygen permeability has not yet been achieved. However, the ingredients for such smart contact lens (LCD display, autofocus sensing, control circuitry, battery, and charging circuits) intended for 1 month use are being demonstrated (see Figure 19.9).

On top of challenges for the circuitry and the system components, also true fabrication technology innovations are appearing [10] that allow fabricating interconnections and circuits on flat deformable substrates that need to be curved to get the required optical shape.

Although at first glance one thinks about improving the functions of contact lenses with advanced technology developments, other uses are also under development such as the integration of tear analysis sensors that enable the measurement of the glucose levels in tear fluid [12] (Google, Medella, etc.).

(a) (b)

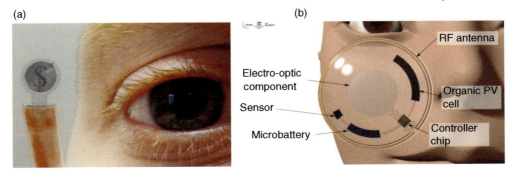

Figure 19.9 (a) LCD display integrated in a contact lens [10]. (b) System components vision for smart contact lens [11].

19.3.6
Fitness and Stress

Let us go further back in time. Lars is now 30 years old. He has a stressful job and has a hard time handling stress. He developed the habit of eating too many sweets and is now overweight. Current fitness trackers are able to confront him with his level of activity, but do not reflect his improving/worsening cardiorespiratory condition. A textile-integrated health patch (see Figure 19.10) not only gives Lars frictionless technology but also provides him with a longitudinal personalized cardiorespiratory fitness assessment.

Quantification of Lars' stress level however is another challenge. How can technology help? Unfortunately, there is not a single sensor for stress evaluation; there is no such device or marker as a stress thermometer. When Lars' brain registers stress, it sets things in motion and stress hormones are released. You could measure them in the blood but that is not convenient, nor real-time and certainly not unobtrusive. These hormones trigger measurable changes that we can measure: Lars' heart rate increases and associated variability decreases, his

Figure 19.10 Textile-integrated health patch for frictionless wearable technology.

perspiration increases, his pupils dilate, and his respiration changes. And these autonomic nervous system responses can be measured. Current stress-sensing wearables typically monitor heart rate and/or maybe his skin conductance. For monitoring Lars's stress level, a truly personalized sensor is needed, because studies are revealing that mental stress expresses itself differently from person to person, so several markers need to be followed up, such as heart rate, heart rate variability, skin conductance, muscle tension (EMG), activity level, social behavior, and so on.

Long-term stress leads to health problems such as depression, burnout, and cardiovascular disorders. Research on the development of objective, continuous, and personalized stress measurement techniques with wearable as well as efficient stress management techniques is still in its infancy. However, more than just hardware is required here, a frame for motivating people toward continuous awareness and behavior change is also required. Stress management is possible only through the implementation of behavioral science strategies-based personalized machine learning algorithms.

For stress management, context-aware machine learning algorithms are needed to determine the suitable combination of many (up to 20) physiological parameters. In a large trial with over 1200 subjects, this approach is being tested and validated. In the next generation, just-on-time messaging will ensure that information from behavioral scientists and psychologists (i.e., the right message) is given at the right time (based on multisensory information as well as contextual information). This aims to keep Lars engaged with the aim to reduce the risk of, for example, burnout.

19.3.7
Pregnancy

Even before Lars was born, wearable sensors could play a key role. Lars' mom would really benefit from preventive health tools while being pregnant. Pregnancy monitoring could be a smart patch, which would provide continuous monitoring of both biochemical markers from peripheral blood or interstitial fluids and electrical (ECG) signals from both mother and fetus. Such wearable sensor could create a comprehensive view of both maternal and fetal health (see Figure 19.11). These wearables will capture an extremely rich and comprehensive data set on maternal and fetal health and may ultimately pave the way for evidence-based clinical algorithms to predict and manage pregnancy complications such as preterm birth, preeclampsia, and gestational diabetes.

19.3.8
Advanced Computing Needs Only Grow

Popularized by wristbands to track one's activity, this area is rapidly expanding due to the development and usage of advanced sensors that can measure much more than the number of steps walked. More and more users expect small

Figure 19.11 Wearable device for monitoring maternal and fetal parameters. (*Source:* Bloom Technologies.)

devices that take accurate measurements while lasting very long on a single-battery charge. Moreover, they want to track progress on their smart watch, smartphone, or Web sites, and get personalized recommendations to improve their lifestyle. It is, therefore, no longer possible to talk about separate applications that run on an ultralow power wristband, a smartphone, and the desktop, but rather one application that spans multiple devices, taking measurements from sensors, showing the results on a smartphone, aggregating the measurements of all users and applying large-scale analytics in backend data centers, personalizing the knowledge, and alerting the user. There is, therefore, a large and growing need for novel distributed software applications that are also enabled by the explosion of the cloud computing paradigm, which itself is enabled by ongoing advanced nanoelectronics technology development – following Moore's law.

19.4
Conclusions

It is now recognized that the monitoring and prevention of diseases and follow-up of patients are two key components for improving the *quality of life*. The improvement in quality of life and the reduction of hospitalization bring associated cost-savings at both the individual and society levels. On the other hand, similar technologies can be used as personal trainers for the well-being of active people. Smart wearable systems for health and well-being are specially designed to simultaneously measure many physical and physiological parameters. Captured data can be analyzed in real time to provide the physiological status of the person wearing the device. Such body parameters are useful for recording the physiological status of healthy people as well as that of people with a chronic disease or frailty.

In smart systems dedicated to health care and wellness applications, sensors collect physical, chemical, and biochemical data to enable interpretation and monitoring of a person's physiological status, in relation to the actual environmental and social context.

The devices are to monitor specific body signals, intermittently or continuously and linked to an individual. Therefore, they are on or near the body or embedded in clothing. They should be comfortable and unobtrusive. This set upper limits to their dimensions and weight. Typically they are on the order of millimeters to centimeters. The devices either store their data internally and are read out regularly or they are wirelessly linked to a central hub (e.g., smartphone), which enables to share the data externally through cloud computing. The devices should be autonomous, implying there should be sufficient energy available to operate the device during the entire lifetime of the application – ranging from a few minutes to a few months. Therefore, this stresses the fact that the ongoing nanotechnology development is a strong enabler for these requirements.

References

1 OECD (2015) Health spending (indicator). doi: 10.1787/8643de7e-en (accessed October 28, 2015).

2 http://www.reforminghealthcare.eu/economist-report/challenges-facing-healthcare-systems/demography-and-disease-load (accessed January 17, 2017).

3 White, C. (2007) Health care spending growth: how different is the United States from the rest of the OECD? *Health Aff. (Millwood)*, **26** (1), 154–161.

4 Goonewardene, S.S., Baloch, K., and Sargeant, I. (2010) Road traffic collisions-case fatality rate, crash injury rate, and number of motor vehicles: time trends between a developed and developing country. *Am. Surg.*, **76**, 977–981.

5 Lobodzinski, S.. (2013) ECG patch monitors for assessment of cardiac rhythm abnormalities. *Prog. Cardiovasc. Dis.*, **56** (2), 224–229.

6 ec.europa.eu/digital-agenda/en/alliance-internet-things-innovation-aioti (accessed November 4, 2015).

7 http://www.hopkinsmedicine.org/news/media/releases/johns_hopkins_teams_with_belgian_based_research_organization_to_expand_health_care_applications_for_silicon_nanotech.

8 Deshpande, P. (2013) Sensors in medicine, London, March www.sensor100.com/sensmed2013.

9 Anand, I.S. *et al.* (2011) Design of the multi-sensor monitoring in congestive heart failure (MUSIC) Study: prospective trial to assess the utility of continuous wireless physiologic monitoring in heart failure. *J. Card. Fail.*, **17** (1), 11–16.

10 De Smet, J., Avci, A., Joshi, P., Schaubroeck, D., Cuypers, D., and De

Smet, H. (2013) Progress toward a liquid crystal contact lens display. *J. Soc. Inf. Disp.*, **21**, 399–406.

11 De Smet, J. (2014) Ph.D. dissertation, UGENT. ISBN 978-90-8578-663-4.

12 Liao, Y.-T., Yao, H., Lingley, A., Parviz, B., and Otis, B.P. (2012) A 3-μW CMOS glucose sensor for wireless contact-lens tear glucose monitoring. *IEEE J. Solid-State Circuits*, **47** (1), 335–344.

20
Smart Energy

Moritz Loske

Fraunhofer Institute for Integrated Circuits IIS, Networked Systems and Applications Department, Nordostpark 84, 90411 Nürnberg, Germany

Energy is a cross-cutting topic in today's and future society. Production, economy, and modern life are depended on a reliable and cost-efficient energy supply, especially in industrial nations. Smart energy systems, like smart grids and decentralized microgrids and energy management systems emerge to encounter today's societal challenges.

This chapter gives an overview about smart energy as a nanoelectronic-enabled sector and its societal challenges. Starting with the evolution of the energy supply caused by environmental and societal ambitions and the resulting need for smart energy systems, the viable technical measures are discussed. Finally, nanoelectronics as well as micro- and power-electronics as key enabler for smart energy are addressed.

20.1
Energy Revolution – Why Energy Does Have to Become Smart?

Since the industrialization started in the eighteenth century, a reliable and affordable energy supply is the basis for modern economy and society [1]. From this century, the combustion of fossil fuels like coal and oil and later nuclear power were the primary energy sources. The resulting environmental issues and the shortage of fossil resources led to a rethinking of the energy supply and a more rational use [2].

The future goal of the energy revolution is a carbon neutral, sustainable and efficient, but also affordable energy supply based on renewable energies. With the 20-20-20 agenda, the European Union aims to rise the share of renewable energies for the overall energy consumption up to 20% until the year 2020 [3]. The decentralized energy generation and the intermittent in-feed of renewable energies lead to instabilities and thus great challenges for the energy supply. In addition, new mobile applications and self-sustaining energy systems, for example, sensor nodes for monitoring purposes emerge. This development veers toward the total electrification and the "all-electric society" [2]. The International

Nanoelectronics: Materials, Devices, Applications, First Edition. Edited by Robert Puers, Livio Baldi, Marcel Van de Voorde, and Sebastiaan E. van Nooten.

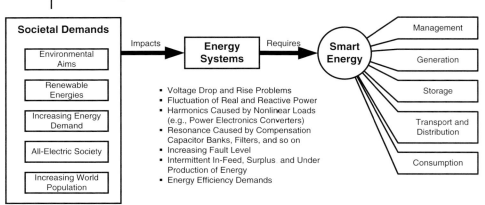

Figure 20.1 Smart grid energy revolution.

Energy Outlook says that until the year 2040 the net-generation of electricity will almost be doubled compared to 2010 to a value of 39.0 trillion kWh per year [4]. Minimizing global consumption of electrical energy is therefore undoubtedly one of the major goals [5]. According to the World Energy Outlook, the total consumption of electric energy could be more then halved by the end of the next decade by improving power efficiency [6].

Figure 20.1 illustrates the cause-and-effect chain of the energy revolution. Today's society is affected by (1) environmental aims resulting in the integration of (2) renewable energies, (3) the increasing demand of energy caused by (4) the "all-electric society," and (5) the growing world population. These demands have direct impact on the supply of energy and on energy systems. These energy systems have to face the following new and inevitable challenges:

- The challenges for the energy grid are intermittent in-feed, surplus, and under production of energy that results in voltage drop and rise problems, fluctuation of real and reactive power, resonance caused by compensation capacitor banks, filters, and an increasing fault level [7].
- The challenges for small and local energy systems and devices are harmonics caused by nonlinear loads (e.g., power electronics converters) and increasing energy efficiency demands [7].
- The challenges for mobile and self-sustaining applications and devices are the increasing energy efficiency demands and advanced functionality and miniaturization goals [2,5].

To handle these challenges caused by the societal demands, energy has to become smart in manners of a smart energy management, generation, storage, transport, and distribution and consumption. Smartness is achieved on one hand by advanced software and algorithms, and on the other hand by applying nanoelectronics, microelectronics, and power-electronics to the devices and the system components. The focus of this chapter thereby lies on semiconductor technologies. The combination of specialized integrated circuits (IC) and, for example, high power switches and converter offer a large grade of interactivity and control capabilities. These measures will lead to a certain intelligence of the system components and an overlaying intelligence of the central energy management – altogether smart energy.

The term smart energy and the elements necessary to achieve it are dealt with in the following sections.

20.1.1
Smart Energy and Systems

The term smart energy comprises all intelligent technologies of energy generation, energy storage, transmission distribution, consumption, and control. Therefore, the entire value chain from the energy generation to the energy consumption is addressed. To make energy smart, these components have to interact with each other via an overlaying intelligent energy management [8].

A more simplified definition of smart energy would be the intelligent and on demand supply of applications and devices with energy, whereas unconsumed energy is stored for later use or provided for other applications. The involved system components interact by utilizing communication technologies.

The general objectives of smart energy are on one hand to increase efficiency, reliability, and sustainability of energy supply and on the other hand to minimize costs. Additionally, smart energy enables new and advanced applications.

The term smart energy often refers to electric and electronic systems. The smart grid is the most relevant and familiar example for a smart energy system. But smart energy must be seen in a larger context than just the energy grid [9]. There are other energy systems of different size and applications, for example, local energy management systems, home and building automation, and micro-energy grids. Even the energy supply and management of mobile and self-sustaining systems are representatives of smart energy systems. These are, for example, self-sustaining sensors and sensor networks and other cyber physical systems (CPS) or applications of electric-mobility like buses or trains that recuperate energy by braking. Smart energy systems can therefore be seen as systems of different size and functionality within the energy value chain. Selected examples of smart energy systems are introduced in Section 20.2.

All smart energy systems feature the same characteristic properties: interconnection between system components and advanced control strategies in combination of an efficient energy generation, storage, transmission and distribution, and also consumption. The difference of an energy system to a smart energy system is

- interconnection of system components,
- dynamic and extensive controllability of the application-process, and
- interaction with costumers and users (demand response, energy market, etc.).

Smart Energy

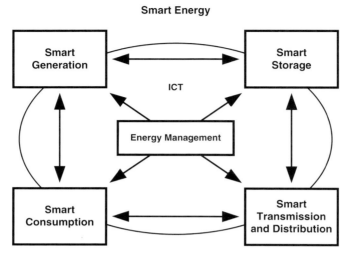

Figure 20.2 Smart energy effect matrix based on "Wirkungsmatrix des Smart Grid" [10].

All smart energy systems can be described using the effect-matrix, which is described in detail in Section 20.1.2.

20.1.2
Smart Energy Effect-Matrix

The main goal of smart energy is to efficiently balance the dynamic demand with the fluctuating energy generation using a flexible and advanced energy management. To make energy smart, each element of the energy supply chain has to become smart. The effect-matrix in Figure 20.2 gives an overview on the smart energy elements and their interaction.

20.1.2.1 Smart Generation
First, the energy has to be generated in a sustainable and economic manner based on renewable energy sources. To run modern applications and devices commonly, electric power is required. This power has to be generated using primary energy sources. A steady increasing share of the consumed energy is generated from renewable energy sources like photovoltaic panels on house roofs, windmills, or hydroelectric power stations. New mobile applications like self-sustaining sensor systems emerge, which demand a local and direct energy generation decoupled from the energy grid connection. Already since 1770, with the invention of the self-winding watch self-powered applications are known [11]. Energy harvesting technologies are used to harvest small-scale electric energy form environmental and ambient energies. Against the trend of decentralization, still large-scale energy is generated in large power plants like offshore wind parks, solar parks, or hydropower stations.

20.1.2.2 **Smart Storage**

As renewable energy sources have the characteristic of a fluctuating generation and in-feed, the energy has to be buffered or stored in smart energy storages for short- or long-term storage. Due to this fluctuation, smart energy storage systems are one of the key technologies to balance and buffer the generation and consumption of renewable energy. Many approaches are already available, like lithium-ion batteries, power-to-gas transformation, heat storage systems, or local hybrid storage systems. There are also ambitions to utilize the capacity of electric vehicles as temporary and mobile energy storage systems. Power-to-gas has a great potential, as gas has a high energy density and can be easily stored, transported, and retransformed into electric power, not restricted on the location of production or in-feed.

20.1.2.3 **Smart Transmission and Distribution**

Another important element of smart energy is the energy transmission and distribution. Often energy is generated away from the place of usage and utilization. Thus, on one hand an efficient and flexible energy transmission is essential for smart energy. On the other hand local and decentralized, and sometimes self-sustaining applications emerge, which require a smart local distribution of energy directly form generation to consumption.

20.1.2.4 **Smart Consumption**

An energy-efficient consumption is essential, either in small-scale households and mobile applications or in large-scale applications in industry and goods production. Smart energy consumption is an important element of smart energy, since it means the energy efficient and sustainable use of a valuable commodity created with great technological effort. To achieve that, intelligent and efficient applications and algorithms together with energy efficient and power saving devices and hardware are required.

20.1.2.5 **Energy Management**

To finally enable these elements to form smart energy systems, the elements have to intelligently interact by means of an overlaying energy management. An interconnection of the elements is thus indispensable. Energy management is based on data acquisition, on status estimation, and a reliable information and communication technology (ICT) [12–14]. To realize smart energy and smart energy systems, the smart energy elements have to be equipped with communication interfaces to connect the components to the internet of energy [15].

A smart energy management knows the variation of energy consumption and demand of the day, as well as a prognosis about the expected energy generation. Energy management provides the control structure to mediate between the smart energy elements within a smart energy system of a certain application and size. It has the function of balancing the energy generation and demand and to offer a reliable and efficient energy supply.

20.2
Applications of Smart Energy Systems and their Societal Challenges

In this section, common and specific examples of smart energy systems are introduced. The most important representative is the multi-energy smart grid followed by high voltage energy transmission and energy distribution systems. Besides the holistic and comprehensive approach of the multi-energy smart grid, also microenergy grids are discussed. Energy harvesting systems and mobility applications complete the brief insight into smart energy systems. Further, but not discussed, examples of smart energy systems are battery management systems, street lighting applications or mobile devices and applications.

20.2.1
Multi-energy Smart Grid

From the beginning of the electrification, the energy generation is altered from a decentralized to a centralized solution. With respect to the societal aspects (cf. Section 20.1) this trend is reversed back again to a decentralization of the energy supply and local generation. The system is changing from a quasistationary top-down architecture to a dynamic top-down and bottom-up energy system [16]. To maintain and even increase energy efficiency despite of this alteration, conventional energy grids have to be transformed into smart grids. An integrated approach is needed that includes rural, urban, and local energy supply (sub-)systems and which bridges the gap between renewable and conventional energies [2]. Smart cities are a novel and future application of smart energy systems. As urban areas will increase in the next years, the energy supply of cities is a major challenge of the future. When today about 50% of the world population lives in urban areas, the United Nations Department of Economic Affairs predicts a rapid increase to 70% in the year 2050 [17]. As this topic is of high significance for the future, it is comprehensively discussed in Chapter 23 of this book.

An additional trend is the combination and direct interaction of different energy grids primary electricity, also gas and heat [2]. So-called multi-energy smart grids are energy networks that intelligently integrate the actions of all generators, consumers, storages, and utilities connected to it, with the ambition to assure a sustainable, economic, and secure energy supply [10,18,19] and to optimize the use of system-immanent capacities and resources [16].

Figure 20.3 depicts a theoretical structure of a multienergy smart grid, combining the energy carriers electricity, gas, and heat to a joint network.

On the higher voltage or transmission level on the left side of the diagram wind power (WP) and gas turbine power plant are exemplified for a centralized energy generation. On the lower voltage or distribution level, photovoltaics (PV) and a microcombined heat and power plant (µCHP) as common representatives for local generation are shown. Additionally, a battery (BAT) as short-time storage system for energy buffering purposes and a controllable load (CL) for demand response application are displayed. The connection between the

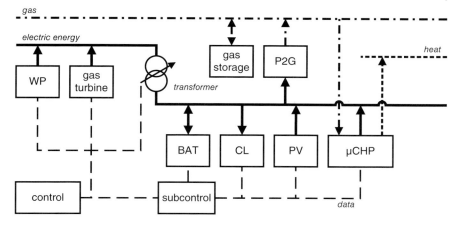

Figure 20.3 Schematic of a mulitenergy smart grid.

transmission and distribution grid is realized with a controllable bidirectional transformer.

The link from the electric energy grid to the gas energy grid is established via a power to gas (P2G) conversion plant. A link in the other direction, from the gas energy grid to the electric energy grid, is established via a μCHP. The μCHP also supplies energy to the heat energy system (e.g., district heating). The control-structure in the example is done hierarchically. It consists of the subcontrol(s) for the distribution grid and the overlaying control structure.

Distributed intelligence is an important aspect of multienergy smart grid. Individual systems are self-sustaining and are dedicated to an overlaying energy management. The grid is structured hierarchically in order to communicate with the next overlaying energy management system, which means the distributed and local intelligence yields into a central intelligence and form a stable and secure integral system [20]. In case of the outage of a subsystem caused by fault or manipulation, the overall system stays stable and secure. Out of this set up, a digitally monitored and self healing energy system emerges that reliably delivers electricity, gas, or heat from generating sources, including challenging renewable sources, to the points of consumption [5].

To sustain those energy systems, a bidirectional communication between the distributed system elements is required. For that existing information, networks will be utilized to exchange information between power plants, storage systems, and users to keep the system efficient and manageable [2]. The reliable power delivery is optimized by facilitating a two-way communication across the grid, enabling end user energy management, minimizing power disruptions, and transporting only the required amount of power in order to achieve the efficient interplay between global and local technical optimization (distributed grid management) [21]. The result will be lower cost for the utility-provider and the customer, more-reliable power distribution, and reduced carbon emissions [5]. The accurate forecast of the energy demand and also the generation (e.g., weather prognosis for PV or WP) as well as the state estimation of the smart grid are essential to efficiently control the energy flow and to offer a reliable and sustainable energy supply. For that, the energy-grid and relevant devices and components have to be equipped with sensors, actuators, intelligent controllers, and an appropriate information and communication technology for the interaction. Especially, sensor systems and their networking allow grids to become smart and aware of bottlenecks and fault protection, and help to automate the grid control and functioning [2]. A key trend and future technology is the virtualization of power plants as well as energy storage by virtually combining distributed renewable energies and respectively energy storage systems. Such virtual power plants (VPP) and storages aim to optimally integrate these energy units into the energy grid [2]. The key advantages of smart grids are as follows:

- Optimal use of ressources and capacities
- Equitable power distribution
- Control of consumption overloads
- Optimized use of power
- Power efficiency
- Reduction of peak demand
- Intelligent monitoring and control [2,10,22]

Besides efficient energy generators and consumers, also energy storage systems play an important role to realize self-sustaining distributed smart energy systems. Small stationary battery systems buffer the local energy fluctuation of buildings and midsize storage systems (e.g., redox-flow batteries) cover load peaks of urban industrial production or punctual overproduction of waste incineration and biogas plants. Another possibility to store energy for longer periods of time is the binding of energy in carrier mediums, like gas or liquids. Examples for such technologies are power-to-gas plants: Energy surplus from renewable energy sources, for example, great sunshine hours is used to produce hydrogen H_2 via water electrolysis and subsequently produce methane in a methanisation reactor using carbon dioxide (e.g., from biogas plants) and the renewable generated hydrogen. Power-to-gas has a great potential, as gas has a high energy density and can easily be stored, transported, and retransformed into electric power, not restricted on the location of production or in-feed. These processes have to be optimized and be made more efficient. Another possibility to store energy is a chemical hydrogen storage system, which stores hydrogen in a liquid organic hydrogen carrier (LOHC). The advantage of these technologies is, that the energy can be stored in tanks and easily be transported to the place of usage (e.g., gas network). The electric power can locally be generated by CO_2-neutral combustion. Means of energy transmission and distribution are considered separately in Section 20.2.2.

20.2.2
High Voltage Transmission and Distribution Systems

Energy is often generated remotely from the place of demand and consumption. Thus, the energy has to be transmitted over long distances, for example, from

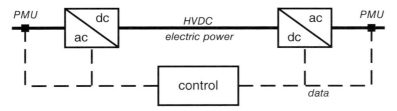

Figure 20.4 Schematic of a high-voltage DC transmission system.

offshore wind parks in the North Sea to heavy industries in the south of Europe. To minimize power losses in the transport medium, high voltage technologies are applied. The main technological segments for the high voltage transmission (HVT) are the high voltage DC (HVDC) transmission and flexible AC transmission Systems (FACTS) [7,22–24].

A promising technology for large distance transmission is the high-voltage direct-current technology, which is already in use for some offshore links and intercontinental circuits. A recent example is the realization of the link to the BorWin3 offshore wind park with a transmission power of 900 MW. Besides the transmission of large amounts of energy, HVDC systems can also be used for the connection of two AC networks with asynchronous frequencies or different frequency control philosophies [2]. In Figure 20.4 a, DC coupling link between two ac-networks is illustrated. Besides the power electronics components for the AC–DC conversion also nanoelectronics for measurement, for example, phasor measurement units (PMU), data processing and control are employed. Further applications of HVDC are point-to-point transmission and multiterminal systems (dc-networks), transmission delivery to large urban areas and underground/submarine cable and offshore transmission. The two so far applied conversion methods for HVDC are the classical current-source converter (CSC) technology based on thyristors and the self-commutated voltage-source converter (VSC) technology based on pulse-width modulation using high power transistors (IGBTs). The greatest impact on the overall energy loss of the HVDC transmission is contributed by the converter valves (thyristors or IGBTs) with about 50% using CSC and 70% using VSC technology [22–24].

To manage the power flow within AC-grids, FACTS are deployed. These systems comprise the control of the reactive power (from 100% inductive to 100% capacitive), of the voltage in manners of magnitude and sign, of the phase angle and of the frequency [2]. Losses are curtailed by ensuring the reduction of reactive power through inductance and capacitance components [22].

There are different kinds of implementations available, for example, the static Var compensator (VCS), which uses thyristors to achieve a fast control of the reactive power injections or absorption. Another approach is the static voltage source (SVS), where the output voltage, frequency, and phase angle is controllable [7]. The main advantage of FACTS is the enhancement of the grid capacity and the reduction of losses.

Once the energy is transmitted it must be distributed to the end user and the place of demand. Like it is for transmission system, the focus here lies on the electronic and smart control of the power flow. It is believed that by 2030 most of the electrical power flow between generation and distribution is managed by power electronic devices [2,18]. So far power electronics, mainly converters, are used in distribution grids only to interface the power generation (e.g., photovoltaic) to the grid. But like it is feasible in the case of FACTS, power electronics can also be used to control and support the distribution grid, which is then called flexible ac-distribution system (FACDS) [7].

20.2.3
Microenergy Grid

As preliminary pointed out the energy supply shifts to decentralized and local energy systems. Besides the multienergy smart grid, also microenergy grids (μ-grid) are of high interest in the energy revolution.

An example for microgrid as a smart energy system could be the energy management system of a building or office block. According to Catrene Scientific Committee Working Group [2], buildings consume in the range between 39 and 48% of total primary energy and are therefore an attractive target for energy efficiency measures and intelligent energy management. This could mean saving of energy through intelligent control and efficient and sustaining system components. Buildings will be equipped not only with renewable energy systems like combined heat and power (CHP) generation in microblock heating stations, or PV facilities but also with energy storage systems. These decentralized stations can be used for the balance of volatile energy portions like solar or wind power [2].

Figure 20.5 shows the principle of a microenergy grid. It consists of a micro combined heat and power plant (μCHP) for electric power and heat generation, a PV facility for power generation, a battery system for buffering and storage purposes, prosumer (combined producer and consumer) in form of electric vehicles (EV) (e.g., a company fleet) and heat and electricity consumer. The holistic control and energy management of the μ-grid is the major task of a smart energy

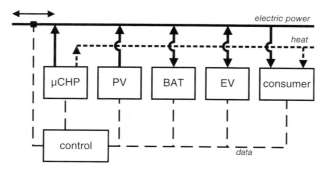

Figure 20.5 Building energy management system.

system. The challenge of these systems is the local balancing of the generation and the demand, to be self-sustaining or even generate energy surplus for infeed to the overlaying energy grid (cf. Section 20.2.1)

The μ-grid energy management knows the variation of energy consumption of the day as well as a prognosis about the expected PV energy generation. With this knowledge, a smart charging and discharging of the battery is possible, which enables grid stabilization over the day. On the technical side such a smart energy management system requires intelligent software algorithms and communication interfaces.

20.2.4
Energy Harvesting Systems

Energy harvesting systems are becoming a much asked for technology. The key applications are wireless and self-sustaining sensor network and wearable and mobile devices. Figure 20.6 shows the principle of an energy harvesting system. Electric energy can be harvested from different ambient conditions. There are many physical quantities that can be used for this purpose, like mechanical motion or vibration, solar radiation, thermal gradients, and radio-frequency. The energy is generated in so called energy transducers [25–27]. Energy harvesting transducers like thermo-, piezo-, or electrodynamic generators provide a high dynamic range of output voltages and currents [2]. Thermogenerators provide very low voltage levels like a few millivolts but high currents. Piezoelectric materials produce higher voltage levels up to 50 or 100 volts, but very low currents in the microampere range. The fluctuating and dynamic power is electronically rectified and buffered, respectively, stored in a battery. This can be typically Li-ion batteries or capacitors, depending on the application. In some cases, the application is directly powered by the energy transducer. The power management electronics like voltage converters and charge circuits for batteries must cope with this dynamic range. Power management integrated circuits (PMIC) are used to handle these tasks. The harvested energy is used to power the applications, like the measurement of ambient conditions (e.g., temperature) and the transmission within a wireless communication network.

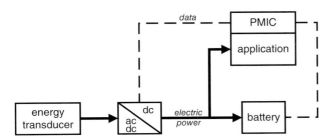

Figure 20.6 Principle of an energy harvesting application.

To enable energy harvesting based applications, especially the sleep modes of microcontrollers should be implemented in an energy-efficient way. A lot of sensor applications remain in sleep mode most of their time. Thus, ultralow power clock circuits will be necessary to keep the overall power consumption at a minimum [2].

The microscale energy transducer technologies need to be improved to enhance the amount of energy harvested and so to enable advanced applications [2,25].

20.2.5
Mobility

Applications of smart energy systems in the field of mobility mainly relate to electric vehicles, like buses, electric cars, or even electric bicycles. But also already established mobility solutions like trams, trains, or metros can be considered as smart energy system.

Private and public transportation, besides freight traffic, is one of the origins of environmental pollution with a share of 11.3% of the worldwide hydrocarbon combustion [28]. Therefore, to reduce CO_2 emissions, the alternative of electric traction is growing in importance [5]. To minimize emissions, electric mobility is in the focus of government programs. Germany, for example, strives for a share of one million electric vehicles on German roads in 2020 [29].

Figure 20.7 shows the principle of an energy system aboard an electric vehicle. The system consists of a battery, for example, lithium-ion cells, the electric engine, an AC–DC-converter, and the control-unit. For a simple application, basically the following three states are relevant:

- *Charging*: In charging mode, the electric energy flow is from the public energy grid, through the converter to the battery, where the energy is stored for driving.
- *Driving*: In driving mode, the electric energy flow is from the battery to the electric engine, where the energy is transformed into traction.
- *Recuperation*: In recuperation mode, the energy flow is from the electric drive to the battery. In the case of braking the electric engine performs as generator. The recuperated energy is then stored in the battery to power driving.

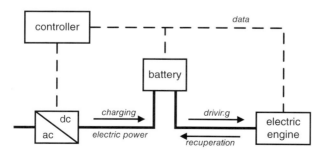

Figure 20.7 Energy system of an electric vehicle.

The energy flow is managed in the control unit, which is connected to each system component. More complex and advanced energy systems could be electric vehicles with additional combustion engine, for example, hybrid cars or electric vehicles that offer a bidirectional energy flow to support the public energy with battery capacities at times of high demand [5].

Electrical traction requires advanced semiconductor technologies for power conversion stages in dc–dc and ac–dc-converters, plug-in battery chargers, and power-inverters stages [5].

20.3
Nanoelectronics as Key Enabler for Smart Energy Systems

Semiconductors are indispensable in today's high-technological countries. On one hand to enable an efficient use of energy in production and industry and on the other hand ensure a reliable public energy supply. Power-electronics together with micro-/nanoelectronics and semiconductor integrated circuits (IC) are essential to assure the energy supply of the future and are key enablers to realize smart energy systems and to pave the way for the energy (r)evolution. The discussed smart energy systems in Section 20.2 require novel components based on advanced semiconductors technologies [2,5]. The estimated energy savings potential is more than 25% of the current electricity consumption in the EU countries [30]. The greatest future energy saving will come from the adoption of semiconductor technologies, which are forecasted to consume around 50% less electricity in 2030 [5].

This section gives a brief overview about the semiconductor key products for smart energy systems and their technological requirements.

20.3.1
Key Products for Smart Energy systems

The intermittent in-feed of renewable energy, the bi-directional energy flow, and the dynamic energy demand lead to stringent requirements concerning the energy supply system. The performance of the energy systems is depending on the quality and capabilities of the electronic components, which are crucial for efficient control and conversion of electrical energy [18,31].

Key products for smart energy systems are on one hand power-electronic components like converters and power switches and on the other hand micro- and nanoelectronics for control and energy management purposes and energy harvesting applications. The key products for smart energy systems are listed as follows:

- Efficient thyristors, IGBTs, and circuit breakers (CB) are the key enabling technologies for HVDC networks [23].
- For the connection to the grid, more efficient and adaptive power converters and power-electronics, like switches are required.

- Efficient stabilization and control of AC flow requires fast and efficient power-electronic components. The most popular FACTS equipment types are based on thyristors, GTOs, or IGBTs [2,24].
- For energy harvesting applications, microscale energy transducers are key products.
- Control and energy management applications demand specialized integrated circuits and control algorithms.

20.3.2
Technological Requirements and Challenges

The technological requirements of the key products can be divided into power-electronics and micro-/nanoelectronics.

20.3.2.1 Requirements of Power-Electronics

Advanced power-electronics (PE) in respect to performance, reliability, efficiency, and packaging are required [18,30]. PE have to operate at high temperatures above 200 °C with a switching frequency of more than 10 kHz and voltage ratings above 10 kV [18], and have to offer highly efficient performance without compromising the price and size of the device [2]. The reliability of the power-electronics is paramount to ensure the energy supply.

One of the major technological requirement is the increase of the power density. The power density can be defined by the relation (Eq. (20.1)) of the power handling capability and the device volume.

$$\text{Power Density}^\uparrow = \frac{\text{Power handling capability}^\uparrow}{\text{Volume}_\downarrow} \quad (20.1)$$

To support high transmission capacities, power-electronics for HVDC must demonstrate high efficiency and high loading capacities with power handling capabilities of several hundred kilovolts both during nominal and overload operation [18]. The evolution of power semiconductors has arrived at a level where packing density restricts the achievable performance of the final device. A package for a power semiconductor has to be able to remove the heat, provide security insulation against the heat sink, conduct current, and has to be electromagnetically and thermomechanically reliable [30]. A more sophisticated thermal design of the packaging to realize reliability and by this longer life times are required. Additionally, miniaturization and efficient package design reduce system costs and losses and enable the integration of more functions [18].

To meet miniaturization and performance goals and to overcome the limitations of conventional silicon technology, new materials with superior electrical properties are being introduced. Wide-bandgap semiconductors like silicon-carbide (SiC) or gallium nitride (GaN) are gaining momentum in the market and are regarded as the future material for power grid applications [22,30]. To improve efficiency in converters, the usage of silicon carbide (SiC) Schottky diodes will minimize switching losses and improve thermal performance.

Cold-bypass switches are an example of device evolution by improving traditional Schottky bypass diodes through reducing power losses in both forward and reverse polarity [5].

20.3.2.2 Requirements of Micro-/Nanoelectronics

Power-electronic devices such as rectifiers, converters, or switches have to be controlled and the energy flow in the energy system has to be sustainably managed. For that highly efficient microelectronics, integrated circuitries and sensors are required. The main future challenges of microelectronics are enhancement of the performance and efficiency and the high integration level, and thus enable higher application processor capabilities [2]. Specialized integrated circuits (IC) and system-on-a-chip (SoC) for energy harvesting are necessary to handle the functionary requirements. The voltage levels of the electronic circuits or regulators have to be scaled down and current consumption must be decreased to assure performance with small amounts of energy. Especially, for low power applications (e.g., energy harvesting, sensor networks) the power supply voltage will be decreased in the future. To enhance the performance, the on-chip clock frequency has to increase. Table 20.1 highlights the estimated future trend.

To facilitate better functionality in demanding applications, the microelectronic devices have to become less power consuming and thus enable longer operating duration. Since ICs are the key source for heat generation, an efficient power management and selection of operating parameters, together with a smart design is required [33]. Integrated power management systems enable energy efficiency by replacing conventional methods with energy saving devices and regulating power supply methods to convert the fluctuant and intermittent environmental and renewable energy (e.g., vibration, sunlight) into a stable power source [33,34].

The microscale energy transducer technologies need to be improved to enhance the amount of energy harvested and so to enable advanced applications [2,25]. Major challenges are:

- reduce leakage currents
- lower start up voltages to several mV
- currents in the low μA range should be processable

Future key trends are customized power management solutions, which include unique MOSFETs, PMICs, and drivers. Digital power management will have higher importance in the market [33].

Table 20.1 Prognosticated decrease of supply voltage and increasing on-chip clock [32].

Year	2015	2017	2019	2021	2023
Power supply voltage (V)	0.83	0.80	0.77	0.74	0.71
On-chip clock (GHz)	5.95	6.44	6.69	7.53	8.18

20.4
Summary and Outlook

The answer of the energy revolution toward handling the societal demands and impacts is the introduction of smart energy systems. After explaining the reasons for the alternation of the energy supply the terms smart energy and smart energy systems are introduced and defined. For a better understanding of the smart energy elements, an effect-matrix is introduced in Section 20.1.2. Section 20.2 highlights some common and specific examples of smart energy systems and their sociatal challenges. Besides the discussion of obligatory smart energy grids, also energy transmission and distribution systems and energy harvesting and mobility applications are pictured. Semiconductor technologies are key enablers for smart energy. In Section 20.3, a brief overview about key products and their technological requirements is given.

Smart energy was defined as the intelligent interaction of technological advanced system components via a communication infrastructure, and intelligent control of the energy flow to balance energy generation and demand. It was shown that advanced semiconductor technologies, like power-, micro-, and nanoelectronics are essential to realize smart energy and thus further push the energy revolution. Three major challenges of these technologies can be constituted: the components size, the power density, and the power consumption.

The reduction of size is motivated by devices that can offer efficient gains and cost-reduction through miniaturization. Nanoelectronics opens up new opportunities as designers can incorporate increased functionality in a limited space. But also power-electronics for FACTS and HVDC-systems need to be reduced in size. So far the installation of such systems claims a lot of space, as many components are required to fulfill the power capability requirements. Miniaturization is closely related to the increase in power density.

Increasing the power density leads to more efficient components, in manners of size and performance. Thus, power- and nanoelectronic devices with higher performance could be realized in the same available space. And vice versa, components with same performance could be realized with smaller packaging.

Additionally, the reduction of power consumption and on-state losses should be in the focus of the ongoing research. Pure semiconductor switches with minimal on-state losses and new wide band-gap power semiconductor devices, for example, with silicon carbide (SiC) or gallium nitride (GaN) need to be developed [2].

As a final remark it can be stated: What actually makes energy smart is the application that brings all system components in a context and not only the advanced technology of the components itself. Both sides, advanced technologies and the energy-management with control algorithm are equally important.

There is still a lot of effort necessary to be put into the research and development of adequate components and tailor made energy systems, to enable a secure, reliable, and efficient energy supply entirely based on renewable energy sources.

References

1 Deane, P. (1995) *The First Industrial Revolution*, 2nd edn, Cambridge University Press.

2 Catrene Scientific Committee Working Group (2014) Semiconductor technologies for smart cities, December.

3 European Union (2009) Directive 2009/72/EC.

4 U.S. Energy Information Administration (2013) International Energy Outlook 2013: With Projections to 2040, July.

5 Papa, C. (2012) The role of semiconductors in the energy landscape. 2012 IEEE International Solid-State Circuits Conference (ISSCC), pp. 16–21.

6 Internation Energy Agency (2008) World Energy Outlook 2008.

7 Kechroud, A., Myrzik, J.M.A., and Kling, W. (eds) (2007) Taking the experience from flexible AC transmission systems to flexible AC distribution systems. doi: 10.1109/UPEC.2007.4469031

8 DATACOM Buchverlag GmbH (2013) ITWissen. Das große Online-Lexikon für Informationstechnologie. Available at www.itwissen.info/.

9 Lund, H., Andersen, A.N., Østergaard, P.A., Mathiesen, B.V., and Connolly, D. (2012) From electricity smart grids to smart energy systems – a market operation based approach and understanding. *Energy*, **42**, 96–102.

10 Verband der Elektrotechnik Elektronik Informationstechnik e.V. (2010) Die Deutsche Normungsroadmap. E-Energy/Smart Grid, March 28.

11 Leonov, V. (2011) Energy harvesting for self-powered wearable devices, in *Wearable Monitoring Systems* (eds A. Bonfiglio and D. De Rossi), Springer, pp. 27–49.

12 Union of the Electricity Industry–EURELECTRIC (2012) Public Consultation on Use of Spectrum for more efficient energy production and distribution, April.

13 Ancillotti, E., Bruno, R., and Conti, M. (2013) The role of communication systems in smart grids: architectures, technical solutions and research challenges. *Comput. Commun.*, **36**, 1665–1697.

14 Department of Energy (2010) Communications Requirements of Smart Grid Technologies, October 5.

15 Bundesverband der Deutschen Industrie e.V. (2008) *Internet der Energie. IKT für Energiemärkte der Zukunft*, BDI-Drucksache.

16 Birkner, P. (2013) Intelligente Verteilnetze. *Gwf – Gas + Energy*, **154**, 654–658.

17 The United Nations (2010) Urban and Rural Areas 2009. Available at www.unpopulation.org.

18 Frost & Sullivan (2012) Advances in Power Electronics Enabling Future Smart Grid. Advanced Power Electronics Driving Next-Generation Power Grid, D4CE-TI.

19 Smart Grids (2013) European technology platform for the electricity networks of the future. Available at www.smartgrids.eu/.

20 Wu, F.F., Moslehi, K., and Bose, A. (2005) Power system control centers: past, present, and future. *Proc. IEEE*, **93**, 1890–1908.

21 BDEW Bundesverband der Energie- und Wasserwirtschaft e.V. (ed.) (2013) BDEW roadmap. Realistische Schritte zur Umsetzung von Smart Grids in Deutschland, February 11.

22 Frost & Sullivan (ed.) (2012) Analysis of the global semiconductors market for smart grids. Demand for Efficiencies and Power Give Rise to a Growing Opportunity for Semiconductors, NAF8-26, December.

23 Franck, C.M. (2011) HVDC circuit breakers: a review identifying future research needs. *IEEE Trans. Power Deliv.*, **26**, 998–1007.

24 Bahrman, M. and Johnson, B. (2007) The ABCs of HVDC transmission technologies. *IEEE Power Energy Mag.*, **5**, 32–44.

25 Spies, P., Pollak, M., and Rohmer, G. (2007) Energy harvesting for mobile communication devices. 29th International Telecommunications Energy Conference, pp. 481–488.

26 Pirisi, A., Grimaccia, F., and Mussetta, M. (2012) An innovative device for Energy

Harvesting in smart cities. 2012 IEEE International Energy Conference, pp. 39–44.

27 Lu, C., Raghunathan, V., and Roy, K. (2011) Efficient design of micro-scale energy harvesting systems. *IEEE J. Emerg. Select. Top. Circuits Syst.*, **1**, 254–266.

28 Dr. Ing. Metz, N. (1999) Worldwide CO2-distribution from 1980 to 2010. 20th Internationales Wiener Motorensymposium, May 6–7.

29 Die Bundesregierung (2009) Nationaler Entwicklungsplan Elektromobilität der Bundesregierung, August.

30 Catrene (2013) Integrated power and energy efficiency. Power Device Technologies, Simulations, Assembly and Circuit Topographies Enabling High Energy Efficiency Applications.

31 U.S. Department of Energy (2005) Power Electronics for Distributed Energy Systems and Transmission and Distribution Applications, ORNL/TM-2005/230, December.

32 ITRS (2013) International Technology Roadmap For Semiconductors. 2013 Ed.

33 Frost & Sullivan (2013) Next Generation Power Management For Electronic Device (Technical Insights). Enabling Energy Efficient and Greener Electronics with Cutting Edge Power Management Technology, D51E-TI, December.

34 Ha, D.S. (2011) Small scale energy harvesting - principles, practices and future trends, in 14th International Symposium on Design and Diagnostics of Electronic Circuits & Systems (DDECS) (IEEE), p. 9.

21

Validation of Highly Automated Safe and Secure Vehicles

Michael Paulweber

AVL List GmbH, Global Research & Technology Management, Hans-List-Platz 1, 8020 Graz, Austria

21.1
Introduction

Automated, connected, or autonomous systems, for example, highly automated vehicles are the next major technology field and challenge. This new generation of systems extend embedded automation systems by two new innovative components: a view of the surrounding world is continuously updated within these systems. This virtual world (or cyber world) is used to perform very difficult control tasks, for example, automated driving of vehicles in complex traffic situations and/or finding the energy and emission optimal trajectories when driving electrified trucks from one location to another. The second component is typically a connection to back-end system in the cloud, which combines data from many systems from the present and the past. Therefore, these automated, connected, or autonomous systems are called automated cyber-physical systems. Automated cyber-physical systems enable on the one hand a huge number of new applications, business models, and major commercial opportunities for high-tech companies in many industry sectors and, on the other hand, help overcome burning societal problems.

Recent studies have identified an estimated annual market value for autonomous systems in transport (automotive, aerospace, rail, maritime) of €82 billion and of €2.8 billion in agriculture. Other studies predict that the global market for smart machines, including autonomous systems and cyber-physical systems, is expected to grow to $15.3 billion by 2019, with a 5-year compound annual growth rate (CAGR) of 19.7% [1]. The worldwide market for unmanned ground vehicles is even expected to have a CAGR of 30% by 2019 [2]. The market growth for partially and fully automated vehicles is forecasted from around $42 billion in 2025 to $77 billion in 2035 [3]. These numbers show the vast potential of this technology. Due to this high market potential, technological giants such as Google, Tesla, and Apple as well as large industrial conglomerates such as Rio Tinto are investing heavily in the automated cyber-physical system domain, and there are also many initiatives taken in Europe in this area.

Nanoelectronics: Materials, Devices, Applications, First Edition. Edited by Robert Puers, Livio Baldi, Marcel Van de Voorde, and Sebastiaan E. van Nooten.

21.2
Societal Challenges

Automated cyber-physical systems offer solutions to many of today's grand societal challenges. Highly automated driving functions will, for example, increase traffic safety, reduce traffic jams, increase passenger comfort, and also enable disabled or elderly people to live independently. Automated vessel navigation systems will avoid collisions and groundings, protect the environment, reduce bunker consumption, and improve the maritime traffic flow as well as intermodal transportation. Unmanned automated vehicles can perform heavy and dangerous tasks in harsh environments, such as maintenance services on airplanes or underwater constructions. The applications and benefits of automated cyber-physical systems are countless.

Automated cyber-physical systems will help to reach several goals, which do contradict partially each other. Therefore, complex optimization procedures have to be executed in real-time to find optimal solutions. Exemplary goals are as follows:

- *Reaching the "vision zero" target for road safety*: Ninety percent of all fatalities in road transport are caused by human errors. Automated functions in vehicles will help to reduce these fatalities to nearly zero. Reduction in fatalities is even more important, as the aging society wants to keep its level of individual mobility. Studies have shown that the percentage of fatalities is high in novices in traffic; it then gets lower and starts to rise again significantly in those of 80 years of age [4]. As more and more senior citizens will reach this age, smart electronic systems are required to support these citizens in their mobility in a safe way. Mobility is one of the biggest values of the western countries. This can be seen in the fact that many families, after buying a house or an apartment for living, spend the second largest amount of money on vehicles. As people get older, they want to keep the personal mobility they have been used to all their life. In order to keep the trend of fatalities decreasing, it is necessary to support elderly people when driving cars even at the age of 80 and older. Without advanced driver assistance systems, elderly people have to stop driving or have an increased risk of fatal accidents. A statistics from Ref. [4] shows this trend, which can be reversed by the use of advanced driver assistant systems.
- *Traffic management to take better advantage of the more and more precious space in megacities:* Today's traffic jams in megacities are caused by the limited space in this city. Vehicles looking for a parking space increase this problem additionally. As it is not possible to increase the space reserved for vehicles (streets, parking places, parking structures), it is necessary to make better usage of this precious resource. Communication between vehicles and infrastructure to find the best route as well as adapt the speed to the environmental conditions (traffic light sequence, traffic density, etc.) can reduce the burden for citizens by traffic jams. Automatic parking assistance systems or even parking chauffeurs, which know the next open parking place and can find it without a human interaction, will reduce stress for the vehicle passengers as well as

environment pollution due to lower times for getting from the final destination to an available parking spot. Additionally, automatic parking functions without passengers inside the vehicle can store more vehicles in a parking structure than in a conventional parking garage.

- *Reducing CO$_2$ and toxic emissions from transportation systems:* Vehicle-to-vehicle (V2V) and vehicle-to-infrastructure (V2I) communication together with precise maps and geo-positioning information in cars will allow finding the most energy-efficient trajectories from current position to the desired final destination. This "connected vehicle" functionality will need on the one hand a powerful and reliable communication link from vehicles to the outside world and, on the one hand, a computing power to optimize continuously the trajectories for the set values of the vehicle (throttle position, braking, battery management, energy distribution, transmission management etc.). This can lead to higher emission reductions than possible with the most advanced combustion engine improvements.

21.3
Automated Vehicles

It is commonly accepted that the automated driving functionality will not come overnight to our vehicles, trucks, off-road machines, ships, or airplanes. In some areas automated functions are already in use since many years, such as autopilots in airplanes, while in other areas only the first starts have been made to introduce automation.

SAE defined five different levels, which are commonly used to describe the degree of automation in vehicles (see Figure 21.1) [5,6].

	Level	Name	Narrative description	Execution of steering and acceleration/deceleration	Monitoring of driving environment	Fallback performance of dynamic driving task	System capability (driving modes)
Automated driving system monitors the driving environment	5	Full automation	The full-time performance by an automated driving system of all aspects of the dynamic driving task under all roadway and environmental conditions that can be managed by a human driver	System	System	System	All driving modes
	4	High automation	The driving mode-specific performance by an automated driving system of all aspects of the dynamic driving task even if a human driver does not respond appropriately to a request to intervene	System	System	System	Some driving modes
	3	Conditional automation	The driving mode-specific performance by an automated driving system of all aspects of the dynamic driving task with the expectation that the human driver will respond appropriately to a request to intervene	System	System	Human driver	Some driving modes
Human Driver monitors the driving environment	2	Partial automation	The driving mode-specific execution by one or more driver assistance systems of both steering and acceleration/deceleration using information about the driving environment and with the expectation that the human driver performs all remaining aspects of the dynamic driving task	System	Human driver	Human driver	Some driving modes
	1	No automation	The full-time performance by a human driver of all aspects of the dynamic driving task, even when enhanced by warning or intervention systems	Human driver	Human driver	Human driver	n/a

Figure 21.1 SAE levels of automated driving.

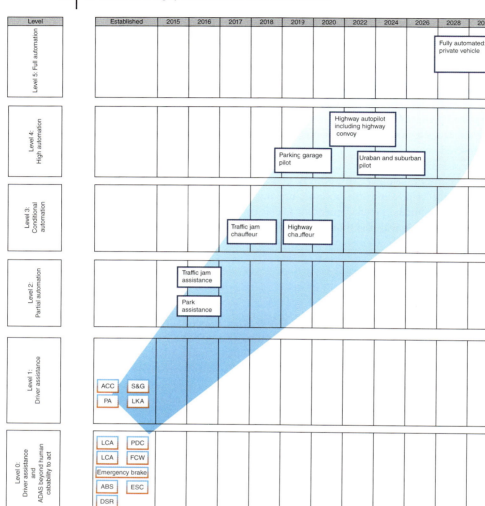

Figure 21.2 Deployment path for automated driving in passenger cars [5].

OEMs already offer vehicles with several automated driving functions of different levels to the market. Many vehicle manufacturers promise to deliver more automated driving functionalities in the next years. An automotive expert group from the ERTRAC industry association predict the introduction for the next levels in automation according the roadmap depicted in Ref. [5] (Figure 21.2).

The functionalities in use are as follows [5]:

- *LCA (lane change assistant) (SAE level 0):* This monitors the areas around the car and warns the driver in case of dangerous situations by means of flashing warning lights (e.g., in the external mirror of the vehicle).

- *PDC (park distance control) (SAE level 0):* It helps the driver at parking maneuvers in indicating the distance to outside objects.
- *LDW (lane departure warning) (SAE level 0):* It warns the driver, when it leaves the road lane by means of warning lights or vibrations of the steering wheel.
- *FCW (front collision warning) (SAE level 0):* It indicates to the driver a too low distance of vehicles or other objects in front by acoustic or optical signals.
- *PA (parking assistant) (SAE level 1):* It partially takes over during parking maneuvers, the driver still has to operate the accelerator and braking pedal, the automated driving systems operates the steering wheel and gives advice via optical and acoustical signal to the driver.
- *ACC (adaptive cruise control) (SAE level 1):* It is a combination of cruise control and a distance control. Whenever the distance from the ego-car to objects in front of it gets too low, the automated driving system decelerates the car. Older versions operate up to a lower speed limit before it signals the driver to take over at very low speeds again, newer version of ACC can operate up to standstill.
- *LKA (lane keeping assistant) (SAE level 1):* It helps the driver to keep the lane. It performs small lateral movements in order to keep the car on the correct lane. Typical systems in use still require the driver to keep his hands on the steering wheel.

21.4
Key Requirements to Automated Driving Systems

Automated driving systems are complex combinations of advanced sensors, high-performance real-time computing power, communication technology, advanced software programs, and optimized actuators (see Figure 21.3). The key requirements to these automation platforms are provision of necessary computational resources for complex algorithms (up to cognitive algorithms) and acceptable dependability of the complete system including the vehicle. Dependability shall comprise safety, security, and reliability [7].

The vehicle with the integrated automated driving system must structurally comply with the regulatory framework. In addition, it is necessary to provide an intuitively easy-to-use human–machine interface. Vehicles equipped with automated driving systems must blend itself into the outside traffic environment, which means it has to adapt to the commonly used driving practice of different countries in order to not to invoke accident due to unexpected behavior of the automated vehicle for other nonautomated traffic participants.

The basic structure of automatic driving systems is shown in Figure 21.3. Many sensors such as positioning sensors, radar sensors, LiDAR sensors, video cameras, ultra-sonic sensors, speed sensors, acceleration sensors, and so on are used to generate an image of the outside world. This is supported using data from accurate maps and data communicated from other vehicles

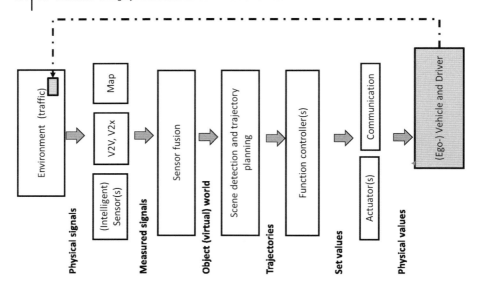

Figure 21.3 Structure of automated driving system control unit.

(vehicle-to-vehicle communication as well as vehicle-to-infrastructure communication, or in short V2X).

As many sensors have difficulties under specific environmental conditions as rains, fog, show, night, glaring, and so on, complex sensor fusion algorithms are used to combine data from different sensors to generate more reliable signals. Modern control units of automated driving systems associate probability values with every sensor value taken, which is used in the sensor fusion algorithms as well is in the subsequent trajectory generation procedures. All sensor signals together with data from V2X communication are used to continuously update an image of the external world. The ego-car itself is an object within this virtual world. This is indicated by the dot–dashed line in Figure 21.3.

The subsequent trajectory planning is done in several steps: strategic planning to generate the navigation data, tactical planning to initiate maneuvers lane changing, and reactive planning to ensure, for example, no obstacles are hit [7].

In order to find the optimal trade-offs between fast and smooth travel to the desired destination under utmost safety conditions and achieve the lowest energy consumption and emissions, additional data received from infrastructure traffic computers (V2I communication) as well from other vehicles (V2 V communication) are used to optimize the trajectory not only for the current time but also for the forthcoming future. This requires complex environmental models simulating all traffic, weather conditions, vehicle dynamics, and so on, which are used in optimization algorithms in order to find the optimal trajectories (see Figure 21.4). This means that the control units use the virtual environment models together with projections into the future behavior of the outside objects (using big data analysis on historical data such as weather data, traffic data, etc.)

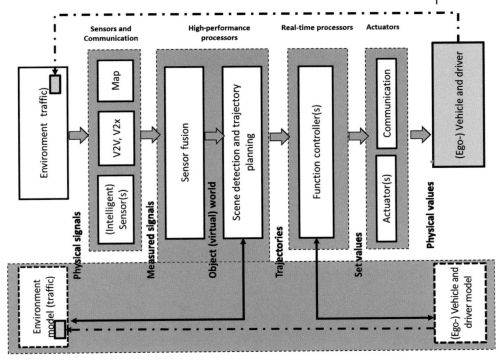

Figure 21.4 Extended structure of automated driving system control unit.

to also find optimal control strategies in the future. This can be compared with the experience of advanced drivers when deciding which the best route is or which maneuver will not only lead to the minimal time when moving from A to B but also ensure optimal safety. Advanced automated driving systems can find multidimensional optima that are more complex than most of the human drivers can find. An example is the best trade-off between shortest time and lowest energy consumption under optimal safety conditions.

All this requires very performant hardware components for control units in automated driving systems. The main hardware components are indicated in Figure 21.4.

Additional requirements for these control units are as follows (see also Ref. [7]):

- Highly dynamic resources (dynamic real-time environment)
- Very large resource demands with respect to communication bandwidth
- Memory and mass storage
- Raw CPU power
- In-vehicle update of parameter data, models, and algorithms
- Low hardware costs.

Especially the in-vehicle software and parameter update are essential for control units in automated driving system as the developers of these system face the problem that they never know all environmental conditions that can occur and may lead to safety-critical situations. Therefore, it is commonly accepted that continuous evaluation of all occurring situations in automated vehicles for potential safety threads, which were not foreseen at design time of the software, is necessary. If potential problems (as well as other HW/SW errors) are detected, software updates are immediately generated by the OEM and automatically downloaded in a safe and secure way to automated vehicles concerned. The new TESLA models have already this functionality in use [8].

To achieve adequate confidence/adequate assurance levels, suitable strategies – in the form of claim, argument, and evidence – need to be developed and mapped into the regulatory context.

21.5
Validation Challenges

Even if it possible to develop automated driving systems, demonstrating the reliability, safety, and robustness of the technology in all conceivable situations, for example, in all possible traffic situations under all potential road and weather conditions, has been identified as a key challenge (e.g., see Refs [5,9]) for the automotive domain and is today the main roadblock for product homologation and certification and thus commercialization of automated driving systems. Recent scientific publications predict that more than 100 Mio km of road driving would be required for the thorough validation of an automated car [10,11]. Only if these extensive tests have been done, it can be statistically proven that the automated vehicle is as safe as a manually driven vehicles.

Taking further into account the high number of vehicle variants and software versions, it becomes obvious that new approaches are required to validate control units for automated vehicles within a reasonable time period at reasonable costs. The complexity of the validation of automated driving systems has various reasons:

- Many new sensors such as video sensors, radar sensors, LiDAR sensors, ultrasonic sensors, vehicle-to-vehicle and vehicle-to-infrastructure communication, GPS sensors, data from digital maps, and so on have to be integrated. Most of the sensors can give only values with a certain probability, which has to be taken into account in further calculations. Many of these sensors are intelligent subsystems, which have significant built-in data processing capabilities. Validation systems have to cope with streams of object lists generated by these sensors.
- Vehicles with automated driving systems communicate with the surrounding environment. Other vehicles driven by human drivers change their path based on the reaction of the automated vehicle. Therefore, it is also necessary to take

into account different behaviors of human actors in different countries (e.g., South Italy versus Sweden) when validating these automated driving systems.

- Outside environment conditions influence the performance of these sensors. This requires tests under many different weather conditions. If this is done using real vehicles, the necessary time is enormous – therefore, testing in virtual validation environments is mandatory.
- Security breaches in vehicle communication systems may lead to wrong images of the outside world, which creates unwanted trajectories and might even result in accidents. Therefore, extensive security validation is very important.

Vehicles with automated driving system have to be safe under any environmental conditions and scenarios. Therefore, the most intuitive validation approach is the validation on the road. When trying to perform validation road testing in short periods of time, it is impossible to ensure that all possible environmental conditions (hot, cold, low or high altitude, snow, rain, fog, ice, etc.) occur within the time frame of the road testing.

Additionally, many validation sequences require dangerous maneuvers, which are safety critical to test drivers piloting vehicles with new automated driving system versions. The exchange of validation on the road by validation in a virtual world helps to reduce the testing effort dramatically. Eliminating these parts of test time on the road, where no safety-relevant events occur, in the virtual testing can reduce the testing time already significantly. Additional time savings are possible when using advanced design-of-experiment methods [12].

21.6
Validation Concepts

In order to reduce the validation time even further, new concepts are developed [13,14]. An example of the required workflow is shown in Figure 21.5. Test vehicles record data from sensors of automated driving systems, environment, and road data and vehicle data (e.g., engine speed, vehicle velocity, vehicle geoposition). These data are analyzed and it is decided whether safety-critical scenarios occurred, which were not considered when developing and validating the system. These scenarios are immediately analyzed and if required, the automated driving systems software is adapted and downloaded to the effected vehicles as soon as possible.

Sophisticated design-of-experiment algorithms are used to generate the minimal number of test scenarios needed to achieve the required reliability level.

The last validation step is an executor of generated test sequences in different test environments throughout the development process of the systems. It starts with tests in model-in-the-loop (MIL), continues with software-in-the-loop testing during development of software components and hardware-in-the-loop (HIL) tests to power-in-the loop validation (xIL), and finally tests on proving ground and on the road. It is obvious that reusable test procedures for all test environments will bring big cost-savings.

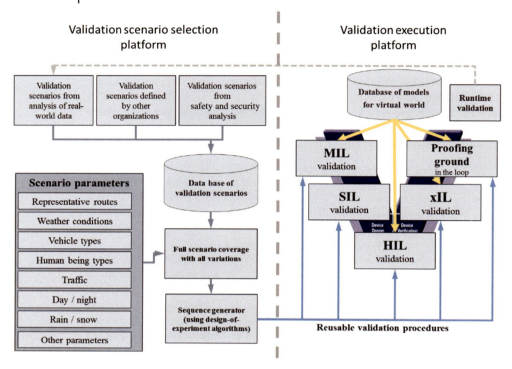

Figure 21.5 Validation tool chain for automated driving systems [13].

21.7
Challenges to Electronics Platform for Automated Driving Systems

The main challenge in developing and validating control units for automated driving systems is the unprecedented level of complexity caused by the sheer endless number of driving situations these systems have to handle, the necessity to deal with sensors that are not reliable under all environmental conditions (e.g., cameras cannot see very well under heavy rain), the enormous need for computing power to perform the optimization of trajectory determination using complex real-time adapting virtual world models, and the need for reliable communication links to the outside world (V2X) in the most implemented concepts.

This leads to three different challenges for hardware and software development, but have to be solved together:

1) The hardware development is very much driven by the later production costs of these systems. Thus, it is mandatory to create hardware platforms that cover a broad application range and can therefore take advantage of economy of scale. Hardware R&D projects must always consider the whole value chain required to produce the indented systems. The development of cost-effective and reliable sensors to create a real-time image of the outside world

within the virtual world models in the control units still poses many challenges.

The requirement to provide hardware platforms for time-critical real-time control algorithms, reliable communication to other vehicles, and the infrastructure as well as real-time optimization in complex virtual worlds to create optimal trajectories lead to the necessity for very powerful multicore processors with dedicated 3D calculation support, built-in communication, and general-purpose real-time software capabilities.

2) Software, on the other hand, has to take the burden of providing the required flexibility. As it is not possible to define upfront all requirements for automated driving systems, it is therefore mandatory to establish mechanisms that allow an efficient, fast, and smooth (in-vehicle) update path for the software, whenever new knowledge about the required scenarios and environment conditions of the system become available. Additionally, the software also has the task to support future variants of vehicles using the same hardware platform. Here a flexible, supreme, and stable software architecture is of utmost importance.

3) This new complexity also needs new agile development processes as they are already implemented successfully in new companies as TESLA [8].

21.8
Conclusion

Automated driving systems will contribute to solve a number of important and increasing societal problems. The development of these systems comes with a new level of complexity. To handle this, it is necessary to work closely together in the development of mechanical, electrical, electronics, and software components. On the other hand, it is also important to understand the different goals of hardware (production and economy-of-scale orientation) and software (agile and flexible) during the R&D process. New validation methods are necessary to ensure a timely validation of these automated driving systems.

References

1 Research, B. (2014) Global market for smart machines expected to reach $15.3 billion in 2019: autonomous robots moving at 22.8% CAGR. BCC Research, May 27. Available at http://www.bccresearch.com/pressroom/ias/global-market-smart-machines-expected-reach-$15.3-billion-2019 (Accessed October 11, 2015).

2 Newswire, P. (2015) World unmanned ground vehicle market to see 30% CAGR to 2019. PR Newswire, April 29, 2015. Available at http://www.prnewswire.com/news-releases/world-unmanned-ground-vehicle-market-to-see-30-cagr-to-2019-501689211.html (accessed October 11, 2015).

3 Group, B.C. (2015) The road to autonomous driving, Boston Consulting Group, January 8, 2015. Available at http://de.slideshare.net/TheBostonConsultingGroup/

the-road-to-autonomous-driving (accessed September 17, 2015).

4 Oxley, J., Fildes, B., Ihsen, E., Charlton, J., and Day, R. (1997) Differences in traffic judgements between young and old pedestrians. *Accid. Anal. Prev.*, **29** (6), 839–847.

5 ERTRAC (2015) ERTRAC roadmap automated driving. ERTRAC, July. Available at http://www.ertrac.org/uploads/documentsearch/id38/ERTRAC_Automated-Driving-2015.pdf (accessed July 31, 2015).

6 Winner, H., Hakuli, S., and Wolf, G. (2011) *Handbuch Fahrerassistenzsysteme: Grundlagen, Komponenten und Systeme für aktive Sicherheit und Komfort*, Vieweg+Teubner.

7 Watzenig, D. (2015) Challenges in the evolution of highly automated system in transport applications. Artemis Technology Conference, Torino.

8 Vance, A. (2015) *Elon Musk: Tesla, SpaceX, and the Quest for a Fantastic Future*, HarperCollins Publishers Inc., New York.

9 Winner, H. and Wachenfeld, W. (2013) Absicherung automatischen Fahrens. 6. FAS Tagung, Munich, Germany.

10 Winner, W. and Wachenfeld, W. (2013) Absicherung automatischen Fahrens. 6. FAS Tagung, München.

11 Winner, H., Wachenfeld, W., Maurer, M., Gerdes, C., and Lenz, B. (2015) Die Freigabe des autonomen Fahrens, in *Autonomes Fahren. Technische, rechtliche und gesellschaftliche Aspekte*, Springer.

12 Horn, M., Watzenig, D., Paulweber, M. *et al.* (2016) Validation of highly automated safe and secure systems, in *Automated Driving – Safer and More Efficient Future Driving*, Springer International Publishing AG.

13 Müller, B. and Robert Bosch AG (2015) Challenges in demonstrating safety for highly automated systems. Safetrans Industrial Day 2015, Renningen.

14 Yu, L., Lei, Y., Kacker, N., and Kuhn, R. (2013) ACTS: a combinatorial test generation tool. IEEE 6th International Conference on Software Verification and Validation, Neumunster Abbey, Luxembourg.

22
Nanotechnology for Consumer Electronics

Hannah M. Gramling[1], Michail E. Kiziroglou[2], and Eric M. Yeatman[2]

[1]*University of California, Berkeley, Department of Mechanical Engineering, Berkeley, CA 94703, USA*
[2]*Imperial College, Faculty of Engineering, Department of Electrical and Electronic Engineering, South Kensington Campus, London SW7 2AZ, UK*

22.1
Introduction

Advances in the cost and physical scale of computing power have enabled devices geared at a consumer market to carry considerable onboard processing capability. In turn, low cost has spurred mass adoption of handheld and smaller computing products. Many devices that started as box-like, single purpose devices – like landline phones, stereos, and televisions – have since evolved into lightweight, low-profile, and cross-functional products. We have already entered an age of ubiquitous smart consumer devices. Smartphones, nominally a replacement for landlines of old, offer the widest range of functions and services, though function-specific products with onboard computing are surging in popularity. Wearables, including fitness monitors, as well as smart sensors and controls for the home, such as automated thermostats, have seen shipment numbers more than triple year-to-year [1].

Innovations at the nanoscale will play a significant role in achieving future consumer devices: nanostructured systems offer electronic performance beyond what is possible with current very large-scale integrated circuits; even devices that do not require significant computing power stand to benefit from improved efficiency achieved using nanostructured materials. Moreover, by enabling the possibility of devices that are low cost and simultaneously flexible, transparent, and lightweight, nanostructured solutions are appealing to nearly any device system. For instance, the future of display technologies is in flexible, low profile, portable displays, many of which will be specifically designed for virtual reality. Interesting optoelectronic properties available at the limit of material thickness, including near-transparency and a direct bandgap, could be pivotal in realizing low-cost, high-performance virtual reality.

Nanoelectronics: Materials, Devices, Applications, First Edition. Edited by Robert Puers, Livio Baldi, Marcel Van de Voorde, and Sebastiaan E. van Nooten.
© 2017 Wiley-VCH Verlag GmbH & Co. KGaA. Published 2017 by Wiley-VCH Verlag GmbH & Co. KGaA.

What is nanotechnology? For the purposes of this chapter, we will use the definition, provided by the US National Nanotechnology Institute, that nanotechnology must involve one of the following aspects:

1) Research and technology development . . . in the length scale of 1–100 nm.
2) Creating and using structures, devices, and systems that have novel properties and functions because of their small size.
3) Ability to control or manipulate on the atomic scale.

Consumer devices both present and future rely on a few underlying technologies: most will contain some combination of processing electronics, energy storage, communications hardware, sensors, and displays. The future of processing electronics is covered in depth in other chapters. For other enabling technologies for future consumer electronics – communications, energy storage, displays, and sensors – we discuss the challenges associated with each, and explore ways in which nanoscale solutions may solve these problems and introduce new device possibilities.

First, we briefly introduce two-dimensional materials, one of the fundamental nanoscale elements that will influence the future of pervasive consumer electronics and sensors.

22.1.1
2D Materials and Flexible Electronics

The considerable appeal of nanoscale electronics for consumer applications stems from their potential to be both flexible and transparent without sacrificing performance.

Two-dimensional materials – single-to-few atom-thick layers which are covalently bonded in the plane – span ballistic conductors, such as graphene, direct bandgap semiconductors, including molybdenum disulfide (MoS_2), and dielectrics (hexagonal boron nitride, hBN). Also called van der Waals (vdW) materials, these materials may be stacked in sequence to form *heterostructures* of multiple two-dimensional materials, each with different properties, enabling familiar device technologies including transistors and diodes [2,3]. Not only are such all-two-dimensional-material devices merely nanometers thick, thus transparent and flexible, they also offer performance gains relative to conventional metal oxide semiconductor field effect transistors (MOSFETs), including ON/OFF current ratios of $\sim 10^6$ [2] and reduced short channel effects, such as leakage current, that plague silicon integrated circuit FETs due to ever-shortening channel lengths. For this reason, among others, two-dimensional materials are hailed as the future of transistors and a possible way forward for Moore's Law [4,5]. However, manufacturing challenges stand in the way of expansive industrial adoption of vdW materials.

By leveraging phenomena such as quantum tunneling that are uniquely available to low-dimensional materials, vdW material transistors may offer up to two orders of magnitude lower power consumption than silicon transistors [6].

Diminished cooling demands [7] compound the advantages of vdW transistors relative to silicon-on-insulator transistors.

Not only is energy efficiency desirable from a sustainability perspective but also for consumer devices, less power-hungry electronics mean greater life between charges and opportunity for a smaller onboard energy storage system. Thermal management has become a limiting factor in the performance of modern chips; by greatly reducing the need for heat sinks and onboard thermal distribution in consumer electronics, energy-efficient vdW electronics can enable new device form factors.

Further reading on two-dimensional materials may be found in Part Three of this book.

22.2
Communications

Communication links, particularly wireless links, are becoming essential for an ever-larger range of consumer devices. Smart phones, laptops, and tablets generally include a range of high data-rate wireless protocols, such as wifi, cellular, and bluetooth; other devices have wireless controllers or peripherals, or internet links for remote monitoring and control. Since almost all these devices are battery powered, low power consumption is a critical requirement, and the communication links are often one of the most power hungry functions. For many devices, size is also highly constrained, and as with all consumer devices, the ability to manufacture very large quantities at low unit cost is essential. Trends in wireless communications for consumer electronics include the use of higher frequency bands to increase bandwidth, the integration of increasing numbers of wireless protocols within single systems, multiple antenna systems for space division multiplexing, and the use of wireless optical (or infra-red) links. Nanotechnology is already inherent in communication modules through the ubiquitous use of low cost, highly functional silicon integrated circuits. However, nanoscale devices are also showing promise in other components of these systems.

All radio transmitter circuits, and all but the most primitive receivers, require oscillator circuits. For consumer devices, the radio carrier frequencies, and thus the required oscillator frequencies, are mainly in the 1–2.5 GHz range. High performance depends on these circuits having a high quality factor Q, which is essentially a measure of how precisely the oscillation frequency is defined. To achieve high Q, oscillators often incorporate quartz crystal mechanical resonators; however, these cannot be monolithically integrated, and thus add to cost, size, and complexity. It has for some time been recognized that MEMS (microelectromechanical systems) devices could provide the combination of high Q performance and integration potential. A well-known review by Nguyen *et al.* [8] proposed a range of functions in radio transceivers where micromachined components could offer advantages; besides resonators for frequency references, these included switches, filters, high Q inductors, varactors (variable capacitors), and antennas. High Q MEMS resonators have been successfully developed, but their size is excessive compared to the electronic components with which they are integrated, and achievable frequencies are limited. Further miniaturization to the NEMS (nanoelectromechanical systems) scale inherently reduces these limitations. By 2005, NEMS resonator frequencies, using silicon carbide structures, were reaching 1 GHz [9], well above what has been achieved with equivalent MEMS devices. However, achievable Q values were shown to drop with miniaturization, because of increased surface–volume ratio and consequent damping effects, and this was identified as a key challenge for NEMS.

Later, resonators based on suspended graphene diaphragms were demonstrated [10]. The graphene structure provides strong electromechanical coupling, and an oscillator Q of ~4000 was achieved, although at a frequency of only 50 MHz. (A radio oscillator is a resonator incorporated into an amplified feedback loop, and typically has much higher Q than the resonator itself). The device's structure also allows oscillator tuning, and a tuning range of 14% was demonstrated. Bartsch *et al.* [11] reported monolithic integration of a NEMS tuning fork resonator with a FET receiver circuit. They demonstrated electromechanical demodulation of frequency modulated signals at >100 MHz, using the high Q (800) of the NEMS oscillator at near atmospheric pressure, with low power consumption (<1 mW).

A specific communication capability that has added greatly to the functionality of consumer electronics is the GPS (global positioning satellite) receiver for location determination. Early portable GPS receivers, developed for military applications in the 1980s, weighed over 20 kg; orders of magnitude size reduction have been achieved largely through two advances: the clock speed of silicon microprocessor units increased to the point where the GPS signal processing could be done digitally, allowing the use of silicon integrated circuits, and replacement of the GaAs MMIC (monolithic microwave integrated circuit) analogue components with Si also became possible [12]. Further miniaturization and power reduction will depend on the same advances as for the other radio receivers and digital processors in consumer electronics, and thus are likely to benefit from advances in nanoscale components and materials. For example, performance improvements in GPS radios have been demonstrated by the use of MEMS resonators [13], although frequency stability was insufficient for practical application. The use of NEMS devices in this application would decrease size and power consumption, and could make temperature control more practical, helping to overcome the frequency stability problem.

Antenna sizes are largely determined by operating wavelength, and for current and anticipated radio frequencies these are not going to be at the nanoscale. However, nanoscale materials can offer performance benefits in antenna structures, particularly for antennas on flexible substrates. One of a number of reports of the use of carbon nanotube (CNT) materials in patch antennas is found in [14]. They developed CNT bundle-based patch antennas for wearables. The advantage is reduced weight, as well as increased flexibility, compared to metal structures. Significant polarization dependence was seen according to the

Flexible GNP paper

Figure 22.1 Photograph of a 47 mm diameter sheet of graphene nanoplatelet paper (GNP) held in tweezers, after annealing and compression processes, illustrating its flexibility [15].

orientation of the nanotubes, which may offer some additional design flexibility. The antennas were for X-band operation (10 GHz); this is well above most present-day requirements, but future consumer systems are likely to exploit ever-higher frequencies. Gain was still less than for an equivalent copper antenna, but this was attributed to the limited CNT thickness that could be achieved (about 5 um) compared to the Cu thickness (17 um) and the electromagnetic skin depth at the operating frequency.

A related application is RF (radio frequency) shielding. The use of nanomaterials impregnated into flexible substrates can provide shielding between different RF modules, or screening from external interference, in a format compatible with wearables and other flexible devices. In Ref. [15], the use of graphene loaded paper, shown in Figure 22.1, was reported to achieve high conductivity for flexible RF shielding material. Good shielding effectiveness was measured (≈55 dB) for frequencies up to 18 GHz.

In addition to the peripheral components and systems discussed above, nanotechnology is also showing promise for RF circuits based on nonconventional transistor technology. For example, CNT-based thin film transistors (TFTs) can be useful for flexible circuits. In Ref. [16], semiconductor-enhanced CNTs are used to produce high performance TFTs with cut-off frequencies as high as 170 MHz, indicating the potential for at least some RF applications. More recently, flexible graphene transistors with cut-off frequencies up to 3 GHz were reported [17], and these were used to implement a mechanically flexible, primitive frequency modulation transmitter operating at 2.5 GHz. The reported system, illustrated in Figure 22.2, also used a graphene speaker for audio output at the receiver.

Finally, pressure on spectral availability and bandwidth is encouraging the exploration of alternative systems for wireless communications in consumer devices, such as visible or infrared communications. Nanostructures and devices

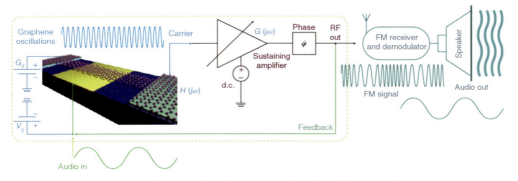

Figure 22.2 Schematic of graphene transistor-based modulator in radio transmitter module [17].

can enable many components for such systems. For example, Luzhansky *et al.* [18] reported a mid-wavelength infrared (MWIR) system for low weight, free-space communication. The source was a quantum cascade laser, a device in which nanostructuring is an essential requirement. The device demonstrated improved tolerance to atmospheric turbulence compared to conventional near-infrared systems.

As nanostructured elements become more integrated in communication systems, we can expect to see continued advantages in reduced power consumption through more efficient high Q oscillators, reduced weight using improved shielding and antenna materials, new operational frequency bands, and new modes of implementation, such as in flexible devices.

22.3
Energy Storage

Some amount of onboard energy storage is critical to any stand-alone powered device. Traditionally, batteries have been the exclusive means of onboard energy storage, and their high specific energy density enables long operation times between charges. However, batteries are subject to degradation at each charge/discharge cycle due to chemical modification of the electrodes. Recently, supercapacitors have attracted interest for their ability to deliver power densities of tens of kilowatts per kilogram, and tens of thousands of charge/discharge cycles without degradation, both orders of magnitude higher than batteries provide. Supercapacitors are potential storage solutions for applications in which the device largely sits idle but is subject to occasional and rapid power demands, for instance for incident sensing and data transmission. Such modes of operation are typical of Internet-of-Things devices, and have been enabled by recent advances in reducing the sleep-mode power consumption of communication circuits. While improvements to energy density and power density are boons to any application, the improvements in both that have been achieved using nanostructured devices are enabling new applications in the consumer device sphere that require energy densities and power densities previously inaccessible in a small form factor device. Recently, a boom in wearable electronic devices aimed at consumers – driven largely by the fitness market, and followed by accessory-scale mobile devices – has brought a commensurate demand for high-performance portable energy storage. In many devices, onboard energy storage need only power the device for a day, and may be rigid; in contrast, wearable devices require not only high energy density but also low profiles (to minimize invasiveness), flexibility (for ruggedness, form factor, and packaging considerations), and transparency (e.g., for displays). By offering inherent mechanical flexibility, limited optical absorption, and extraordinary surface-area-to-volume ratio, low-dimensional and nanostructured materials are attractive choices for electrodes and supercapacitors in consumer device-scale energy storage [19–21].

Nanoscale innovations have been able to contribute to battery performance with improved electrode materials. When considering electrode materials, the ability to host large numbers of ions is critical. In this regard, its high surface-area-to-volume ratio serves graphene well, as it is capable of hosting twice as many Li+ ions as bulk graphite [22]. However, in-use degradation of graphene cripples its feasibility as an electrode material. Ions must shuttle between electrodes in order to make the battery useful, and it is here that graphene suffers. Reduced graphene oxide – a form of graphene made by chemically expanding bulk graphite to facilitate exfoliation, and reducing the final graphene oxide product – undergoes a chemical reduction upon the initial implantation of lithium ions, becoming passivated and restricting ions from traveling between electrodes [23,24]. The ion-induced degradation limits the efficiency of the electrode, negating the value of its ion-hosting capacity. Thus, it is not believed that graphene electrodes will find widespread use [25], though many attempts are underway to architect the material in such a way that the issue of surface passivation is avoided [25].

Nanostructured materials benefit Li batteries by offering reversible intercalation and high charge/discharge rates due to their high surface area to volume ratio; Liu *et al.* conducted a thorough review of the use of carbon nanotubes in electrode applications [26]. To this end, Kim *et al.* [27] achieved a maximum discharge capacity of 1882 mAh/g using multiwalled carbon nanotubes in combination with silicon.

Silicon itself has a high theoretical charge capacity, but has found little use in Li battery anodes as it experiences a 400% volume change during lithiation that was mechanically untenable. Structuring silicon into nanowires allowed researchers to leverage silicon's theoretical charge capacity in actual application, reaching a maximum discharge capacity of 3193 mAh/g, without suffering the mechanical degradation that cripples bulk silicon [28]. A review of the benefits available by nanostructuring silicon and the mechanics of lithiation and delithiation in nanostructures may be found in Ref. [29].

Supercapacitors are fundamentally structured differently from batteries in such a way that enables rapid charge and discharge, which is equivalent to high power density. While batteries rely on chemical insertion and removal of ions from a material, supercapacitors take advantage of the creation of an electric double layer on a charged surface, meaning ions are only loosely, electrostatically adhered rather than chemically adhered. Though this mechanism limits the stored energy density of supercapacitors, it also allows supercapacitors to endure orders of magnitude more charge/discharge cycles than batteries without suffering material degradation. Successful supercapacitor systems must offer:

- high specific capacitance (hundreds of F/g),
- high specific power density (at least tens of W/g), and
- many recharge cycles (tens of thousands) with negligible loss of capacitance,

in addition to, ideally, being fabricable from readily available materials in low-cost manufacturing processes. A sample of properties offered by different supercapacitor and battery systems is given in Tables 22.1 and 22.2.

Ubiquitous and high-performance consumer electronics will need to leverage the improved energy densities which nanostructured materials are providing to both battery and supercapacitor systems. Hybrid storage systems could let device makers combine the benefits of long lifetime between charges in batteries and many charges per lifetime in supercapacitors. Moreover, both storage platforms will likely offer flexibility and facile mass production in the near future, for example, by printing methods as reviewed by Refs [39–42]. Challenges will come in finding cheaper and less-toxic materials for batteries, increasing the energy density of supercapacitors while reducing leakage current (about 10%/day [43]), and fine-tuning production techniques. Furthermore, the promise of environmental energy harvesting offers potential for fully self-powered consumer devices, with intermittent energy generation coupled to local storage. Already, consumer-device scale flexible lithium ion batteries have been coupled with onboard energy harvesting for a fully self-powered integrated approach to wearable electronics [44].

Energy harvesting converts ambient sources of energy (predominantly heat, motion, and light) into electric energy for devices. Heat harvesting via

Table 22.1 Characteristics offered by selected nanostructured supercapacitor systems.

Max Capacitance (F/g)	Max energy density	Max power density	Material	Retention	Reference
540	51 Wh/kg at a power density of 205 W/kg		ZnS-carbon textile	94.6% after 5000 cycles	[30]
234.2	0.72 mWh/cm^3	~0.4 W/cm^3	MnO$_2$/G-gel/NF	>98.5% after 10 000 cycles	[31]
267	17 Wh/kg	2520 W/kg	MnO$_2$	92% after 7000 cycles	[32]
222	106.6 Wh/kg	10.9 kW/kg	Single-walled carbon nanotubes with reduced graphene oxide	~99% after 1000 cycles	[33]

Table 22.2 Characteristics offered by selected Li-ion and nanostructured battery systems.

Energy density (Wh/kg)	Charge capacity (mAh/g)	Material	Retention	Reference
100–200	~150	Lithium ion state-of-the-art	—	[34,35]
~120	377	Na–S	—	[35]
—	4277	Silicon nanowires	~81% over 20 cycles	[28]
—	~2500	Multiwalled carbon nanotubes with silicon	~87% over 12 cycles	[27]
—	2725	Interconnected silicon hollow nanospheres	>92%/100 cycles, for 700 total cycles	[36]
60	40	$LiMn_2O_4$ cathode, $LiTi_2(PO_4)_3$ anode, and Li_2SO_4 electrolyte	82% over 200 cycles	[37]
—	—	Porous PDMS electrode, $Li_4Ti_5O_{12}$ anode and $LiFePO_4$ cathode	70% after 300 cycles	[38]

thermoelectrics has seen performance improvements using nanostructured materials [45]; however, because of the small temperature differences present in any (comfortable) consumer device environment, the main applications for advanced thermoelectrics are likely to be in waste-heat recovery applications with large temperature differences, such as in the automotive sector [46]. For motion harvesting, transduction is often achieved using piezoelectric materials, which directly couple mechanical strain and electric field. Nanostructuring is being investigated to enhance properties of these materials [47]; researchers are aiming to improve film thinness, ability to withstand high flexural strain, and defectivity of piezoelectrics. Complex material structuring methods have been deployed to construct flexible piezoelectric metamaterials at the macroscale using macrofiber-composites [48].

Harvesting from light, using photovoltaic cells, has been commercialized for low power consumer devices such as pocket calculators for several decades. Recent research in photovoltaics is mainly concerned with large-area applications, where cost per unit area is critical; for consumer devices, where area is highly constrained, energy conversion efficiency is a more critical parameter. Nanostructures have been widely investigated for increasing the efficiency of photovoltaics, as reviewed in Ref. [49], by increasing both light trapping and photoelectric conversion, achieving efficiencies as high as 37.9% using multijunction devices [50]. State of the art efficiencies are regularly updated by Green *et al.* in [50].

22.4
Sensors

Sensors that are currently integrated in consumer electronics are mainly motion and image sensors. In addition, some portable devices integrate biomedical sensors to measure pulse rate, electrocardiographs, glucose, or blood oxygen, intended for health monitoring but also sports applications. These systems have greatly extended the functionality of personal electronics, introducing convenient services such as personal navigation, activity logging, connected and tagged photography, and location-sensitive information. Nanotechnology has already played a very important role in this development in the on-chip integration of multiple sensor systems, advanced biochemical assay analyses and nanolenses. In return, the consumer electronics market has been driving the booming progress in motion and camera sensors of the last decade, with state-of-art high-performance sensors being regularly found in a range of personal electronic equipment. This large market creates demand for advances supplied by nanotechnology. In the following subsections, motion processing units, portable biomedical sensors, and imaging sensors are discussed along with relevant nanotechnologies, both current and imminent.

22.4.1
Motion Processing Units

Acceleration sensors are the most developed and widely applied type of MEMS sensor. They are based on an inertial mass that moves inside the sensor package in response to acceleration. Beyond the supporting CMOS (complementary metal-oxide-semiconductor) circuit, these require fabrication of planar and vertical profiles and interfaces at the nanoscale. MEMS accelerometers also require high quality nanoscale springs. At the device level, accuracy, sensitivity, noise, dynamic and spectral ranges, and response time are the main performance metrics. Nevertheless, the key element in the successful incorporation of motion sensors to consumer electronics is chip-level integration of multiple sensors into a CMOS system. This provides compactness, reliability, and connectivity benefits, and also enables the combination of multiaxis linear and rotational motion data, and thus more sophisticated motion-based applications. Such applications include detection of free-falling, gestures, display orientation, and tapping; image stabilization; step counting for athletic applications; and dead reckoning (navigation by integrating motion from a known location). System-level integration of such services is critical for efficient power management in portable devices.

Commercial examples of integrated motion processing units (MPUs) are the Invensense MPU-9250, the Bosch BMX055, and the ST LIS2DS12 microchips. All integrate a three-axis accelerometer and a three-axis gyroscope, and the first two also include a 3D orientation sensor. The sensors are integrated with analog-to-digital converters and a digital processing unit to provide interrupts for detection and counting of specific events such as tapping, dropping, stepping, and motion state change. A comparison of the technical features of these implementations is presented in Table 22.3. Predecessors of these models have been used in popular devices including the Apple iPhone 6 and the Google LG Nexus 5. This type of hybrid system-on-chip integration has recently enabled various features and services in consumer electronics, such as indoor navigation by dead

Table 22.3 Comparison of commercial motion processing unit chips.

Brand/model	Bosch/BMX055	Invensense/MPU-9250	ST/LSM6DSL
Year	2013	2014	2015
Packaging	$3 \times 4.5 \times 0.95$ mm, LGA	$3 \times 3 \times 1$ mm, QFN	$2.5 \times 3 \times 0.86$ mm, LGA
MEMS area	—	~ 2.5 mm^2	—
Accelerometer range, sensitivity, RMS noise	± 2 g to ± 16 g, 1 mg/LSB, 150 µg/$\sqrt{\text{Hz}}$	± 2 g to ± 16 g, 62 µg/LSB, 300 µg/$\sqrt{\text{Hz}}$	± 2 g to ± 16 g, 61 µg/ LSB, 90 µg/$\sqrt{\text{Hz}}$
Gyroscope range, sensitivity, RMS noise	125–2000°/s, 0.0038°/s, 0.01°/s/$\sqrt{\text{Hz}}$	250 to 2000°/s, 0.0076°/s, 0.01°/s/$\sqrt{\text{Hz}}$	125–2000°/s, 0.0044°/ s, 0.0045°/s/$\sqrt{\text{Hz}}$
Orientation magnetometer range, sensitivity	X,Y: ± 1.3 mT, Z: 2.5 mT, 0.3 µT/LSB	± 4.8 mT, 0.4 µT/LSB	—
Active/sleep mode power	18 mW/20 µW	9 mW/20 µW	1.2 mW/5.4 µW
Event recognition	Any motion, high magnetic field, low-g/high-g, new data, tap.	Any motion, step detector/ counter, programmable interrupts.	Free fall, motion change, step detector/ counter, tap.
Connectivity	SPI and I^{2}C	SPI and I^{2}C	SPI and I^{2}C

reckoning, gesture recognition, motion-based power saving, stable minidrones, and augmented reality systems.

The implementation of motion sensors in the nanoscale using conventional concepts of operation is challenging because of the requirement for a substantial proof mass. While new device concepts using 1D and 2D materials have been proposed, these are at an early stage of development [51,52]. Further information about emerging nanoscale motion sensor technologies can be found in Chapter 8.

22.4.2
Nanosensors for Biomedical Applications

Portable, personal biomedical devices have been expanding rapidly in usage and variety in recent years, supported by new sensor types and by low-power sensor, processing, and communication electronics [53]. The electronic personal glucometer, in commercial use since 1971, has been particularly successful: it has allowed a profound improvement in the quality of life for hundreds of millions of people globally [54]. The sensors in these systems are typically either electrochemical, based on measuring the electric current of glucose oxidation through glucose oxidase, or optical, mainly using tissue spectroscopy and fluorophore techniques [55]. Extensive research has focused on the improvement of these methods using nanostructures. Nanoparticles can improve the selectivity, stability, and sensitivity of electrochemical sensors by enhancing glucose absorption and improving electron transfer. Gold nanoparticles are the most popular in such studies, and they have also been shown to improve the sensitivity of optical glucose sensors, by serving as fluorescence quenchers. Nanowires, nanotubes, and nanocomposites have also been used in glucose sensors, mainly to improve the interface area and the selective absorption of glucose molecules. Nanotechnology-enhanced glucose sensors are expected in commercial glucose monitoring systems in the next few years. They may be particularly beneficial in continuous monitoring systems, where sensors are required to be minimally invasive. A review of nanotechnology research for glucose sensors can be found in Ref. [56].

Other personal biomedical devices that have been in commercial use include pulse oximeters, portable electrocardiogram systems, optical heart rate counters, breath analyzers, respiratory monitors, and fertility monitoring systems [57]. These systems have been given a dramatic boost by the advances in handheld personal devices, mainly smartphones and tablets, which offer a platform with a pre-existing display, stored power, and onboard computing capability. While systems such as pulse oximeters or electrocardiographs are fully optical or electrical, a number of personal biomedical sensors involve measuring an analyte through an assay procedure. In these cases, nanoparticles and nanostructures already play a critical role: nanoparticles are used to provide an optical output to selective reactions, in order to be measureable by a camera, which is typically available in the host personal electronic device. Gold nanoparticles in the size range 5–200 nm are often used; applications include the detection of antibiotics in food samples [58]. In the near future, such capabilities may become available to consumers as a smartphone-hosted service. An example of a gold nanoparticle-based assay system implemented as a smartphone sensor is the NutriPhone lab-on-a-chip, which can measure the level of vitamin D in a sample. An image of this prototype device describing the process of absorption enhancement by gold nanoparticles is shown in Figure 22.3 [59]. Quantum dots have also been used as a photoluminescence source for improved assay analysis in smartphone-based point of care biomedical sensors [60]. An overview of

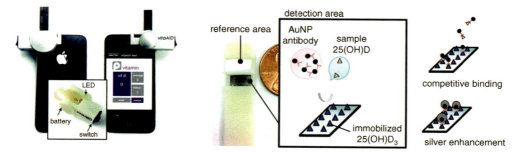

Figure 22.3 The gold nanoparticle based NutriPhone lab-on-a-chip for measuring vitamin D levels using a smartphone. (From Ref. [59].)

the role of nanoparticles in personal diagnostic devices has been recently published in Ref. [61].

22.4.3
Optical Sensors

The development of solid-state image capturing devices, initiated and mainly driven by charge coupled device (CCD) array technology, has led to the availability of low-cost, high-end sensors. Solid-state imaging is now ubiquitous in stand-alone cameras, and integrated into smartphones, tablets, and other personal electronic devices. An image sensor is comprised of an array of cells, each one having a semiconductor channel within which charge is accumulated by the absorption of incident light. The image resolution is defined by the number of rows and columns of cells. In CCD sensors, charge is measured at one side of the array by sequentially forwarding the accumulated charge from one cell to the next, using electrostatic gating. In CMOS image sensors, each pixel is individually addressable and so requires one or more additional transistors as well as further wiring. Both technologies now yield similar performance in terms of image quality, with CMOS being generally lower cost and more compatible with monolithic integration with other electronic functions. Hence, CMOS image sensor technology currently owns a much higher share of the imaging market. The size range of a typical sensor array is in the range of $10–100$ mm^2, with larger arrays being beneficial for better sensitivity, because of larger pixel area, larger depth of field, and field of view specifications. The resolution typically found in high-end commercial devices is in the range of tens of megapixels, corresponding to a pixel pitch of $1–2\,\mu$m. This means that for typical consumer imaging applications, although the feature sizes of components within each pixel are in the submicron range, they can be fabricated by conventional CMOS methods, rather than requiring bespoke nanotechnological processes.

Nevertheless, advances in nanotechnology could soon play a significant role in the evolution of optical sensors for consumer electronics. With the rapid proliferation of applications related to 3D imaging, subpixel optical processing may become highly desirable in next-generation optical sensors especially in combination with high precision motion sensing. With subpixel processing, information about the direction of incoming light can be captured, allowing the reconstruction of a 3D image, in what is called light-field or plenoptic cameras [62]. Advances in nanolens arrays may play an important role; gold nanowire-based nanolenses have already been demonstrated that achieve imaging resolution beyond the Abbé diffraction limit [63].

An important and widely experienced limitation of current image sensor technologies is the poor performance in low light environments. Si-based sensors are not sensitive enough to capture images in dark environments at short exposure times, leading to either dark or blurry pictures. New materials such as monolayer molybdenum disulfide have proven promising for a dramatic increase in sensitivity, with the additional benefit of using only a subnanometer-thin and effectively transparent active area layer [64]. Such materials have not been integrated to device level yet.

Top-down nanoscale fabrication and chip integration techniques will be increasingly important as on-chip sensor integration becomes essential and is expanded to new sensor types such as breath or environmental gas analyzers, indoor localization, and radiation exposure. Silicon nanowires have been identified as a likely successor to state-of-the-art finFET CMOS technology, and nanowire-based sensors, either electrode- or beam-based, could benefit in cost, availability, and compatibility from sharing material and fabrication techniques, just as silicon MEMS benefitted from Si integrated circuit technology in the 1990s. Self-assembly and the fabrication of bulk materials from nanostructures are also expected to play an important role in consumer electronics, especially in optical surface processing layers and transparent conductors for displays, but also in communication components, as discussed elsewhere in this chapter.

22.5
Internet-of-Things Applications

In the Internet-of-Things (IoT) paradigm, a wide range of objects are given intelligence in the form of computation, communication and, typically, sensing capability [65]. IoT will allow such objects to be monitored and potentially operated remotely, either by human users or by automated systems. While "IoT" implies interconnection over the Internet, the concept can be extended to interconnection using other networks, such as private networks in the case of applications with high security demands. For consumer products, the most prevalent IoT products in the market currently are for "smart home" applications. These include Internet-enabled climate control systems and appliances, enabling such functions as control of room temperature and lighting, washing machine operation, or appliance status checks, from any location, using smartphone apps. In fact, IoT functionality will make electronics significant in a much larger range of household objects, including small electric appliances such as toasters and shavers.

The role that nanotechnology can play, for example, in improved energy storage, or novel antennas, discussed in Sections 22.2 and 22.3, will clearly apply in the IoT domain, as will their respective challenges. IoT applications also have particular challenges in terms of form factor, social acceptance, and distributed network communications and data processing.

The range of consumer products for which some IoT functionality is conceivable is vast, but in practice the range of applications will be constrained by the balance of cost, broadly defined, versus perceived benefit. For example, if the use of a product becomes dependent on its IoT functionality, then the IoT components must have high reliability, ruggedness, and long lifetime, as well as being cheap to manufacture and to incorporate into the object. In most IoT applications the electronics will be battery-powered, and therefore long battery lifetime, or the use of battery alternatives such as energy harvesting, will be essential. Minimizing power consumption for the computation and communications components is imperative for long battery lifetime.

One approach to overcoming the power consumption problem, which is suitable for some IoT applications, is that used in radio frequency identification (RFID). RFID tags typically hold stored information about the tagged object that can be interrogated by a reading device. While active RFID tags contain their own power supplies, passive tags gain the required power from the incident RF signal transmitted by the reader, and so have no on-board energy storage requirement. Permanent identity tagging is not a true IoT function, but the RFID platform can be extended to communicate nonstatic information such as the output of an on-board sensor [66]. For example, in Ref. [67], a strain sensor is reported, which uses a resonant L–C (inductor–capacitor) circuit implemented on a flexible substrate, so that the resonant frequency varies to indicate the strain state of the device. Since the tag is entirely passive, no memory is provided and the strain can only be known at the moments when the device is interrogated. Silver nanoink, comprised of silver nanoparticles suspended in an organic solution, is used to achieve the flexible conductors from which the circuit components are fabricated.

More ambitious application of nanotechnology to IoT has also been proposed. In the Internet-of-Nano-Things concept [68], the intelligence is implemented by devices that are themselves at nanoscale. Realization of this concept will require nanoscale sensors, and communication approaches that are compatible with nanodimensions. The latter requirement makes the use of THz frequencies attractive because of the short wavelengths involved, although these are still in the μm rather than nm range. Alternatively, the use of biomolecules as methods of communication between nanodevices [69], particularly within the body, has also been proposed, although this remains far from commercial realization.

22.6
Display Technologies

Display technology has been revolutionized since the late 1990s with the introduction and widespread use of various new types of flat panel displays. These have allowed more compact, higher quality, and more efficient stationary display devices, while simultaneously equipping handheld devices with powerful graphics, paving the way for handheld computing, including various new services such as localized information and augmented reality. The key underlying scientific advancements are the invention of the InGaN quantum well blue LED [70] and bright solid-state lighting, progress in LCD and the semiconductor technology including MEMS microreflector technologies, the development of polymer semiconductors (organic, flexible LEDs) and improved performance of transparent conductors.

22.6.1
Self-Illuminating Displays

Liquid crystal displays (LCDs) employ a liquid layer containing longitudinal molecules that tend to align in a common direction. This type of liquid, called a nematic liquid crystal, is interposed between two patterned surfaces, which guide the crystals to a rotation gradient, thereby rotating the polarization of passing light. By using two orthogonally aligned polarization films, the multilayer is transparent only when the crystals are in the rotational orientation. Hence, the intensity of passing light can be controlled by electrostatic distortion of the crystal rotation. A grid of transparent electrodes allows control of individual pixels. A white LED source is used as backlight, and color is achieved by using multiple cells per pixel and filtering. Other liquid crystal geometries and alignment modes, such as in-plane switching (IPS) and vertical alignment (VA), are described in Refs [71,72]. The IPS-LCD is currently the most widespread display technology for consumer electronics, used in the vast majority of flat panel monitors and televisions as well as smartphone models such as the Google Nexus, the LG G-series, and the Apple iPhone 6.

The light transmission efficiency of LCD displays is below 10%, due to the large number of optical layers required. As energy consumption becomes increasingly important, especially for portable devices, methods for better light management have been proposed. These include the use of polarized light sources to reduce loss from the bottom LCD polarizer, sequential coloring in place of color-filtering, dynamic LED backlight, and reflective polarization using grid polarizers. Nanowire grid polarizers have been proposed and implemented to achieve significant light recycling [73]. Nanowire fabrication techniques that have been used include laser-interference lithography [74], nanoimprint technology [75], and self-assembly, such as block copolymer pattern transfer [76]. Advances in the manufacturability of polarization nanostructures should play an important role in light management for power reduction of displays in the near future.

Organic light-emitting diode (OLED) displays consist of an array of organic semiconductor devices which can be individually lit by application of an electric current, using bottom and top transparent electrode grids to address each pixel. The application of a voltage across an organic semiconductor material results in the injection of electrons from the cathode and holes from the anode, which recombine, emitting a photon. Depending on whether each pixel contains a transistor for switching support or not, OLED displays are subdivided into active matrix (AMOLED) or passive matrix (PMOLED) displays. AMOLED has the advantage of requiring considerably lower current for the same illumination. The photon emission at recombination depends on the spin combination of the two carriers, which leads to a triplet exciton at 75% in three out of four recombinations. These excitons cannot directly decay, delaying transmission and leading to an efficiency of around 25% for simple small molecule organic materials. Implantation of organometallic molecules with heavy metals, such as iridium, assists in the exploitation of phosphorescence from triplet excitons. Commercial OLED devices leverage this technique for red and green emitters, but use a conventional fluorescent small molecule LED for deep blue, due to reliability concerns for blue phosphorescent layers [77]. Commercial OLED flat panel display products include models from most major TV and monitor makers as well as smartphones such as the Samsung Galaxy series.

Nanotechnology is expected to play a significant role in the technology evolution of OLED devices. Nanoscale surface modification of OLEDs has been shown to reduce plasmon-trapping of emitted light, thereby improving the out-coupling efficiency of the device [78]. The reduction of roughness of OLED material layers has been demonstrated by using nanoparticle-based electrospray deposition methods [79]. The use of graphene and other nanomaterials as transparent conductors for OLED devices has also been proposed [80,81]. Transparent conductors are discussed separately later in this section.

Field emission displays (FED) have been also proposed and studied, as a type of illuminating display technology that combines the brightness and contrast benefits of the old cathode ray tube technology with the compactness and low power of modern flat panel displays. In this display architecture, an array of field emission tips is used to emit electrons and excite photons through a color phosphoric panel. This type of display could benefit substantially by employing efficient nanoscale emitting structures such as self-assembled nanoscale tips or carbon nanotubes [82]. Industrial interest in FED technology has declined in recent years, mainly due to the requirement for high vacuum integration and structural reliability issues related to ion impact damage of the nanoscale tips.

22.6.2
Reflective Displays

Reflective displays address the markets of electronic reading, smart windows, dimmable mirrors, large-scale screens, and signing. Electronic readers (eReaders), currently the most widespread application, are typically based on the electrostatic migration of light- and dark microparticles. Such devices are known as electrophoretic displays. Particles of both types are encapsulated in microcontainers. The light reflection from each container is controlled by applying an electrostatic field, which can bring to the surface either the light or the dark particle population. This microencapsulation technique was first introduced by Comiskey *et al.* [83], and is used in the eReader products of E-Ink, currently supporting popular commercial products such as the Amazon Kindle. The main benefit over LCD displays is the absence of backlighting, which means lower eye stress and power consumption. While the "inks" in reflective displays are monochromatic, recent studies have proposed the use of electrophoretic nanoparticles, including organic molecules [84], pigment-silica composite nanoparticles [85], phthalocyanine nanoparticles [86], and block copolymers [87], as color electronic inks. Furthermore, front-lit displays (which enhance reading in dark environments) can operate with a single LED light source, by using nanoimprinted light guides along with an array of holes to evenly disperse light across the screen.

The use of electrochromic materials has also been proposed for flexible reflective displays. Such materials include metal oxides (e.g., WO_3, $Ir(OH)_3$), viologens, and polymers. Electrochromic materials are able to reversibly change their refractive index when they are oxidized or reduced, thereby allowing electronic control of light reflection. A transparent conductor such as indium tin oxide (ITO) is used to control each pixel. The use of mesoporous TiO_2 has been proposed [88] and used in commercial electrochromic products [89] to confine electrochromic molecules, resulting in contrast enhancement at the edges of displayed features. Color electrochromic displays have also been demonstrated, using multilayers of different materials or optical filters [90]. A review of electrochromic materials for display technologies can be found in Ref. [91].

Electrowetting, the modulation of surface tension by the application of an electric field, is also under investigation for reflective displays. In such devices, the intensity of a pixel can be controlled by forcing a light-absorbing liquid to alternately spread over the pixel surface area or to congest at a corner, through surface tension. Despite involving fluid motion, these devices have shown promise for faster switching in comparison with electrophoretic displays [92]. Color electrowetting displays could be implemented by using multilayers of oil-based materials. Controlling the hydrophobic and oleophobic properties of surfaces by surface engineering at the nanoscale is expected to play a critical role in electrowetting devices [93,94]. For example, the use of nanopillars has been shown to enhance the manipulation of liquids in the microscale [95]. An analysis of electrowetting and its applications can be found in [96]. This technology was initially proposed by Philips Research Lab, and subsequently developed through the Liquavista spin-out in 2005. Liquavista was acquired by Amazon in 2013.

22.6.3
Transparent Conductors

A common key technology behind these diverse approaches to making displays is the integration of a transparent conductor material. Such a material is required as an interconnection layer at the front side of the display, in order to allow individual pixel control. In addition, transparent conductors play a vital role in the integration of displays with complementary functionality such as touch-screens. The most common conductor material for these purposes is ITO, which has an absorption spectrum practically identical to that of glass, transparency in the range of 90% for typical film thickness, and an electrical conductivity of 10^4 S/cm [97]. ITO has been studied and used over several decades, providing mature fabrication and integration processes [98]. It holds over 90% of the billion-dollar transparent conductor market [99], with the rest mainly shared between fluorine doped tin oxide (FTO) and aluminum-doped zinc oxide (AZO). The rapidly increasing demand for transparent conductors has led to research for alternative materials, mainly because of low indium reserves and the need for flexible devices.

Figure 22.4 Demonstration of flexibility in a transparent display using metal nano-wire based transparent conductor, Lee *et al.* [105] (left); and comparison of transparency for a graphene based touch screen with 250 Ω/sq. sheet resistance (corresponding to around 10^5 S/cm) against an ITO implementation, Ryu *et al.* [106] (right).

To achieve a combination of high flexibility, transparency, and conductivity, various nanostructures have been considered, including carbon nanotubes, graphene, and metallic nanowires [100,101]. Nanowire conduction lines are usually deposited from a solution, allowing them to be printed or spread onto a surface. This is compatible with existing roll-to-roll production methods that are envisioned for flexible electronic devices. Nevertheless, dry deposition methods do exist, such as preparation of CNT sheets by pulling from vertically grown multi-wall CNTs [102]. CNT transparent conductors exhibit over 80% transparency, and conductivity in the range of 10^3 S/cm, but are difficult to manufacture due to purification and separation challenges [100]. Graphene layers have been recently shown to exhibit similar electrical conductivity and optical transparency performance to ITO, over 10^4 S/cm at 80% transparency [103]. Extensive reviews on the prospects of graphene transparent conductors have been reported recently, including performance comparisons against other materials [80,104]. Examples of touch screen panels using metal nanowire- and graphene-based transparent conductors are shown in Figure 22.4.

Silver nanowires have been demonstrated to be over 90% transparent, with electrical conductivities over 10^4 S/cm [107]. They are considered a very promising alternative to ITO, due to the high conductivity of silver. Conductive lines are fabricated by deposition, as with CNTs, but require annealing to exhibit high sheet conductivity. Such annealing is not always compatible with CMOS integration, and significant effort has been devoted to developing alternative methods for improving nanowire interlinks, such as pressing [108]. In addition, the surface roughness on metallic nanowires is a significant factor, as it can lead to increased reflections. Silver nanowire-based OLED prototypes have already been demonstrated with very promising transparency and durable flexibility performance [109]. A review of the prospects of silver nanowires as a flexible transparent conductive material can be found in Ref. [110]. Various conducting polymers are also under investigation, such as the commercially available PEDOT: PSS with conductivity as high as 10^3 S/cm [111]. An overview of nanotechnology-based flexible conducting materials can be found in Refs [112,100].

The evolution of flat panel displays has already benefitted from various nanoscale optical technologies, such as grid polarizers, reflectors, and color filters, as well as from nanofabrication techniques. Nanotechnologies are of critical importance to the progress of LCD, electrophoretic, and electrochromic displays, all of whose operating principles fundamentally rely on nanoscaled structures. Nanomaterial-based transparent and flexible conductors, such as CNT bundle layers or silver nanowires, are expected to join ITO in display technology in the short term, enabling higher resolution and energy efficiency, denser functionality, and improved transparency and flexibility, which is expected to give rise to vastly different ways for users to interact with handheld electronics.

22.7
Conclusions

Nanotechnology is essential to the continuing advances in integrated electronics: increasing computational power, reducing device scale, and limiting energy consumption. In addition, as we have seen in this chapter, nanostructured materials are providing performance improvements and new functionalities in a range of associated technologies, such as nanomechanical devices, transparent and flexible conductors, display materials, and high capacity energy storage. These developments provide critical support to current trends in consumer electronics – in particular, increased functionality and battery life in smartphones and other handheld devices, rugged and comfortable wearables, wide area and flexible displays, and a wide range of sensing devices.

We anticipate a future where distributed devices, with specialized sensing capabilities and long-lasting energy supplies, are ubiquitous and well integrated into our environments and ways of living. By providing such devices with both the unique functionalities, including flexibility and transparency, and the required performance, nanoelectronics will be critical to realizing such a future.

References

1 IDC (2015) Worldwide Quarterly Wearable Device Tracker, Framingham, MA.

2 Roy, T., Tosun, M., Kang, J.S., Sachid, A.B., Desai, S.B., Hettick, M., Hu, C.C., and Javey, A. (2014) Field-effect transistors built from all two-dimensional material components. *ACS Nano*, **8** (6), 6259–6264.

3 Das, S., Gulotty, R., Sumant, A.V., and Roelofs, A. (2014) All two-dimensional, flexible, transparent, and thinnest thin film transistor. *Nano Lett.*, **14** (5), 2861–2866.

4 Ferrari, A.C. (2014) Science and technology roadmap for graphene, related two-dimensional crystals, and hybrid systems. *Nanoscale*, **7** (11), 4598–4810.

5 Kim, S.J., Choi, K., Lee, B., Kim, Y., and Hong, B.H. (2015) Materials for flexible, stretchable electronics: graphene and 2D materials. *Annu. Rev. Mater. Res.*, **45** (1), 63–84.

6 Ionescu, A.M. and Riel, H. (2011) Tunnel field-effect transistors as energy-efficient electronic switches. *Nature*, **479** (7373), 329–337.

7 Islam, S., Li, Z., Dorgan, V.E., Bae, M.H., and Pop, E. (2013) Role of joule heating on current saturation and transient behavior of graphene transistors. *IEEE Electron Device Lett.*, **34** (2), 166–168.

8 Nguyen, C.T.C., Katehi, L.P.B., and Rebeiz, G.M. (1998) Micromachined devices for wireless communications. *Proc. IEEE*, **86** (8), 1756–1767.

9 Ekinci, K.L. and Roukes, M.L. (2005) Nanoelectromechanical systems. *Rev. Sci. Instrum.*, **76** (6), 061101.

10 Chen, C., Lee, S., Deshpande, V.V., Lee, G.-H., Lekas, M., Shepard, K., and Hone, J. (2013) Graphene mechanical oscillators with tunable frequency. *Nat. Nanotechnol.*, **8** (12), 923–927.

11 Bartsch, S.T., Rusu, A., and Ionescu, A.M. (2012) A single active nanoelectromechanical tuning fork front-end radio-frequency receiver. *Nanotechnology*, **23** (22), 225501.

12 Stotts, L.B., Karp, S., and Aein, J.M. (2014) The origins of miniature global positioning system-based navigation systems [SP History]. *IEEE Signal Process. Mag.*, **31** (6), 114–117.

13 Lovseth, J., Hoffmann, T., Kalyanaraman, S., Reichenauer, A., Olen, V., and Hrncirik, D. (2012) Communication and navigation applications of nonlinear micro/nanoscale resonator oscillators, in Proc. of SPIE, **8373**, p. 837305.

14 Keller, S.D., Zaghloul, A.I., Shanov, V., Schulz, M.J., Mast, D.B., and Alvarez, N.T. (2014) Radiation performance of polarization selective carbon nanotube sheet patch antennas. *IEEE Trans. Antennas Propag.*, **62** (1), 48–55.

15 Tamburrano, A., Paliotta, L., Rinaldi, A., Bellis, G.De., and Sarto, M.S. (2014) RF shielding performance of thin flexible graphene nanoplatelets-based papers. 2014 IEEE International Symposium on Electromagnetic Compatibility (EMC), pp. 186–191.

16 Wang, C., Chien, J.-C., Takei, K., Takahashi, T., Nah, J., Niknejad, A.M., and Javey, A. (2012) Extremely bendable, high-performance integrated circuits using semiconducting carbon nanotube networks for digital, analog, and radio-frequency applications. *Nano Lett.*, **12** (3), 1527–1533.

17 Yogeesh, M., Parrish, K., Lee, J., Park, S., Tao, L., and Akinwande, D. (2015) Towards the realization of graphene based flexible radio frequency receiver. *Electronics*, **4** (4), 933–946.

18 Luzhansky, E., Choa, F.-S., Merritt, S., Yu, A., and Krainak, M. (2015) "Mid-IR free-space optical communication with quantum cascade lasers," p. 946512.

19 Zhu, Y., Murali, S., Cai, W., Li, X., Suk, J.W., Potts, J.R., and Ruoff, R.S. (2010) Graphene and graphene oxide: synthesis, properties, and applications. *Adv. Mater.*, **22** (35), 3906–3924.

20 Chen, J., Li, C., and Shi, G. (2013) Graphene materials for electrochemical capacitors. *J. Phys. Chem. Lett.*, **4** (8), 1244–1253.

21 Yang, X., Cheng, C., Wang, Y., Qiu, L., and Li, D. (2013) Liquid-mediated dense integration of graphene materials for compact capacitive energy storage. *Science*, **341** (6145), 534–537.

22 Dahn, J.R., Zheng, T., Liu, Y., and Xue, J.S. (1995) Mechanisms for lithium insertion in carbonaceous materials. *Science*, **270** (5236), 590–593.

23 Raccichini, R., Varzi, A., Passerini, S., and Scrosati, B. (2015) The role of graphene for electrochemical energy storage. *Nat. Mater.*, **14** (3), 271–279.

24 Winter, M., Besenhard, J.O., Spahr, M.E., and Novák, P. (1998) Insertion electrode materials for rechargeable lithium batteries. *Adv. Mater.*, **10** (10), 725–763.

25 Kou, L., Liu, Z., Huang, T., Zheng, B., Tian, Z., Deng, Z., and Gao, C. (2015) Wet-spun, porous, orientational graphene hydrogel films for high-performance supercapacitor electrodes. *Nanoscale*, **7** (9), 4080–4087.

26 Liu, X.-M., dong Huang, Z., woon Oh, S., Zhang, B., Ma, P.-C., Yuen, M.M.F., and

Kim, J.-K. (2012) Carbon nanotube (CNT)-based composites as electrode material for rechargeable Li-ion batteries: a review. *Compos. Sci. Technol.*, **72** (2), 121–144.

27 Kim, T., Mo, Y.H., Nahm, K.S., and Oh, S.M. (2006) Carbon nanotubes (CNTs) as a buffer layer in silicon/CNTs composite electrodes for lithium secondary batteries. *J. Power Sources*, **162** (2), 1275–1281.

28 Chan, C.K., Peng, H., Liu, G., McIlwrath, K., Zhang, X.F., Huggins, R.A., and Cui, Y. (2008) High-performance lithium battery anodes using silicon nanowires. *Nat. Nanotechnol.*, **3** (1), 31–35.

29 Wu, H. and Cui, Y. (2012) Designing nanostructured Si anodes for high energy lithium ion batteries. *Nano Today*, **7** (5), 414–429.

30 Javed, M.S., Chen, J., Chen, L., Xi, Y., Zhang, C., Wan, B., and Hu, C. (2016) Flexible full-solid state supercapacitors based on zinc sulfide spheres growing on carbon textile with superior charge storage. *J. Mater. Chem. A*, **4** (2), 667–674.

31 Zhai, T., Wang, F., Yu, M., Xie, S., Liang, C., Li, C., Xiao, F., Tang, R., Wu, Q., Lu, X., and Tong, Y. (2013.) 3D MnO_2–graphene composites with large areal capacitance for high-performance asymmetric supercapacitors. *Nanoscale*, **5** (15), 6790.

32 Peng, L., Peng, X., Liu, B., Wu, C., Xie, Y., and Yu, G. (2013) Ultrathin two-dimensional MnO_2/graphene hybrid nanostructures for high-performance, flexible planar supercapacitors. *Nano Lett.*, **13** (5), 2151–2157.

33 Jha, N., Ramesh, P., Bekyarova, E., Itkis, M.E., and Haddon, R.C. (2012) High energy density supercapacitor based on a hybrid carbon nanotube-reduced graphite oxide architecture. *Adv. Energy Mater.*, **2** (4), 438–444.

34 Goubard-Bretesché, N., Crosnier, O., Favier, F., and Brousse, T. (2016) Improving the volumetric energy density of supercapacitors. *Electrochim. Acta*, **206**, 458–463.

35 Thackeray, M.M., Wolverton, C., and Isaacs, E.D. (2012) Electrical energy storage for transportation–approaching the limits of, and going beyond, lithium-ion batteries. *Energy Environ. Sci.*, **5** (7), 7854.

36 Yao, Y., McDowell, M.T., Ryu, I., Wu, H., Liu, N., Hu, L., Nix, W.D., and Cui, Y. (2011) Interconnected silicon hollow nanospheres for lithium-ion battery anodes with long cycle life. *Nano Lett.*, **11** (7), 2949–2954.

37 Luo, J.-Y. and Xia, Y.-Y. (2007) Aqueous lithium-ion battery $LiTi_2(PO_4)_3/LiMn_2O_4$ with high power and energy densities as well as superior cycling stability. *Adv. Funct. Mater.*, **17** (18), 3877–3884.

38 Liu, W., Chen, Z., Zhou, G., Sun, Y., Lee, H.R., Liu, C., Yao, H., Bao, Z., and Cui, Y. (2016) 3D porous sponge-inspired electrode for stretchable lithium-ion batteries. *Adv. Mater.*, **28** (18), 3578–3583.

39 Khan, S., Lorenzelli, L., and Dahiya, R.S. (2015) Technologies for printing sensors and electronics over large flexible substrates: a review. *IEEE Sens. J.*, **15** (6), 3164–3185.

40 Mahadeva, S.K., Walus, K., and Stoeber, B. (2015) Paper as a platform for sensing applications and other devices: a review. *ACS Appl. Mater. Interfaces*, **7** (16), 8345–8362.

41 Kim, S.-H., Choi, K.-H., Cho, S.-J., Choi, S., Park, S., and Lee, S.-Y. (2015) Printable solid-state lithium-ion batteries: a new route toward shape-conformable power sources with aesthetic versatility for flexible electronics. *Nano Lett.*, **15** (8), 5168–5177.

42 Gaikwad, A.M., Arias, A.C., and Steingart, D.A. (2015) Recent progress on printed flexible batteries: mechanical challenges, printing technologies, and future prospects. *Energy Technol.*, **3** (4), 305–328.

43 Merrett, G.V., Weddell, A.S., Lewis, A.P., Harris, N.R., Al-Hashimi, B.M., and White, N.M. (2008) An empirical energy model for supercapacitor powered wireless sensor nodes. Proceedings of the 17th International Conference on Computer Communications and Networks, pp. 1–6.

44 Pu, X., Li, L., Song, H., Du, C., Zhao, Z., Jiang, C., Cao, G., Hu, W., and Wang,

Z.L. (2015) A self-charging power unit by integration of a textile triboelectric nanogenerator and a flexible lithium-ion battery for wearable electronics. *Adv. Mater.*, **27** (15), 2472–2478.

45 Heremans, J.P., Dresselhaus, M.S., Bell, L.E., and Morelli, D.T. (2013) When thermoelectrics reached the nanoscale. *Nat. Nanotechnol.*, **8** (7), 471–473.

46 Orr, B., Akbarzadeh, A., Mochizuki, M., and Singh, R. (2015) A review of car waste heat recovery systems utilising thermoelectric generators and heat pipes. *Appl. Therm. Eng.*, **101**, 490–495.

47 Briscoe, J. and Dunn, S. (2015) Piezoelectric nanogenerators – a review of nanostructured piezoelectric energy harvesters. *Nano Energy*, **14**, 15–29.

48 Sodano, H.A., Park, G., and Inman, D.J. (2004) An investigation into the performance of macro-fiber composites for sensing and structural vibration applications. *Mech. Syst. Signal Process.*, **18** (3), 683–697.

49 Yu, R., Lin, Q., Leung, S.-F., and Fan, Z. (2012) Nanomaterials and nanostructures for efficient light absorption and photovoltaics. *Nano Energy*, **1** (1), 57–72.

50 Green, M.A., Emery, K., Hishikawa, Y., Warta, W., and Dunlop, E.D. (2015) Solar cell efficiency tables (Version 45). *Prog. Photovoltaics Res. Appl.*, **23** (1), 1–9.

51 Lu, Qianbo, Bai, Jian, Lian, Wenxiu, and Lou, Shuqi (2015) A novel scheme design of a high-g optical NEMS accelerometer based on a single chip grating with proper sensitivity and large bandwidth. 10th IEEE International Conference on Nano/Micro Engineered and Molecular Systems, pp. 248–253.

52 Hurst, A.M., Lee, S., Cha, W., and Hone, J. (2015) A graphene accelerometer. 28th IEEE International Conference on Micro Electro Mechanical Systems (MEMS), pp. 865–868.

53 Chandrakasan, A.P., Verma, N., and Daly, D.C. (2008) Ultralow-power electronics for biomedical applications. *Annu. Rev. Biomed. Eng.*, **10**, 247–274.

54 Clarke, S.F. and Foster, J.R. (2012) A history of blood glucose meters and their role in self-monitoring of diabetes mellitus. *Br. J. Biomed. Sci.*, **69** (2), 83–93.

55 Vaddiraju, S., Burgess, D.J., Tomazos, I., Jain, F.C., and Papadimitrakopoulos, F. (2010) Technologies for continuous glucose monitoring: current problems and future promises. *J. Diabetes Sci. Technol.*, **4** (6), 1540–1562.

56 Scognamiglio, V. (2013) Nanotechnology in glucose monitoring: advances and challenges in the last 10 years. *Biosens. Bioelectron.*, **47**, 12–25.

57 Choi, S. (2015) Powering point-of-care diagnostic devices. *Biotechnol. Adv.*, **34** (3), 321–330.

58 Peng, J., Wang, Y., Liu, L., Kuang, H., Li, A., and Xu, C. (2016) Multiplex lateral flow immunoassay for five antibiotics detection based on gold nanoparticle aggregations. *RSC Adv.*, **6**, 7798–7805.

59 Lee, S., Oncescu, V., Mancuso, M., Mehta, S., and Erickson, D. (2014) A smartphone platform for the quantification of vitamin D levels. *Lab Chip*, **14** (8), 1437–1442.

60 Petryayeva, E. and Algar, W.R. (2016) A job for quantum dots: use of a smartphone and 3D-printed accessory for all-in-one excitation and imaging of photoluminescence. *Anal. Bioanal. Chem.*, **408** (11), 2913–2925.

61 Petryayeva, E. and Algar, W.R. (2015) Toward point-of-care diagnostics with consumer electronic devices: the expanding role of nanoparticles. *RSC Adv.*, **5** (28), 22256–22282.

62 Hiramoto, M., Ishii, Y., and Monobe, Y. (2014) Light field image capture device and image sensor. US 20140078259 A1, March 20, 2014.

63 Casse, B.D.F., Lu, W.T., Huang, Y.J., Gultepe, E., Menon, L., and Sridhar, S. (2010) Super-resolution imaging using a three-dimensional metamaterials nanolens. *Appl. Phys. Lett.*, **96** (2), 023114.

64 Lopez-Sanchez, O., Lembke, D., Kayci, M., Radenovic, A., and Kis, A. (2013) Ultrasensitive photodetectors based on monolayer MoS_2. *Nat. Nanotechnol.*, **8** (7), 497–501.

65 Whitmore, A., Agarwal, A., and Da Xu, L. (2015) The Internet of Things–a survey of topics and trends. *Inf. Syst. Front.*, **17** (2), 261–274.

66 Tan, Z., Daamen, R., Humbert, A., Ponomarev, Y.V., Chae, Y., and Pertijs, M.A.P. (2013) A 1.2-V 8.3-nJ CMOS humidity sensor for RFID applications. *IEEE J. Solid-State Circuits*, **48** (10), 2469–2477.

67 Kim, J., Wang, Z., and Kim, W.S. (2014) Stretchable RFID for wireless strain sensing with silver nano ink. *IEEE Sens. J.*, **14** (12), 4395–4401.

68 Balasubramaniam, S. and Kangasharju, J. (2013) Realizing the Internet of nano things: challenges, solutions, and applications. *Computer*, **46** (2), 62–68.

69 Akyildiz, I.F., Brunetti, F., and Blázquez, C. (2008) Nanonetworks: a new communication paradigm. *Comput. Networks*, **52** (12), 2260–2279.

70 Crawford, M.H., Wierer, J.J., Fischer, A.J., Wang, G.T., Koleske, D.D., Subramania, G.S., Coltrin, M.E., Karlicek, R.F., and Tsao, J.Y. (2015) Solid-state lighting: toward smart and ultraefficient materials, devices, lamps, and systems, in *Photonics*, John Wiley & Sons, Inc., pp. 1–56.

71 Kim, K.-H. and Song, J.-K. (2009) Technical evolution of liquid crystal displays. *NPG Asia Mater*, **1**, 29–36.

72 Yang, D.-K. and Wu, S.-T. (2014) *Fundamentals of Liquid Crystal Devices*, 2nd edn, John Wiley & Sons, Inc.

73 Wu, S.-T., Ge, Z., Tsai, C.-C., Jiao, M., Gauza, S., and Li, Y. (2009) 45.4: Invited paper: enhancing the energy efficiency of TFT-LCDs. *SID Symp. Dig. Tech. Pap.*, **40**, 677–680.

74 Hoon, K.Sang., Joo-Do, P., and Ki-Dong, L. (2006) Fabrication of a nano-wire grid polarizer for brightness enhancement in liquid crystal display. *Nanotechnology*, **17**, 4436.

75 Kim, D. and Baek, K.H. (2013) Method for manufacturing nano wire grid polarizer. Google Patents.

76 Jung, Y.S., Lee, J.H., Lee, J.Y., and Ross, C.A. (2010) Fabrication of diverse metallic nanowire arrays based on block copolymer self-assembly. *Nano Lett.*, **10**, 3722–3726.

77 Mertens, R. (2015) *The OLED Handbook*, Lulu.com.

78 Kim, D.-H., Kim, J.Y., Kim, D.-Y., Han, J.H., and Choi, K.C. (2014) Solution-based nanostructure to reduce waveguide and surface plasmon losses in organic light-emitting diodes. *Org. Electron.*, **15**, 3183–3190.

79 Ju, J., Yamagata, Y., and Higuchi, T. (2009) Thin-film fabrication method for organic light-emitting diodes using electrospray deposition. *Adv. Mater.*, **21**, 4343–4347.

80 Li, N. (2014) Using graphene as transparent electrodes for OLED lighting. US Department of Energy Solid State Lighting R&D Workshop. Tampa, FL, USA.

81 Wu, J., Agrawal, M., Becerril, H.A., Bao, Z., Liu, Z., Chen, Y., and Peumans, P. (2010) Organic light-emitting diodes on solution-processed graphene transparent electrodes. *ACS Nano*, **4**, 43–48.

82 Lee, N.S., Chung, D.S., Han, I.T., Kang, J.H., Choi, Y.S., Kim, H.Y., Park, S.H., Jin, Y.W., Yi, W.K., Yun, M.J., Jung, J.E., Lee, C.J., You, J.H., Jo, S.H., Lee, C.G., and Kim, J.M. (2001.) Application of carbon nanotubes to field emission displays. *Diam. Relat. Mater.*, **10**, 265–270.

83 Comiskey, B., Albert, J.D., Yoshizawa, H., and Jacobson, J. (1998) An electrophoretic ink for all-printed reflective electronic displays. *Nature*, **394**, 253–255.

84 Oh, S.W., Kim, C.W., Cha, H.J., Pal, U., and Kang, Y.S. (2009) Encapsulated-dye all-organic charged colored ink nanoparticles for electrophoretic image display. *Adv. Mater.*, **21**, 4987–4991.

85 Yin, P., Wu, G., Qin, W., Chen, X., Wang, M., and Chen, H. (2013) CYM and RGB colored electronic inks based on silica-coated organic pigments for full-color electrophoretic displays. *J. Mater. Chem. C*, **1**, 843–849.

86 Li, D., Le, Y., Hou, X.-Y., Chen, J.-F., and Shen, Z.-G. (2011) Colored nanoparticles dispersions as electronic inks for electrophoretic display. *Synth. Met.*, **161**, 1270–1275.

87 Yin, P.-P., Wu, G., Dai, R.-Y., Qin, W.-L., Wang, M., and Chen, H.-Z. (2012) Fine encapsulation of dual-particle electronic ink by incorporating block copolymer for electrophoretic display application. *J. Colloid Interface Sci.*, **388**, 67–73.

88 Weng, W., Higuchi, T., Suzuki, M., Fukuoka, T., Shimomura, T., Ono, M., Radhakrishnan, L., Wang, H., Suzuki, N., Oveisi, H., and Yamauchi, Y. (2010) A high-speed passive-matrix electrochromic display using a mesoporous TiO_2 electrode with vertical porosity. *Angew. Chem., Int. Ed.*, **49**, 3956–3959.

89 Vlachopoulos, N., Nissfolk, J., Möller, M., Briançon, A., Corr, D., Grave, C., Leyland, N., Mesmer, R., Pichot, F., Ryan, M., Boschloo, G., and Hagfeldt, A. (2008) Electrochemical aspects of display technology based on nanostructured titanium dioxide with attached viologen chromophores. *Electrochim. Acta*, **53**, 4065–4071.

90 Yashiro, T., Okada, Y., Naijoh, Y., Hirano, S., Sagisaka, T., Gotoh, D., Inoue, M., Kim, S., Tsuji, K., Takahashi, H., and Fujimura, K. (2013) Flexible electrochromic display. International Display Workshops, pp. 1300–1303.

91 Mortimer, R.J. (2011) Electrochromic materials. *Annu. Rev. Mater. Res.*, **41**, 241–268.

92 Hayes, R.A. and Feenstra, B.J. (2003) Video-speed electronic paper based on electrowetting. *Nature*, **425**, 383–385.

93 Daub, C.D., Bratko, D., Leung, K., and Luzar, A. (2007) Electrowetting at the nanoscale. *J. Phys. Chem. C*, **111**, 505–509.

94 Feng, X.J. and Jiang, L. (2006) Design and creation of superwetting/antiwetting surfaces. *Adv. Mater.*, **18**, 3063–3078.

95 Krupenkin, T.N., Taylor, J.A., Schneider, T.M., and Yang, S. (2004) From rolling ball to complete wetting: the dynamic tuning of liquids on nanostructured surfaces. *Langmuir*, **20**, 3824–3827.

96 Frieder, M. and Jean-Christophe, B. (2005) Electrowetting: from basics to applications. *J. Phys. Condens. Matter*, **17**, R705.

97 Granqvist, C.G. and Hultåker, A. (2002) Transparent and conducting ITO films: new developments and applications. *Thin Solid Films*, **411**, 1–5.

98 Chopra, K.L., Major, S., and Pandya, D.K. (1983) Transparent conductors – a status review. *Thin Solid Films*, **102**, 1–46.

99 IDTechEx (2015) Transparent conductive films (TCF) 2015–2025: forecasts, markets, technologies.

100 Hecht, D.S., Hu, L., and Irvin, G. (2011) Emerging transparent electrodes based on thin films of carbon nanotubes, graphene, and metallic nanostructures. *Adv. Mater.*, **23**, 1482–1513.

101 Hu, L., Hecht, D.S., and Grüner, G. (2010) Carbon nanotube thin films: fabrication, properties, and applications. *Chem. Rev.*, **110**, 5790–5844.

102 Zhang, M., Fang, S., Zakhidov, A.A., Lee, S.B., Aliev, A.E., Williams, C.D., Atkinson, K.R., and Baughman, R.H. (2005) Strong, transparent, multifunctional, carbon nanotube sheets. *Science*, **309**, 1215–1219.

103 Koh, W.S., Gan, C.H., Phua, W.K., Akimov, Y.A., and Bai, P. (2014) The potential of graphene as an ITO replacement in organic solar cells: an optical perspective. *IEEE J. Sel. Top. Quantum Electron.*, **20**, 36–42.

104 Xu, Y. and Liu, J. (2016) Graphene as transparent electrodes: fabrication and new emerging applications. *Small*, **12**, 1400–1419.

105 Lee, J., Lee, P., Lee, H., Lee, D., Lee, S.S., and Ko, S.H. (2012) Very long Ag nanowire synthesis and its application in a highly transparent, conductive and flexible metal electrode touch panel. *Nanoscale*, **4**, 6408–6414.

106 Ryu, J., Kim, Y., Won, D., Kim, N., Park, J.S., Lee, E.-K., Cho, D., Cho, S.-P., Kim, S.J., Ryu, G.H., Shin, H.-A.-S., Lee, Z., Hong, B.H., and Cho, S. (2014.) Fast synthesis of high-performance graphene films by hydrogen-free rapid thermal chemical vapor deposition. *ACS Nano*, **8** (1), 950–956.

107 De, S., Higgins, T.M., Lyons, P.E., Doherty, E.M., Nirmalraj, P.N., Blau, W.J., Boland, J.J., and Coleman, J.N. (2009) Silver nanowire networks as flexible, transparent, conducting films: extremely high DC to optical conductivity ratios. *ACS Nano*, **3**, 1767–1774.

108 Tokuno, T., Nogi, M., Karakawa, M., Jiu, J., Nge, T.T., Aso, Y., and Suganuma, K. (2011) Fabrication of silver nanowire transparent electrodes at room temperature. *Nano Res.*, **4**, 1215–1222.

109 Ok, K.-H., Kim, J., Park, S.-R., Kim, Y., Lee, C.-J., Hong, S.-J., Kwak, M.-G., Kim,

N., Han, C.J., and Kim, J.-W. (2015.) Ultra-thin and smooth transparent electrode for flexible and leakage-free organic light-emitting diodes. *Sci. Rep.*, **5**, 9464.

110 Daniel, L., Gaël, G., Céline, M., Caroline, C., Daniel, B., and Jean-Pierre, S. (2013) Flexible transparent conductive materials based on silver nanowire networks: a review. *Nanotechnology*, **24**, 452001.

111 Stöcker, T., Köhler, A., and Moos, R. (2012) Why does the electrical conductivity in PEDOT:PSS decrease with PSS content? A study combining thermoelectric measurements with impedance spectroscopy. *J. Polym. Sci. Part B Polym. Phys.*, **50**, 976–983.

112 Yao, S. and Zhu, Y. (2015) Nanomaterial-enabled stretchable conductors: strategies, materials and devices. *Adv. Mater.*, **27**, 1480–1511.

Part Seven
From Device to Systems

23
Nanoelectronics for Smart Cities

Joachim Pelka

Fraunhofer Group for Microelectronics, Business Office, Anna-Louisa-Karsch-Str. 2, 10178 Berlin, Germany

23.1
Why "Smart Cities"?

By 2050, 70% of the world's population will live in cities. And, as cities grow and expand their services, management, operations, and governance across the city become increasingly complex. To manage this increasing complexity, it becomes necessary to make cities "smart."[1] Technological, organizational, and systemic innovation is required to deal with the most urgent questions. The goal is to overcome the discrepancies between individual and societal needs for mobility, consumption, and quality of life on one hand and resource conservation, climate protection, and sustainability on the other.[2]

Following an analysis of Frost & Sullivan,[3] *smart cities* are cities built on "smart" and "intelligent" solutions and technologies that will lead to the adoption of at least five of the eight following smart parameters – smart energy, smart building, smart mobility, smart health care, smart infrastructure, smart technology, smart governance and smart education, and smart citizen (Figure 23.1). Having this definition in mind, a *smart city* can be basically defined as a city *that uses information and communication technologies* to make its critical infrastructure and its components and utilities more interactive and efficient, and making citizens more aware of them.[4] This means, making cities "smart" requires sensors, signal processing, communication, and storage, which need not only software but also the nanoelectronics-based hardware.

---

1) http://www.ibm.com/smarterplanet/us/en/smarter_cities/overview/.
2) Tobias Hegmanns in Bosshard *et al.*, *Sehnsuchtsstädte*, transcript Verlag, Bielefeld, 2013.
3) *Strategic Opportunity Analysis of the Global Smart City Market*, Frost & Sullivan, August 2013.
4) *Smart Cities Study: International Study on the Situation of ICT, Innovation and Knowledge in Cities*, The Committee of Digital and Knowledge-Based Cities of UCLG.

Nanoelectronics: Materials, Devices, Applications, First Edition. Edited by Robert Puers, Livio Baldi, Marcel Van de Voorde, and Sebastiaan E. van Nooten.
© 2017 Wiley-VCH Verlag GmbH & Co. KGaA. Published 2017 by Wiley-VCH Verlag GmbH & Co. KGaA.

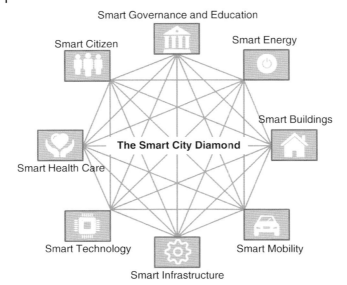

Figure 23.1 Key areas to define smart cities. (Adapted from Frost & Sullivan,[3] icons by Freepik, www.flaticon.com.)

Within a *smart city*, urban innovation will therefore be based on an additional nanoelectronics-based infrastructure – the information network, which will become as essential as the power grid or the water supply. To be precise, for realization of such a new infrastructure,[5] nanoelectronics and smart systems are needed for secure generation, transmission, processing, and storage of information as well as for handling data within applications. In addition, energy is needed to operate everything. Thus, energy efficiency and power semiconductors will also play an important role.

This short introduction already sketches the main challenges to be addressed in future for *smart cities* by the semiconductor business:

- Information
  - Generation of information
  - Communication and information processing
  - Storage of information
  - Use of information
- Energy
  - Efficiency, reduction of losses
  - Power supply
  - Smart grid
- Cross-cutting aspects
  - Safety/security
  - Applications

5) www.smartsantander.eu/.

23.2
Infrastructure: All You Need Is Information

The evolution of *smart cities* is still at its very beginning. Even the attribute "smart" is quite often not clearly defined and each *smart city* or *smart region* today has its own specific requirements: What is regarded as "innovative and smart" in one city or region may already be state of the art in another one.

Making a city "smart" always starts with digitization of what is called *urban processes* (Figure 23.2). In a "smart" city, *urban processes* become part of the most advanced version of an enhanced city planning by adding a feedback loop for information. Whereas in former times all planning work was based on manually captured data, modern information and communication technologies (ICT) allow for a continuous, complete, and effective feedback loop. With close interaction made possible by this feedback loop, a *smart city* will become a huge "system of systems," as described in Mühlhäuser and Encarnação.[6] Addressing

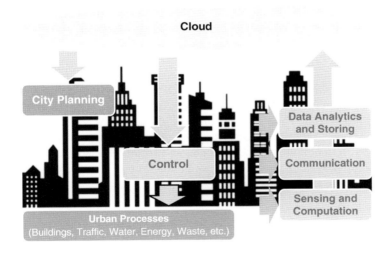

Figure 23.2 What makes a city smart? The flow of information.

6) E. Mühlhäuser and J. Encarnação, Integrierende IKT für die Stadt der Zukunft, ACATECH, 2014.

"Integrative ICT for the City of the Future" from a software viewpoint, a *smart city* is regarded as a huge "cyber physical system," which is a system comprising *information generation* (sensing), *information processing*, and *communication*.

Sensors are used to monitor traffic, pollution, flow of goods, energy consumption, and many other parameters. Sensors are either spread over the whole city area such as "Smartdust"[7] or are part of personal, mobile devices such as cellphones or cars (crowdsourcing).[8] All these sensors are linked and communicate with each other. Captured information is transmitted into the *Cloud* by different types of networks. Wireless communication is used via short distances like in (local) sensor networks, whereas wired and optical communication networks are the backbones of medium and wide area communication. Information is processed and stored in the *Cloud*, and is used for modeling of all processes in a city. The results of the modeling processes are used twofold: for control (e.g., control of traffic flow based on pollution data) and for future city planning.

The three basic elements *sensing*, *processing*, and *communication* are based on micro/nanoelectronics and microsystems and cannot be realized without high-performance hardware components. RFID tags, wired and wireless sensor nodes, local servers, and further automatic identification and data capture (AIDC) devices constitute basic monitoring and communication capabilities in *smart cities*. They will turn devices and systems into smart objects,[9] which are equipped with perception capabilities, embedded intelligence, and high level of autonomy. These smart objects are able to communicate and are interconnected. This means that at the end of the day, applications of *smart cities* will be driven by the *Internet of Things* (IoT). A *smart city* will be a main use case of the IoT.

As a consequence, this will lead to a huge number of connected sensors and devices and a tremendous amount of data to be collected, processed, transmitted, and stored. Referring, for example, to the Bosch Sensor Swarm approach,[10] experts expect 7 billion people to be served by the Internet by 2017 with 1000 linked sensors per person. These sum up to 7 trillion sensors (Figure 23.3), a great part of which will serve as infrastructure in *smart cities*, monitoring everything: air quality, traffic, flow of goods, people, and so on. Hewlett Packard[11] expects a resulting growth of Internet data communication by a factor of 1000 by 2018, just to handle the sensor data traffic.

7) https://en.wikipedia.org/wiki/Smartdust.

8) https://de.wikipedia.org/wiki/Crowdsourcing.

9) N.L. Fantana, *et al.* (2013) IoT applications: value creation for industry, in *Internet of Things: Converging Technologies for Smart Environments and Integrated Ecosystems* (eds. O. Vermesan and P. Friess), River Publishers.

10) http://www-bsac.eecs.berkeley.edu/scripts/show_pdf_publication.php?pdfID=1365520205.

11) T. Perković *et al.* (2009) Multichannel protocols for user-friendly and scalable initialization of sensor networks. *Security and Privacy in Communication Networks*. Springer, Berlin, pp. 228–247.

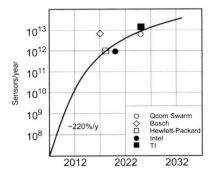

Figure 23.3 Forecast of the increase in the number of sensors produced each year on a global level until 2037. (*Source:* Fairchild Semiconductors.[11])

Within a *smart city*, urban innovation will therefore require an additional (nanoelectronics based) infrastructure – the information network – that will become as essential for an urban environment as the power grid or the water supply of today. Such a high-performance information backbone has to be able to handle this enormous data traffic and has to be as reliable as today's conventional infrastructures. For realization of such a new infrastructure,[12] nanoelectronics and smart systems are needed for generation, transmission, processing, and storage of information as well as for all applications using these data.

The functionality of this new information infrastructure will rely on some fundamental semiconductor technologies (Figure 23.4):

- *Cyber-physical systems (CPSs) – smart sensors with data preprocessing and communication means based on zero-energy sensor nodes/systems:* For a long time, the simple sensors of the past have become smart by combining different types of sensors, by enhancing them with data processing capabilities, and by integrating energy harvesting and smart power supply. Today, a smart sensor is turned into a cyber-physical system by adding communication means. A formerly simple sensor is becoming a complete, miniaturized system in a single package.
- *Increased miniaturization/packaging/lifetime of electronic components, modules, and systems:* Using sensors everywhere is not only a challenging task for miniaturization but also for packaging and reliability. Further miniaturization requires heterogeneous integration techniques as well as using the third dimension (chip stacking). Packages must shield highly miniaturized, complete systems from harsh environments and guarantee reliable functionality over many years.
- *Sensor and communication networks with a focus on high data rates @ low power and security issues (reliability, resilience, manipulation):* Adding communication means requires the use of not only mixed-signal technologies for integration of RF modules but also additional technologies for integrating

12) www.smartsantander.eu/.

Cloud

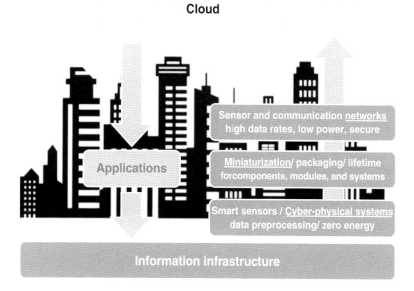

Figure 23.4 The information infrastructure needed to make a city smart.

power supplies: energy harvesters, converters, and batteries. On the load side, ultralow- or even zero-power technologies and efficient power management (sub)systems are becoming increasingly important to ease the need for sufficient power. However, one of the most pressing challenges will be security. Besides the software-related aspects of authorization, authentication, data integrity, and so on, hardware integrity also has to be ensured. Otherwise, software-based security measures might be cheated by manipulated hardware. PUF technologies (PUF: physically unclonable functions) will help to guarantee hardware integrity of the *smart cities'* ICT backbone.

The International Technology Roadmap for Semiconductors (ITRS) has already included this shift in paradigm in its latest big update, which goes far beyond the old "scaling"-based Moore's law. The recently set up ITRS 2.0, now called IRDS (International Roadmap for Devices and Systems), introduces a new system-oriented approach and is being extended to cover topics like heterointegration, heterocomponents, system integration, and outside system connectivity, as well as application-specific aspects.[14] IRDS is now addressing the whole technology basis for cyber-physical systems and the Internet of Things. Additionally, *smart city* applications are leading to another extension: Energy supply is becoming a more and more important issue (Figure 23.5).

Figure 23.5 Internet of Things – a driving force for smart cities and semiconductors.[13],[14]

23.3
Nothing Will Work Without Energy

   This roughly sketched information infrastructure for *smart cities* will consume a significant amount of energy when generating, processing, transmitting, and storing information (Figure 23.6). Therefore, in addition to the above-mentioned requirements in terms of high data rates and secure communication, *power supply* will become one of the biggest issues – Solutions such as energy harvesting and self-powered sensors must exist before a trillion-sensor universe can become reality.

   On the other hand, energy supply in a *smart city* will not only be challenged by a sensor swarm but also has to solve energy supply for buildings, traffic, production, and so on. One of the main energy consumers will be the ICT infrastructure. In 2006, information technology in the United States consumed as much as 1.5% of US electricity.[15] It is predicted that by 2020, 14% of the electrical energy will be consumed by the ICT sector (corresponding to 8% of primary energy).[16] For the operation of Germany's ICT infrastructure, about 40 TWh of

13) M. Graef and P. Gargini (2014) ITRS 2.0, SEMICON Europa 2014, TechArena.
14) M. Badaroglu, *System Integration*, SEMICON Europa 2014, TechArena.
15) D. Miller (2014) *Low energy optoelectronics for interconnects*, INC10, Washington, DC, May.
16) M. Pickavet, W. Vereecken, S. Demeyer, P. Audenaert, B. Vermeulen, C. Develder, D. Colle, B. Dhoedt, and P. Demeester (2008) Worldwide energy needs for ICT: the rise of power-aware networking. *2nd International Symposium on Advanced Networks and Telecommunication Systems (ANTS'08)*, December.

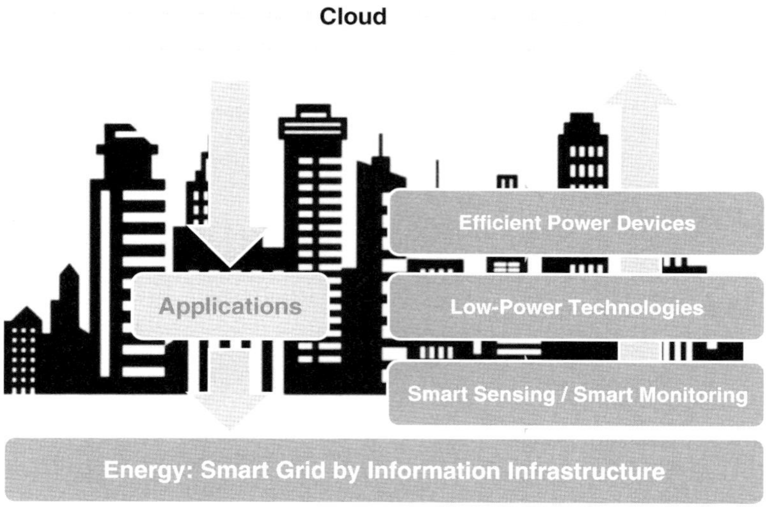

Cloud

Figure 23.6 Energy makes everything happen.

electric energy will be necessary in 2015, of which 90% will be evoked by wireless communications.[17]

In total, the sophisticated energy supply of cities has to face a variety of challenges, such as the steady increase of renewable energies and their intermittent infeed to the grid, the distributed generation and the transmission of energy over long distances to urban areas, and finally the efficient distribution to the customers. In order to guarantee an efficient handling of energy, nanoelectronics and power electronics are needed to build the next-generation smart energy grid:

- *Ultralow power electronics*
  The issue of energy consumption, especially of the ICT infrastructure, has to be tackled by using ultralow power microelectronics and optics for sensing, processing, and communication. It is also necessary to have a closer look at wireless communication technologies, where a big potential for improvement can be seen in terms of energy-efficient algorithms, telecommunication systems, circuits, and devices.
- *Power devices*
  The second big challenge is power distribution, which needs highly efficient switches and converters besides the necessary ICT for monitoring and control. Most promising developments for future energy supplies result from the application of wide-bandgap materials: silicon carbide (SiC) and gallium nitride (GaN), but also from the improvement of existing silicon (Si) technology and in advanced circuit design.

17) VDE Kongress Smart Cities, 2014.

- *System aspects*
From a system point of view, focus will lie on the reduction of on-state losses and on the increase of efficiency of semiconductor components. Besides highly efficient power devices, low-power integrated circuits and microelectronics for monitor and control purposes are required; for example, specialized ICs and systems-on-chip (SoCs) for battery management systems and energy harvesting are necessary to handle the functionary requirements. The voltage levels of the electronic circuits, such as system-on-a-chip (SoC), integrated circuits (IC), or regulators have to be scaled down and power consumption must be decreased to ensure performance with small amounts of energy. Another more general field of potential improvement is the miniaturization and the increase in power density, as well as an efficient and sustainable electronic control of the generation, transmission, distribution, and consumption of energy in future cities. Miniaturization and integration, which play an essential role in nanoelectronic components for energy and battery management for local and distributed systems, go hand in hand with the requirement to increase the power density. Overall, the efficiency of the power and micro/nanoelectronics is a major challenge in energy supply.
- *Energy harvesting*
Energy harvesting transducers such as thermo-, piezo-, or electrodynamic generators provide a high dynamic range of output voltages and currents. Thus, power management electronics such as voltage converters and charge circuits for batteries must cope with this dynamic range. Integrated power management is indispensable for converting the fluctuant and intermittent environmental and renewable energy (e.g., vibration, sunlight) into a stable power source.[18]

23.4
Application: What Can Be Done with Information

As already described, the digitization of the urban processes will result in an information infrastructure[19] that will be the basis for a broad variety of "smart" applications within a *smart city*. Besides monitoring, controlling, and planning of urban processes, the most important applications will be as follows:

- Smart buildings
- Smart mobility and transport
- Smart production and logistics.

18) D.S. Ha (2011) Small scale energy harvesting: principles, practices and future trends. *14th International Symposium on Design and Diagnostics of Electronic Circuits & Systems (DDECS)*, p. 9.
19) I. Schieferdecker and W. Mattauch (2013) ICT for smart cities: innovative solutions in the public space, in *Computation for Humanity: Information Technology to Advance Society* (eds J. Zander and P.J. Mosterman), CRC Press.

Once a *smart cities* information infrastructure is installed, the realization of a lot of other visionary ideas will become possible.

23.4.1
Smart Buildings

As one of the major challenges, a *smart city* of the future will be built on inter-acting buildings and infrastructure that are designed for autarkic energy supply by, for example, solar energy and a complete use of building envelopes for energy generation (Figure 23.7). Disposal of excess heat is achieved by high energy efficiency in combination with wind inflow, smart use of daylight for illu-mination, and smart waste water systems with heat recovery. Heat islands within buildings as well as within metropolitan areas no longer exist. Optimized aerody-namics is used for ventilation. The building envelope is used not only for energy harvesting but also as a sink for emissions, which are collected and disposed of by natural processes (e.g., moss). Buildings not only consume energy but are also an important resource for urban energy generation and for energy storage.

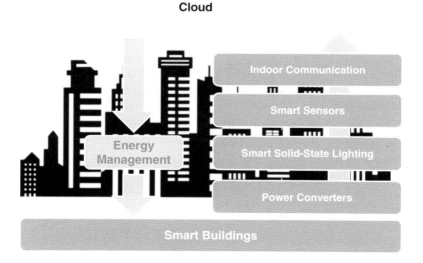

Figure 23.7 Buildings will become smart especially under the energy saving aspect.

Buildings are becoming part of the smart grids for electrical and thermal energies.

Sustainability roadmap concepts for cities are necessary, and role models for sustainable buildings are important for raising awareness and garnering wide acceptance for innovative solutions from all city stakeholders. Smart *solid-state lighting* (SSL) will be a key technology for reducing energy losses from lighting by increasing the efficacy of the light sources themselves and by making the lighting "smart," which means light will be generated only where it is needed and in the quantities that are needed. Nanoelectronics will also be essential for all monitoring and control systems in order to ensure energy efficiency. Sensors and sensor nodes are key to feeding the "smart building cockpit" with the necessary data; local servers will deliver computing power and data storage and will take care of controlling all heating and cooling devices, ventilation, lighting, and communication in a smart building in accordance with the guidelines obtained from the urban processes (Figure 23.8).

With respect to smart buildings, there are the following challenges where the future development of micro/nanoelectronics and smart systems will be responsible for achieving high energy efficiency:

- *Solid-State Lighting*
  For solid-state lighting, four major issues can be identified. The first is today's nonavailability of SSL devices with (very) high luminous output, as needed for outdoor lighting. The second is the necessary increase of efficiency for low-cost power converters, a major prerequisite for competitiveness. As a third challenge, the potential of control of color temperature should be addressed

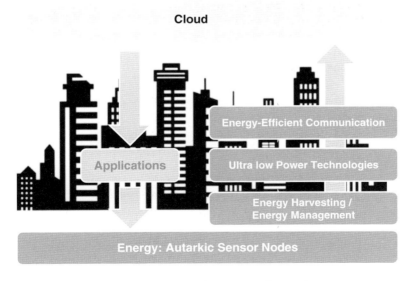

Figure 23.8 Besides the "smart grid," autarkic sensor nodes are also necessary.

for indoor lighting, which requires the integration of control functions into current lighting infrastructure (retrofit) or introduction of new lighting systems. Finally, the manufacturing process must be made more flexible to meet the demands of a market of individually shaped luminaires without replaceable light sources.

- *Sensors for Smart Home and Building Energy Management*
  For smart home, ambient assisted living (AAL) and energy management applications sensors are needed that, for example, allow determining the presence and activity of people inside the building. Other sensors are needed to measure air quality, temperature, humidity, and other measurands of the technical building equipment. These sensors need to be low cost and low power and have to be intelligent, that is, they can be tailored to privacy requirements in hardware and to transmit only information really needed.

- *Power Converters*
  Increased deployment of renewable energy harvesting equipment in smart buildings demands highly efficient power converters and storage closely related to the *smart grid*. More detailed information on this challenge can be found in Chapter 20.

- *Indoor Communication*
  High data rate applications will call for further improvement of standards such as Ethernet and 802.11. For low data rate applications such as energy management, extremely low power, highly secure, robust, and easy-to-deploy wireless communication systems have to be developed. For all wireless communication systems, the problem of coexistence of many networks should be addressed by developing tools and procedures to monitor communication, manage resources (i.e., bandwidth), and resolve problems in high traffic situations.

23.4.2
Mobility and Transport

In a *smart city* of the future, nobody can imagine the solid lines of cars in a traffic jam of today, which produce a lot of harmful exhaust. Traffic is completely aligned with the needs of the population. It is highly efficient in terms of sustainability and quality of life. Traffic management means no longer modifying the cities according to the traffic density but controlling traffic according to the real needs of city and people, which is quite a challenge when it comes to sensing, collecting, and processing all the necessary data (Figure 23.9).

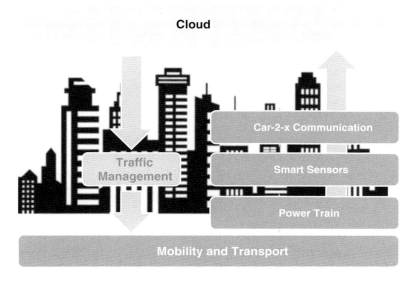

Figure 23.9 Mobility and transport in a smart city are based on the same infrastructure elements as a smart city itself: sensors, communication, and energy. Crowdsourcing will help manage traffic in a highly efficient way.

In the future, urban mobility will be based on shared vehicle concepts. Smart reservation and booking and a dynamic distribution of cars have reduced the required parking space by a factor of 10. This is supported by innovative parking lot concepts including high rack-type car parks and parking space management systems that guide vehicles to the next free parking space on the shortest available track. The total number of cars has been reduced by cooperative use. One shared car is now replacing 20 individual vehicles.

Thanks to reliable communication concepts including car-2-car and car-2-environment communication, accidents have almost gone "extinct." Autonomously driving cars are the basis for successful personal rapid transport (PRT). Downtown areas are open only to battery- or fuel cell-powered electric vehicles; combustion engines can be found in vintage cars only. Traffic management controls traffic flow by means of dynamic toll systems and achieves a smooth and undisturbed traffic stream. A lot of alternative concepts have replaced big parts of today's individual traffic. Noise has been significantly reduced by electric drives; personal traffic and transport of goods have been decoupled.

The need for sustainable mobility concepts becomes obvious when looking at global carbon dioxide emissions caused by transport. Today, successful measures are mostly connected to large infrastructure projects, which are shaped by decades of development, policies, and local framework conditions. Tomorrow, nano-electronics-based solutions will take care of controlling the traffic flow according to traffic density, weather conditions, and air pollution. Electric cars, which have become more or less "chipsets on four wheels," are fully packed with sensors that capture all necessary data for autonomous driving from their environment and that feed information back to the urban processes infrastructure (*crowdsourcing*). Doing so requires sufficient computing power on board of the vehicles as well as secure broadband wireless communication resources for data exchange with other vehicles, the environment, and the infrastructure (Figure 23.9).

In order to accelerate the transition to "smart mobility," several challenges to the development of nanoelectronics have to be addressed:

- *High-voltage power semiconductors:* The power train of a fully electrical car has to rely on highly efficient power semiconductors. Most importantly, semiconductor-based power devices allowing for high switching frequencies in drive inverters and power converters are essential for the achievement of small, lightweight, and reliable systems. Silicon carbide and gallium nitride are promising technologies and power devices such as these are already entering the market. However, further work is required toward both higher reliability for automotive applications and the increase of breakdown voltage. The true potential of these wide bandgap semiconductors – operation beyond 6.5 kV – is not yet commercially exploited. However, this corresponds to voltage classes encountered in electric trains, where grid voltages may be as high as 25 kV.
- *High-frequency power semiconductors:* There is also demand for RF power transistors with cutoff frequencies beyond 10 GHz. These devices could be facilitated to manufacture RF amplifiers capable of transmitting data at higher data rates and with more parallel participants. This requirement arises from the larger number of wireless sensors, traffic data, and wireless applications associated with autonomous driving and demand-driven mobility. Similarly, car radar applications require RF devices operating at 77 GHz and beyond.
- *Smart sensors and sensor systems:* Highly integrated sensor systems with dedicated functionality have to be developed. Autonomous driving will be based mainly on sensors within the vehicle. Traffic monitoring and control will rely on information collected from both the vehicles and the sensors mounted on streets, traffic signs, buildings, and so on.
- *Car-2-x Communication:* Car-2-car and car-2-environment communication are prerequisites for autonomous driving and enable new applications such as cooperative exchange of information for the avoidance of accidents. All such applications depend on reliable and robust radio communication between vehicles and between vehicles and environment or infrastructure, respectively. Because the reliability of the wireless connection directly depends on the type and quality of the radio communication, better understanding and realistic simulation of the dynamics of the wireless communication open the way to adaptive design and accurate but low-cost evaluation of future communication systems. Car-2-x communication becomes an add-on to conventional sensor systems such as RADAR and video sensors and requires reliable hardware and secure communication.

23.4.3
Production and Logistics

In a *smart city* of the future, innovative production and logistics schemes would be the basis for a smooth flow in transportation and handling of goods, commerce, and service and guarantee the provision to people of all products of vital importance (Figure 23.10). However, a reorientation of design and operation of urban manufacturing locations and logistics networks is necessitated by future economic, ecological, and societal boundary conditions and is closely related to the challenges of *mobility and transport* and to *urban processes*.

Manufacturing locations in a *smart city* are designed to be "urban compatible," which means designed for minimal environmental stress and pollution. For example, urban manufacturing locations are usually within walking distance from the living areas – living and working can be arranged easily. Thus, manufacturers and the city are operating in a symbiotic manner. Waste heat, excess energy, and recycled raw materials will be exchanged between urban supply

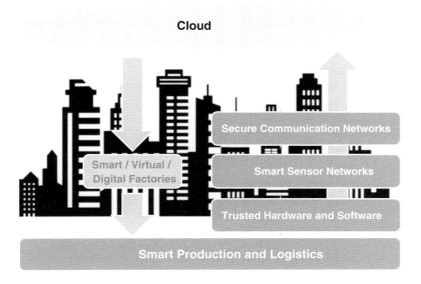

Figure 23.10 In connection with smart mobility management, the infrastructure of smart cities will allow bringing production back into the cities. Smart, digital factories will increase quality of life significantly by reducing commuting.

systems, disposers, and production plants. Due to the close cooperation of manufacturers and urban providers, materials flows do not suffer any longer from the decentralization. The efficiency of infrastructure is further increased by collaborative use of urban resources.

Flows and processes for the transportation of goods have changed. The core issue of the innovative logistics strategy is distribution of goods, intermodal traffic systems, and efficient use of infrastructures by flexible and collaborative pooling of commodity flows. Innovative solutions for (automated) delivery, distribution, express services, and other "last mile" services combine a variety of individualized services of commerce and production.

The *smart city* is much more interleaved with production and logistics than today. It provides, plans, and monitors urban infrastructure and even parts of logistics and production themselves. Basic system conditions are changing due to the growing population and density of metropolitan areas, new production and consumption patterns, and demographic change. Customized products and e-commerce lead to more individual manufacturing and smaller quantities of goods resulting in higher delivery frequencies and atomization of shipments. It is crucial to not just optimize individual technologies or product components but also analyze value chains in a systematic and holistic way. Starting from product design, the complete value chain of production starts with raw materials and ends with the delivered product. Since value is and should be added in this chain, the initial goal of improving production is based on the ideal of maximizing the added value and minimizing the waste.

Manufacturers must transform their operations by automating, executing, and managing the performance of uniform global business processes across their value chain. This means end-to-end integration of processes spanning supply chain, production, maintenance, distribution, quality, and labor operations – regardless of where these facilities and operations are physically located.[20] Naturally, wireless sensor and information networks provide a means to fulfill this goal by connecting the physically distributed system components into a cooperating system.

Reliable and high-quality data are the key to successful performance. It is almost impossible to imagine well-functioning processes without proper data. It has been claimed that most of the disasters in modern industry are in some way caused by improper or missing data. Data quality has to be improved at sensor level by developing self-learning, context-aware, and intelligent sensor networks that evaluate the measured data with other data sources. Problems related to missing data will be solved by developing *ad hoc* and disposable sensors that can be used to gather critical information. Through information management plans, critical data and information from the production viewpoint have to be defined. Quality of information fed into systems by humans will be improved by user-interface technologies. Using

---

20) *Guide to Resource Efficiency in Manufacturing.* Available at http://blogs.ec.europa.eu/orep/guide-to-resource-efficient-manufacturing/.

(sensor) data provided by others makes one vulnerable to errors in these data when generated or transported, especially when the data are used, transformed, and upgraded by several parties in the chain. Design for auditability provides the means to get insight into the data/information chain, to reason with uncertainties of measurements.

Advances in the state of the art are carried out by overcoming the different technological challenges in data quality weakness. Intelligent decision-making systems in industrial applications have a real challenge not only in terms of their basic functional performance but also to prove themselves in the absence of real implementations that validate the reliability and usefulness of these systems.

All this is based on ICT-related technologies, on software, as well on hardware components. The relevant technological requirements and, therefore, the challenges on nanoelectronics and smart systems can be summarized in two categories:

- *Secure data generation, transmission, processing, and storage:* In order to benefit from big data, distributed machines, manufacturing processes, and workers should be connected seamlessly to facilitate real-time information exchange, prompt responsiveness, better decision-making, and finally optimization of a highly distributed complex system. Information and communication technologies play a crucial role here. Both transmission and storage of massive sensing data pose new challenges to nanoelectronics. Information should be conveyed promptly and reliably, and should also be stored safely and be retrieved quickly.
- *New models and methods for decision making and prediction:* Different from conventional industrial systems, the more complex and cooperative systems should be described by new models to have information correlation, dynamic description, and behavior prediction. Data mining of massive data sets across greatly varied platforms and formats involves methods at the intersection of artificial intelligence, machine learning, statistics, predictive analytics, and database systems.

Thus, ICT-related technologies are the key enabler for improving a production system at three levels[21]:

- *Smart factories:* Agile manufacturing incorporating novel sensors and robotics with automatic control, planning, and automation.
- *Virtual factories:* Connected to network, the production procedure and management are linked to and based on physically distributed assets.
- *Digital factories:* Improved management and decision systems are built upon advanced simulation, modeling, and prediction methods.

To successfully integrate this into "smart productions and logistics" of a *smart city*, it is not sufficient to follow the general trend toward industry automation only.

---

21) *Factories of the future PPP: strategic multi-annual roadmap*, prepared by the *Ad Hoc* Industrial Advisory Group.

"Smart production and logistics" has to be more than that and nanoelectronics drives all these applications. Therefore, one has to face the following challenges:

- *Secure sensor and communication networks for controlling of smart, virtual, and digital factories:* Design, maintenance, and managing tasks are "virtualizing" more and more with the support of ICT technologies, and finally employees will work in virtual factories. Remote data monitoring and actuator controlling will be key to that, but data reliability and availability must be imperative. This requires adequate sensor and communication networks.
- *ICT support for logistics:* Industrial production and retail sales are two key industrial market drivers. This market indicates new demands for new logistics facilities, including huge e-fulfillment centers at the heart of national and regional distribution networks, local delivery centers serving individual cities, and returns processing centers. Sensors for monitoring the flow of goods as well as, for example, the environmental conditions in the case of sensitive products such as uninterrupted cold chain, tracking of goods and vehicles, and all the necessary data communication, data processing, and storage behind will be part of the information infrastructure of *smart cities*.

Key requirements for smart production and logistics are a dense sensor network, sufficient bandwidth, and secure transmission and processing of data.

23.5
Trusted Hardware: Not Only for Data Security

In a *smart city*, safety and security has become self-evident in daily life. This results in absolute freedom. Children can play everywhere, airports and stations have become multifunctional buildings of public life, and parks and footpaths are crowded even in darkness. But public safety and security is not guaranteed by security guards alone; every citizen has become part of a comprehensive resilience concept (sustainable security), supported by electronic security systems based on sensors and nanoelectronics, using the available *smart city's* ICT infrastructure.

However, when using ICT for security concepts, when dealing with urban processes, when controlling traffic flow, when collecting and processing thousands of megabytes of sensitive data, data security and data integrity play an essential role. Therefore, all means of data capturing, processing, and transmission must be trustworthy, reliable, and secure. Data must be protected at any time against manipulation, stealing, or misuse. In addition, privacy of information must be ensured.

Therefore, a *smart city* environment has to guarantee that any bit of information is generated by a trusted source, has not been manipulated during transmission and processing, and has been securely stored.

In order to ensure this, the necessary authentication and identification processes can only partially be based on software solutions. Secure and trusted hardware is necessary to guarantee data integrity. This leads to the three major challenges that need support from nanoelectronics hardware:

- *Authentication:* Authentication is the process of determining whether someone or something is, in fact, who or what it is declared to be. In an IoT environment like a *smart city*, device authentication is basically something different from the user-based authentication (for instance, user name and password). Authorized device notion of unique device and copy protection to ensure uniqueness using mechanism as physical unclonable function (PUF) could be an alternative for devices such as sensors and actuators. In other use cases, the usage of gateways with a secure element and the introduction of a trusted third party (TSM) should permit the deployment of strong authentication models, for instance, smart grid and smart governance. Authentication used to link a device to an individual (for liability and service charge) needs to define rules for this type of link update, transfer, and revocation. So authentication mechanism has also to support this kind of features, such as in smart health and smart mobility.
- *Access control:* The ICT infrastructure of *smart city* requires an access control mechanism to access the resources as gateways or devices in order to prevent unauthorized use of these resources. As a consequence, it has to support various authorization models according to the heterogeneity and diversity of the devices or gateways. There is also a need to handle all security operation by itself without human control using new techniques as self-learning to reach a self-management behavior.
- *Privacy:* A significant part of data captured in a *smart city* has to be regarded as personal data. Thus, there is a clear need to support anonymity and restrictive handling. Crypto technologies are used to enable protected data storage processing and sharing. Technologies such as homomorphic encryption are interesting in this perspective. Preservation of geolocation or citizen behavior through anonymization techniques is a key feature. Privacy by design has to be used in order to address identification, authentication, and anonymization. The secure generation of information and the transmission of information in sensor networks require that the data collected by the sensors be processed accurately, and that the data from the sensors to the monitoring, analysis, or control units (which can be stationed at a larger distance) be transmitted securely. The types of sensors in urban centers will be used, for instance, to monitor air quality, traffic density, or gatherings of people in crowded places.

The sensors must be equipped with a (receiver) transmitter unit, which has to meet the following conditions:
- Sufficiently high data rates
- Secure transmission encryption of the measurement data
- Low energy consumption per transmitted bit
- High reliability, lifetime, and robustness
- Low acquisition and operation costs

23.6
Closing Remarks

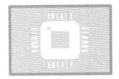

Nanoelectronics will soon be faced with the huge challenges of *smart cities*. This will be mainly caused by the upcoming Internet of Things with its enormous complexity of the necessary control and sensor systems. *Smart cities* will become a prominent application of the IoT. However, the high number of possible scenarios related to the future combination of communication/sensor networks and the networks of mobile systems make it difficult to give precise estimations of what is really needed by a *smart city*, but what will make a city smart will always be based on information, communication, and energy. Besides the already existing infrastructure networks such as energy and water, which have "only" to be made smart in the future, a completely new communication infrastructure has to be set up. This information backbone has to guarantee secure and reliable handling of huge amounts of data necessary to monitor and control an urban region. This new information infrastructure has to form a bridge from the edge (sensor nodes) to the cloud (servers) by smart selection and preprocessing of data in order to reduce the amount of data to a reasonable amount and it has to enable the privacy of citizens. Moreover, such a sensitive infrastructure has to guarantee the data security and integrity, which has to be partially based on unclonable hardware. If these requirements can be fulfilled in a sufficient manner, the application field mentioned here will be the beginning of an exciting future of nanoelectronics.

Acknowledgement

This chapter is based on a study of the CATRENE (Cluster for Application and Technology Research in Europe on NanoElectronics) working group on "Semiconductors for Smart Cities," which was performed in 2014 and published by the CATRENE organization in early 2015 (CATRENE Study on Semiconductor Technologies for Smart Cities, 2015. Available at http://www.catrene.org/web/downloads/communication_catrene/Smart_Cities_Final_Report_2015-01-13.pdf.). The author would like to thank the CATRENE organization and especially its Scientific Committee for guidance and the chapter editors of the working group for their support and their agreement to use parts of the study for this chapter:

Silke Cuno (Fraunhofer FOKUS, Berlin, Germany)	Urban Processes
Guido Dolmans (imec-nl, Eindhoven, The Netherlands)	Production and Logistics
Tobias Erlbacher (Fraunhofer IISB, Erlangen, Germany)	Mobility
Joachim Pelka (Fraunhofer Mikroelektronik, Berlin, Germany)	Editor and overall coordination
Florian Pebay-Peyroula (CEA-Leti, Grenoble, France)	Safe and secure cities
Moritz Loske (Fraunhofer IIS, Erlangen, Germany)	Energy
Ina Schieferdecker (Fraunhofer FOKUS, Berlin, Germany)	ICT
Andreas Wilde (Fraunhofer IIS-EAS, Dresden, Germany)	Buildings

Part Eight
Industrialization: Economics/Markets – Business
Values – European Visions – Technology Renewal
and Extended Functionality

24
Europe Positioning in Nanoelectronics

Andreas Wild

Former executive director of ECSEL Joint Undertaking, Avenue de la Toison d'Or 56-60, 1060 Brussels, Belgium

24.1
What is the "European" Industry

"European" is a description that can be understood geographically, as a location anywhere on the continent; or with a narrower meaning, as a short form for countries belonging to the European Union (EU). In any case, the geographic location of a company is not particularly important for the company itself. In fact, in nanoelectronics, like in many other industries, companies must act globally, positioning their structural elements at different locations to optimize their business conditions; for example, a controlling share of the company is owned by investors in country A; the company is incorporated in country B; the headquarters are established in country C; and the company has operational units or affiliates established as legal entities located in country X. The location is, on the contrary, significant when public institution are defining and implementing policies, since they can exercise their authority only in a well-defined administrative area. They must formulate eligibility criteria for incentive programmes explicitly requiring that the candidates must be located in the area in which that particular institution is competent.

A company can be considered as being "European" even if not all countries in which it is active are located in Europe; the determination depends upon the context and the objectives pursued. For example, if the objective of the policy is job creation, then the location of the operational units with high employment numbers matters most; alternatively, the headquarter location may be the element to consider if the objective is to ensure controlled access to a given technology.

In this overview, the different significations of the expression "European" would not be always explicitly stated, trusting that the meaning will be more or less obvious from the context.

Nanoelectronics: Materials, Devices, Applications, First Edition. Edited by Robert Puers, Livio Baldi, Marcel Van de Voorde, and Sebastiaan E. van Nooten.

24.2
European Strategic Initiatives

24.2.1
The European Commission

The Lisbon European Council from March 23/24, 2000 defined a new strategic goal for the EU: "to become the most competitive and dynamic knowledge-based economy in the world capable of sustainable economic growth with more and better jobs and greater social cohesion."[1] Given the limited financial means at the EU level, the contribution of the European institutions consists in seeding initiatives to enticing strong participation of the member states and – even more important – of the private sector. Leverage is key to success.

To structure the activities in line with the Lisbon objective, the European Commission invited leading representatives from all industries to elaborate a strategic research agenda in their field and constitute European Technology Platforms (ETPs). The most convincing ETPs with clear strategies have been institutionalized in the form of Joint Technology Initiatives implemented by autonomous legal entities called Joint Undertakings.

The European Commission established in 2010–2011 and 2013–2015 two successive High Level Groups on Key Enabling Technologies (KET) that recommended to shift the point of gravity of the public funding toward higher Technology Readiness Levels (TRLs).[2] The European Commission called for the nanoelectronics, recognized as a KET, to start growing again in Europe toward a share comparable with the EU share in the world GDP, fulfilling the vision expressed metaphorically as an "Airbus of Chips."[3] To progress toward an executable plan, the Commission established an Electronics Leaders Group that proposed in 2014 to take a wider view to the value chain, include electronic system houses servicing end users, and generate growth by balancing technology push and market pull, leveraging emerging applications to expand regional demand and drive technology progress.[4] The group also recommended establishing the ECSEL JU as an essential element in implementing the European strategy in nanoelectronics.

24.2.2
ECSEL Joint Undertaking

Positioned as a collaboration between the EU and the member states, the ECSEL (Electronic Components and System for European Leadership) Joint

---

1) http://www.europarl.europa.eu/summits/lis1_en.htm.
2) http://ec.europa.eu/growth/industry/key-enabling-technologies/european-strategy/high-level-group/index_en.htm.
3) http://europa.eu/rapid/press-release_SPEECH-12-382_en.htm.
4) https://ec.europa.eu/digital-agenda/en/electronics-strategy-europe.

Undertaking (JU) is an autonomous organization continuing the JUs ARTEMIS, and ENIAC that functioned in the period 2008–2014.

Among the instruments available to the EU, the JU seems best positioned to contribute to the implementation of the Lisbon agenda in nanoelectronics because it brings together all relevant players in a three-way partnership encompassing the private sector, the member states, and the EU. This results in one of the highest leverage factors among all European programmes. At the same time, it is probably the most difficult instrument to run, because of the inherent complexity of the setup.

The success depends critically upon aligning strategies and priorities, such that each participant – public or private – can obtain a positive return justifying its engagement. This is not trivial. For example, the European principle of equal treatment irrespective of the country of origin collides with the large variability in the national incentive systems that can use exclusively grants, or tax incentives, or revolving loans, or equity investments, and so on, or any combination thereof. It is impossible to find a unique European rule that, in combination with the national systems, would be satisfactory to every participant. To make it function, the common objective must take precedence over the procedures: the players must reach a high level of mutual trust, allow some flexibility in the rules under their control, and accept that a positive return can be beneficial to them even if it is not identical country by country and company by company.

Establishing trust requires time; after almost 3 years since inception, once it achieved a solid basis of trust, ENIAC JU could engage in implementing KET recommendations and enticed the industry to engage in 2012 and 2013 about 2 billion euro in "pilot lines" projects moving technology closer to manufacturing.[5] Although largely insufficient to recover by itself alone the investment gap accumulated in Europe over several years, this action demonstrates the feasibility of impactful policies and is coherent with the positive evolution showed lately by the European companies. Now, ECSEL is making progress along the same lines, being chartered to support capital-intensive actions such as pilot lines or large scale demonstrators at higher TRLs up to level 8.

24.2.3
Combining Instruments

Further support mechanisms are explored at the European level. The framework programme 2014–2020 of the EU encourages synergistic funding of actions from different instruments. For example, innovative projects following objectives that synergistically address both excellence in technology and sustainable regional development can be funded at the same time from the European mechanisms supporting innovation and regional cohesion, respectively. The new ambition visible in Europe both in the private and in the public sector at the national/regional level will surely take advantage of the new opportunities, both in the

5) http://eniac.eu/web/downloads/presentations/2014-04-09__innovpriorities_apcm.pdf, slide 15.

emerging markets and in the public support, and will put forward specific proposals qualifying as "important projects of common European interest," elevating to the next level the multiple sourcing model pioneered by the JU model in order to reach the scale observed in the other regions of the world.

24.3
Policy Implementation Instruments

24.3.1
In The World

In most regions of the world, public authorities developed policies to support nanoelectronics, considered as a critical industry from both economic and strategic viewpoints; for example, following are just a few public announcements:

- The 2015 budget proposed to the US Congress in support of the National Nanotechnology Initiative (not exactly identical with, but including aspects pertinent to nanoelectronics) requests $ 1.5 billion to be awarded through 11 state departments and agencies, with an increased emphasis on technology transfer from research to commercially relevant activities.[6]
- Also in the United States of America, the Advanced Manufacturing Partnership Steering Committee outlined in 2012 recommendations for spurring investment and positioning the United States of America for long-term leadership in advanced manufacturing. This resulted in establishing the National Network for Manufacturing Innovation (NNMI)[7] chartered to create, showcase, and deploy new capabilities and new processes; four out of seven NNMI Institutes that were operational in 2014 address topics relevant for nanoelectronics: "DMDII" on digital manufacturing and design, similar to some extent with the Industry 4.0 initiative; "Power America" on wide bandgap semiconductors; "AIM Photonics" on integrated photonics; and "Flexible Hybrid Electronics."
- India made public announcements and promised tax and customs incentives as well as subsidies up to 20–25% for 10 years on capital expenditure to attract national and international investments, as part of the "Make in India" presidential initiative.[8]
- China represents about 45% of the worldwide demand for chips, but relies on imports to 90%; a task force set by the Government including ministries and private sector leaders has been charged with setting an aggressive growth strategy, and issued in 2014 the *Guideline of the National IC Industry Development*

6) THE NATIONAL NANOTECHNOLOGY INITIATIVE, Supplement to the President's 2016 Budget, https://www.whitehouse.gov/sites/default/files/microsites/ostp/nni_fy16_budget_supplement.pdf.
7) http://manufacturing.gov/nnmi.html.
8) http://www.makeinindia.com/sector/electronic-systems.

Promotion. It foresees public support up to 1 trillion renminbi ($ 170 billion) over the next 5–10 years through a National Industry Investment Fund and provincial-level entities with the objective to sustain till 2020 a 20% compounded annual growth rate for the domestic production.[9]

24.3.2
In Europe

The loss of market share experienced by the companies incorporated in Europe in the last 10 years cannot be explained by the European overall economic situation. The member states of the EU have over 500 million inhabitants and a combined gross domestic product second to none. The EU would be the largest economy of the world if it could be compared with the national states. This is not the case, in principal because the powers and competencies of the European institutions are rather limited: they can address only the topics and only to the extent conceded by the members states.

Europe has elaborate systems to support research and innovation in general that are also applicable to nanoelectronics; the following enumeration includes some actions considered relevant in this context, it is not an exhaustive list.

National Programmes
The majority of the financial means are engaged by the member states. Some of them elaborated national strategies and defined priorities including nanoelectronics, and implement them actively through national programmes, albeit under the scrutiny of the European Commission that assures compliance with the international rules protecting fair competition. Various countries use different incentives, like advantageous loans; equity participation by public entities; financial reassurance schemes for venture capital to stimulate investments and spark growth, and so on.

Intergovernmental Programmes
The EUREKA programme is intended to raise the productivity and competitiveness of European businesses through technology. It assessed project proposals and labels them as conformal with the competition. Since 1985, EUREKA has generated over € 6 billion investments in more than 5700 projects. Besides projects labellized at the central level, the programme also operates clusters on specific topics, one of them being the "Pan-European partnership in micro- and nanoelectronic technologies and applications" (PENTA). It shall start its activities in 2016 and will run for 5 years, continuing the previous programmes in the field that run under the acronyms JESSI (1989–1996), MEDEA and MEDEA+ (1997–2007), and CATRENE (2008–2015).[10]

---

9) http://www.mckinsey.com/insights/high_tech_telecoms_internet/
   semiconductors_in_china_brave_new_world_or_same_old_story.
10) http://www.eurekanetwork.org/sites/default/files/cluster_report_2012.pdf.

Joint Undertaking

ECSEL was granted € 1.2 billion from the EU budget plus commensurate contributions from the member states in order to leverage research and innovation projects with a total volume around € 5 billion. The projects will be engaged in 2014–2020.

European Union Programmes

The EU defines framework programmes over a period of 7 years. The current programme addressing research and innovation is "Horizon 2020," budgeted with € 70 billion over the period 2014–2020.[11] Except for the funding dedicated to specific instruments like the Joint Undertakings Horizon 2020 consists in calls for proposals launched by the European Commission that result in selecting for funding research and/or innovation projects addressing one of three pillars: excellence in science; leadership in enabling and industrial technologies; and societal challenges. Any topic included in the biannual work plans will have an assigned budget; some of them will address topics relevant for nanoelectronics.

The European Structural and Investment Funds (ESIF) are intended to help local areas grow; they shall support investment in innovation, businesses, skills and employment, and create jobs. Among them, the European Regional Development Funds (ERDF) are particularly suitable for exploring synergies, for example, with Horizon 2020, including both the actions of the European Commission and those of the JUs.

At the European level there are also efforts to offer better access to financing. The European Commission initiated a European Fund for Strategic Investments (EFSI) with the goal to mobilize extra private finance in specific sectors and areas; it is estimated to reach a multiplier effect of 1 : 15 of at least € 315 billion in 2015–2017 in real investment in the economy. In addition, the European Investment Bank manages a number of instruments covering various types of loans, guarantees, microfinancing and, in collaboration with the European Commission, venture capital investments through the European Investment Fund.

The EU budget represented in 2014 no more than 2.2% of the combined administrative budget available to the Governments of its member states (Figure 24.1); it consists almost exclusively of budgetary transfers from the member states.[12] It is therefore obvious that the European initiatives can only make a difference if they are embraced by the member states and the private sector.

24.4
Europe's Market Position

However good the strategy, you should occasionally look at the results.[13]

---

11) http://ec.europa.eu/research/participants/portal/desktop/en/home.html.
12) http://appsso.eurostat.ec.europa.eu/nui/submtViewTableAction.do.
13) Winston S. Churchill (Author), Richard M. Langworth (Editor), "Churchill In His Own Words," EBURY PRESS, ISBN 978-0-09-19333-6-4.

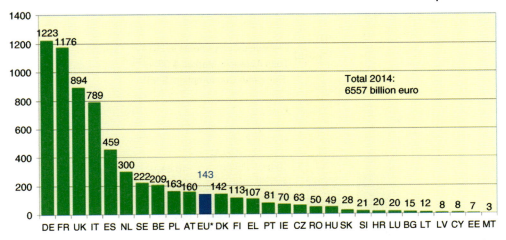

Figure 24.1 The budget of the European Union as compared with the budget of the Member States in 2014.

24.4.1
European Market Share: Consumption

Figure 24.2 shows the world consumption of electronic components per market segments in the main regions of the world. The largest amount is consumed for communications, followed by computing applications; combined, these two segments represent about two-thirds of the total consumption of the world. The remaining one-third is split among consumer, automotive, and a segment described as "industrial-medical-other," plus the Government applications that complete the statistics with just a fraction of one percent (Government applications have a much lower weight in components than in the OEM market).

In terms of the regional consumption,[14] Asia-Pacific has a dominant position, growing toward 60% of the world market. This is not necessarily surprising considering on one side the demography and on the other side the general economic progress in the region.

In 2014, components in value of about $ 38 billion have been consumed in Europe. This represents slightly more than 11% of the worldwide component production.

The European consumption by market segments presents a different pattern than in the rest of the world. The largest segment is automotive, in which the European consumption also ranks as #1 in the world, followed closely by Asia-Pacific. Computing follows closely as second largest segment, the third one being industrial-medical-others, in which the European consumption ranks as second largest in the world behind Asia-Pacific. Europe is a distant follower in the

14) The volume of deliveries "shipped to" or "billed to" an address located in that region.

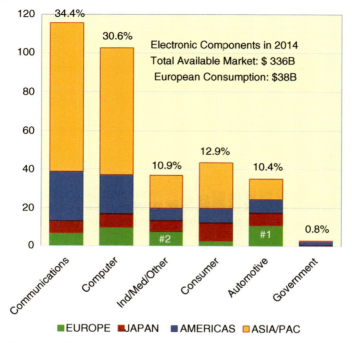

Figure 24.2 The component consumption by market segment in the main regions. Europe ranks as #1 in automotive and #2 in industrial–medical–other.

communication and consumer segments, the picture being completed by the rather insignificant consumption in Government applications.

If all regions would be equally active in electronics, then the consumption would probably follow more or less the demographics. Only about 6% of the world population lives in the EU, so that the European consumption of electronic components is over-proportional with respect to its demographics. Europe has still a robust demand, leads innovation in several areas, and many suppliers consider it a market worth servicing.

24.4.2
European Market Share: Supply

The combined sales of the suppliers incorporated in Europe amount to about two-thirds of the European consumption. From this perspective, Europe is a net importer of electronic components.

Although more than half of the world production of electronic components is consumed in the Asia-Pacific region, the supply did not followed suite. In fact, in the last 25 years, sustained investments and strong innovation both in technology and in business models enabled the companies incorporated in the United

States of America to consolidate their leading position and increase their market share from 38% in 1990 to 54% in 2014.

Over a similar period, the three European leaders[15] (alphabetically, INFINEON, NXP, and STMicroelectronics) combined had a negative evolution, after maintaining or slightly increasing their market share in the first 15 years, they experienced in the last decade considerable market share losses, from 10.4% in 2004 to 5.4% in 2014. This was also the trend of the whole European component industry that only adds about 2% to the share of the leaders. On one side, companies headquartered in Europe have been affected by adverse evolutions of their lead customers in the region, for example, European OEMs missing the smartphone opportunity. On the other side, Europe invested less than the other regions: its weight in the worldwide expenditures in R&D diminished in this period from about 13 to about 7%, while the weight of the capital investments diminished abruptly from more than 10% in 2000, remaining around 3%, flat over the last 5 years.

In spite of the overall unfavorable evolution in the last decade, Europe conserved the following areas of considerable strength:

- In "More Moore," Europe brings decisive contributions in equipment, dominating lithography, and in materials; in addition, it generates important results in fully depleted silicon on insulator (FDSOI) manufacturing process.
- In "More than Moore," Europe is second only to Japan; both the IDMs and the components divisions of vertically integrated OEMs (such as Bosch, Philips, or OSRAM) developed strong capabilities in power devices, MEMS and LEDs, ranking favorably among the sales leaders in O-S-D: 3 European companies are in top 10 and 5 in top 15.
- Europe has a vibrant academic research and hosts Research Technology Organizations (RTOs) with excellent capabilities and global reputation, generating a stream of ideas that fuel progress in all domains, including future innovations going "Beyond CMOS."

Finally, the European leaders seem to have stabilized their businesses and consolidated their positions. It can be forecasted that the 2015 numbers will show a reversal of the downward trend in market share, as illustrated in Figure 24.3. Two European leading companies participated in last years' strong merger and acquisition activities and grew inorganically: Infineon acquired International Rectifier, and NXP merged with Freescale Semiconductor in a $ 40 billion deal. Although European fabless champions have been acquired by US companies, the last merger and acquisition transactions should result in a net increase of the European market share.

In the near future, European suppliers have significant growth opportunities as well as challenges to overcome.

---

15) Here, the "European leaders" are the companies incorporated in Europe ranked among the top 25 sales leaders in the world.

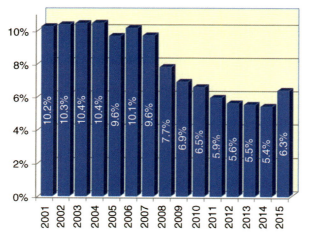

Figure 24.3 Market share of the leading companies incorporate in Europe.

The leading European IDMs are mainly concentrating on the two market segments in which the regional consumption is ranking highest, specifically on automotive and industrial–medical–other. This clear strategic focus had beneficial results, but resulted in servicing only about half of the European demand and one-fifth of the worldwide market. If the European companies maintain this strategic focus, their global growth will depend upon the materialization of the futuristic scenarios forecasting that these two segments should overgrow the market. An alternative strategy – yet less than trivial – could consist in diversifying the product and technology portfolio to address the other half of the regional demand as well as four-fifths of the world market currently not serviced by European suppliers. At this point in time, it looks like the dominant tendency among the European incumbents is to continue on the same path; and no small- and medium-sized companies in sight seem to be growing and capturing share in other market segments.

The European players are already well positioned to compete in emerging markets like Industry 4.0 and IoT building upon their strength in specific components and their relevant system knowhow. European companies have valuable assets in low power computing; sensors and actuators; RF and mixed-signal; and last but not the least, in embedded software and embedded systems, in which the European OEMs and their regional suppliers are credited with a share exceeding 30%. In order to get a fair participation in the value created by the system solutions delivered to the emerging markets, the Europeans must consider investing to fill gaps in their products and technologies portfolio. Furthermore, they must complement their in-house capabilities by forging successful alliances with contract manufacturers and/or IDMs from other regions.

Since the economics of the industry are basically unchanged, it can be safely anticipated that the fragmentation of the value chain will continue, so that the share of the foundries and the fabless companies will increase at the expense of the IDMs. At this point in time, the United States of America is the region with the strongest position in the emerging value chain, the second largest foundry is incorporated in the United States of America, and US fabless companies have a share of almost two-thirds of the market. The Asia-Pacific region ranks second, hosting the largest foundry capacity and fast growing fabless players. European is rather weak and getting even weaker in this area; pure-play foundries incorporated in Europe are small in size and lack 300 mm capacity, while leading fabless companies have been lately acquired by US-based owners: Qualcomm took over CSR, and Intel took over Lantiq, further reducing the European share in the fabless market to 1–2%. The medium- to long-term effect of the European weakness of the foundry+fabless value chain is surely the biggest structural challenge faced by the European policy makers.

24.4.3
European Manufacturing Capacities

Europe has being a reasonably strong contender in technologies up to 200 mm. Europe even assumed technological leadership, greatly contributing to the industrial introduction of 300 mm wafers: the first demonstration of memory devices on 300 mm wafers has been achieved in 2000 by a collaborative project jointly coordinated by Siemens and Motorola in Dresden, Germany. But right after this success, and in the wake of the severe recession of that year, European companies changed their strategy. They exited the memory market, concentrated on diversifying their existing 200 mm infrastructure on specialty technologies, and started using extensively contract manufacturing outside Europe. Following are the European capabilities in 300 mm:

- The "Crolles 2 Alliance," a collaborative project in 2002–2007 involved STMicroelectronics, Philips, and Motorola, established in France a 300 mm facility called sometimes a "mini-fab," contributed to advancing the state of the art, produced prototypes with excellent turn-over time, and also performed manufacturing in relatively small volume. At the end of the Crolles 2 Alliance, STMicroelectronics assumed full ownership.
- In the last years, Infineon developed leadership in power transistor technologies on 300 mm and invested in manufacturing in Villach, Austria, and Dresden, Germany.

Figure 24.4 summarizes the regional capacity as per headquarter location of the fab owner; altogether, the presence of the European companies in 300 mm manufacturing remains minor, below 3% of the world capacity. The figure also shows the capacity build-ups announced by the end of 2014. There were no announcements from Japanese companies, but Japan already had a sizable 300 mm capacity. No announcements have been made by companies headquartered in Europe.

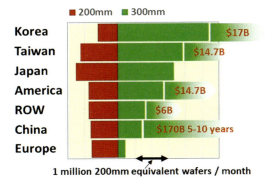

Figure 24.4 Installed capacities as per owner's headquarter location and investment plans announced by the end of 2014.

The picture would not be complete without mentioning companies incorporated in the United States of America that operate considerable 300 mm manufacturing facilities located in Europe:

- Intel, the world's leading IDM, invested $5 billion in the last 4–5 years upgrading its 300 mm facility in Leixlip, Ireland extending its capabilities to the 14 nm node.
- Global Foundries, the second largest foundry in the world, engaged in industrializing the Fully Depleted Silicon on Insulator (FDSOI) technology in Dresden, Germany.

Altogether, the fabs located in Europe, irrespective of owner, represent at this point in time around 10% of the total worldwide capacity, albeit with a strong 200 mm proportion. This percentage is almost in balance with the component consumption in Europe.

24.5
European Perspectives

Clearly, improving the competitive position and the sustainability of nanoelectronics in Europe cannot succeed unless the industry chooses to compete. Public support can incite investments and can accompany efforts, but public interventions cannot replace the private sector engagement.

Europe faces the following few quite obvious choices:

- It can decide to continue doing business as usual, accept its disengagement from the race to miniaturization, build upon its strength in diversified ICs and in O-S-D, and compete on this market that is expected to approach 20% of the electronic component TAM in the following 5 years.
- Europe could also decide that diversification is beneficial, but insufficient, and develop the ambition to re-enter the miniaturization race; in such a case, it

must establish supporting policies to stimulate private investments by players headquartered in Europe or elsewhere, reconnect its very advanced equipment industry with the regional chip manufacturing, and keep in the region critical know-how to secure its controlled access to leading edge technology generations.

- In addition, Europe can decide to further strengthen the links from components to applications, building a coherent ecosystem in which the players at different levels in the value chain will cross-fertilize, duplicating and further developing the global successes of the past like the GSM telephony, to quote just one example to address new opportunities, such as internet technology shortening the time gaps and crossing boundaries among disciplines to penetrate all industries (Industry 4.0!) and interconnect all objects (IoT!).

If the European private sector and policy makers, aware of the particular importance of nanoelectronics, demonstrate the will to engage strongly in the global competition, Europe will be in the position to harness its considerable intellectual and economic assets, build upon its strengths, fill the gaps, and assume a leading role in building the future.

25

Thirty Years of Cooperative Research and Innovation in Europe: The Case for Micro- and Nanoelectronics and Smart Systems Integration

Dirk Beernaert[1] and Eric Fribourg-Blanc[2]

[1]Nano-Tec, Ecosystems Technology and Design for Nanoelectronics, Avenue de Beaulieu 31, 1160 Brussels, Belgium
[2]European Commission, Communications Networks, Content & Technology, A3 – Competitive Electronic Industry, Avenue de Beaulieu, 33, B-1049 Brussels, Belgium

25.1
Introduction

Europe has stimulated the research and development in micro and nanoelectronics and micro-nanotechnologies over the past three decades in the firm belief that these technologies form the basis of new generations of products and manufacturing techniques and will offer large market opportunities. Europe has recognized the strategic nature of these nanoelectronics and micro-nanotechnologies as essential to fulfilling its social and economic targets. Not only is the ECS industry representing an important growing economic sector offering jobs and growth in important European regions; even more, innovation in electronic systems is enabling increased efficiency, new services, and new businesses across all sectors. These tiny components are bringing smartness in all products, enabling intelligence, processing, communication, and networking capabilities in all products, systems, and processes, influencing all parts of society.

Originally, component innovation was mainly driven by computers and communication devices, driving further their miniaturization, cost, and performance. Today, these smart nanoelectronics components and micro/nanosystems are increasingly more embedded in many various systems surrounding us in our daily life, driving an increase in functionality, performance, integration, and complexity. Their applications allow less energy consumption in products and buildings, more efficient and more sustainable generation of energy, enhanced health care and new health applications – for example, predictive medicine, environmental monitoring using wireless sensor technologies, more energy and carbon-friendly production and safer and sustainable transport, shifting into the direction of electric autonomous driving vehicles, better food control to allow ambient-assisted living for elderly, better security, and many more. Nanoelectronic

Nanoelectronics: Materials, Devices, Applications, First Edition. Edited by Robert Puers, Livio Baldi, Marcel Van de Voorde, and Sebastiaan E. van Nooten.
© 2017 Wiley-VCH Verlag GmbH & Co. KGaA. Published 2017 by Wiley-VCH Verlag GmbH & Co. KGaA.

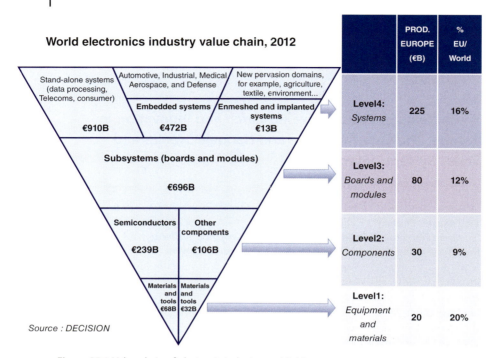

Figure 25.1 Value chain of electronic industry worldwide.

components are at the bottom of this innovation chain enabling access to new digital services, opening new ways for education, personal entertainment, more efficient administration, and increased efficiency and sustainability of the production of goods and services. Increasingly, these smart components are invisibly surrounding us offering the foundations to build new digital products and services poised to disrupt the economy, our society, the way we live, work, and even exist.

European capability to innovate in the markets where it is traditionally strong such as automotive or to build new markets such as in the Internet of Things heavily depends on access to the latest advancements in nanoelectronics and micro-nanosystems. Figure 25.1 shows that a €340 billion components drives a €1400 billion electronic systems market, which in turn elevates €6400 billion services.

European research and development in the field of micro-nanoelectronics and micro-nanotechnology evolved from what was basically a pure electronics component technology, research-driven, miniaturization program (More Moore) into a very multidisciplinary field driven by the integration and optimization of different functionalities, involving a large set of very different materials and requiring a combination of technology and system competencies (More than Moore). Moreover, driven by different application requirements and the need for more solution-oriented approaches, nanoelectronics-enabled processing and computing technologies are directly integrated with sensing and actuating. This requires a mixture of different competencies and technologies, for example, optical, mechanical, bio, and microfluidics to provide the required integrated functionalities.

This increase in complexity, the evolving nature, and increasing importance of integration and application can also be seen in the different Framework Programs for European Research, as it will be demonstrated below.

25.1.1
The European R&D Program in the European R&D Landscape

Unlike some other regions in the world, Europe is not one integrated country, one entity. The European Union with 508 million citizens and a GDP of €13.1 trillion consist of 28 independent Member States. Research and Innovation policy is largely their individual responsibility. Only a few Member States[1] have a program dedicated to research on ICT, nanoelectronics, or micro-nano-technology. Often the research activities are concentrated in specific regions,[2] innovation ecosystems building upon the specific expertise, excellence, or industry needs in that region.

These national and regional initiatives are complemented by two main R&D programs that are based on transnational collaboration: the European Framework for Research and Innovation driven by the European Commission and the cooperation between different MS in the context of the Eureka Framework. Both programs are managing a work program covering different topics and are yearly launching competitive calls. The research and innovation funded in the different European Frameworks of Research represents around 3–4% of the total public resources spending on research in Europe. However, EU funded research in nanoelectronics is around 15% of the total funded research activities in Europe in this field.

Within the European Frameworks for Research and Innovation, projects are submitted toward a call and are selected based on excellence, impact, and on their transnationality (European added value), for example, research projects have to be carried out by consortia, which include participants from three different European Member States; fellowships require mobility over national borders. The Framework Program mainly targets the research challenges that can only be addressed in a cooperative manner at European level. Projects are funded by the R&D budget allocated to the European Union and decided and agreed upon by the European parliament. The funding mainly covers grants to the research actors in order to co-finance research, technological development, and pilot and demonstration projects.

---

1) Major nationally funded research activities in nanoelectronics are going on in Germany, France, the United Kingdom, the Netherlands, Belgium, Italy, Austria, Ireland, and Finland.
2) The Regions of Grenoble, Dresden, Eindhoven–Leuven–Aachen, Cambridge, and others are well known for their competencies and activities in the field of nanoelectronics and micro-nanotechnologies.

The Eureka projects are initiated by industry, labeled by a board and submitted for funding to the individual member states participating in the Eureka cluster. The funding is in charge of the national administration (public authority) of the individual member state of each partner of a project and depends upon their national rules. Funding can cover subsidies for grants or loans.

Both programs have as main objective to stimulate multidisciplinary cooperation and to transfer competence from the research into industrial application for increased industrial competitiveness. The EU framework was traditionally more dedicated to long-term and precompetitive research and Eureka is addressing closer to the market projects; however, this distinction is blurring with an increased industrial and innovation focus on the recent European Frameworks. The current EU framework, Horizon 2020, has as extra objectives to support a set of EU policy initiatives in the areas of energy, mobility, environment, security, and health and contributes to realizing a single European Research Area.

The European Frameworks, due to the competitive nature of their calls and the focus on excellence and cooperation, have attracted many of the best in class European researchers and institutes in their particular field and despite being only around 3% of all research represent a disproportionate impact on Europe's economy,

25.2
Nanoelectronics and Micro-Nanotechnology in the European Research Programs

In the 30 years of European Framework Programs (FP), there was always a dedicated micro and nanoelectronics research and development initiative as part of an overall ICT program. However nanoelectronics research, development, and innovation activities could also be found in other parts of the Commission's Research Framework Programme, such as the NMP (nanotechnology, materials, and nanomanufacturing program), the FET (future and emerging technologies), the IDEAS (physical principles), and the ERC (European Research Council dealing with basic research), the application programs, and CAPACITY (nanoelectronics research infrastructure) programs.

The results of the advanced research at nanoelectronics technology, component, and systems developed in the FPs are further taken up by the more application-oriented programs, for example, the Public–Private Partnership on the electric car or, as of today, in the Internet of Things, or to address major Societal Challenges and/or in other initiatives oriented closer to the market such as the consecutive nanoelectronics research clusters in EUREKA (JESSI, MEDEA, EURIMUS, CATRENE, PENTA) or in the Joint Technology Initiatives. The nanoelectronics and electronics systems Joint Technology Initiatives (JTI) ENIAC and ARTEMIS, combined today in the JTI ECSEL, are developing these advanced research results further to meet industry requirements, including variability, robustness, reliability, and manufacturability (bringing research to higher

TRL: technology readiness levels), or to put them in advanced applications directly going to the market.

Finally, the equipment, production, or design technologies are directly taken up by the European nanoelectronics industrial and regional ecosystems to manufacture and design new products.

Future progress in nanoelectronics will rely increasingly on exploiting specific phenomena at the nano, quantum, and molecular levels. This makes direct cooperation between academics and industry and of nanoelectronics with the nanotechnology program in the Framework essential.

25.3
A Bit of History Seen from an ICT: Nanoelectronics Integrated Hardware Perspective

In order to strengthen the scientific and technological base of European industry, and to enhance its international competitiveness, Europe has started its Framework Programmes for Research and Technological development in 1984. The first three Programmes cover 3 years each, the next three 4 years, and from the 7th Framework, they run for 7 years. From €1 billion/year at the start, the latest Framework, H2020, now allocates more than €10 billion/year for research and innovation or more than €75 billion for the whole of H2020 (Figure 25.2). This represents 8% of the total budget of the Commission and the consecutive increases underline the growing importance the European Commission put on research and innovation. Also, in other policy parts of the Commission, in particular in supporting smart and sustainable growth in the regions and in stimulating SME businesses, increasing emphasis and budget is allocated for research and innovation.

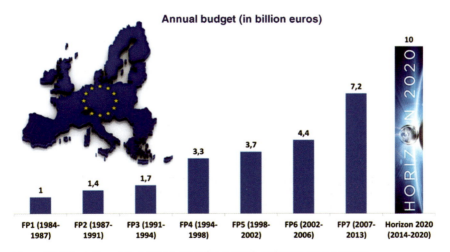

Figure 25.2 Evolution of European budget for research and innovation activities.

25.4
ESPRIT I, II, III, and IV

From 1984 to 1998, research on microelectronics was a part of ESPRIT (European Strategic Programme for Research in Information Technology).

ESPRIT had the following main objectives: (i) promoting EU academic and industrial cooperation in IT, (ii) providing the EU IT industry with the basic technologies necessary to compete in the 1990s, and (iii) contributing to internationally accepted standards.

During *ESPRIT I*, (1984–1987), the microelectronic research was technology driven and classified according to the functionality pursued: high density, high speed, multifunction ICs, and peripheral technologies. A substantial part of the work in high density was targeting application-specific integrated circuits (ASICs) to reach optimized system solutions and to protect proprietary architectures. High-speed semiconductors was targeting telecommunications, mobile radio, and computers. The multifunction ICs included nonvolatile memory cells, analog–digital ICs, optoelectronics, and smart power. Peripherals included high-density magnetic mass storage, magneto-optic drives, and flat panel displays. All these areas involved silicon as well as compound technologies, computer-aided design (CAD), manufacturing equipment, and materials. During ESPRIT I, a total of 49 advanced research projects were launched and all major *European semiconductor companies and institutes* involved in semiconductor area at that time were participating. *ESPRIT II and III* (1987–1994) continued in the same spirit and launched respectively 55 and 79 projects. Projects were covering the same topics with extra emphasis given on packaging, BICMOS, testing, manufacturing science (automation, CAM, etc.), and sensors. In 1989, semiconductor fabrication equipment received particular attention around two axes: lithography with deep involvement of ASML and multiprocess, multichamber equipment. In 1990, a strategic topic on manufacturing science and technology grouped together all major European IC manufacturers. An Open Microprocessor Systems Initiative (OMI) was launched in 1992 with the involvement of the company ARM. In 1990/1991, nine special actions were taken to stimulate awareness and to encourage the uptake of technology by new users, particularly SMEs, with special tailored actions to Italy, Spain, Portugal, and Greece.

The Treaty of the European Union (1993) had decided a large increase of the RTD budget. In 1994, with *ESPRIT IV* (1994–1998), the strategy changed. While in the 1980s and early 1990s Community R&D in Information Technologies followed a technology-push policy, the new focus under ESPRIT IV was on supporting the emerging information infrastructure and the research that should improve competitiveness of IT industry and help enhance the quality of life and environment. The ICT program was to a greater extent led by the needs of users and the markets and stimulated access to information, services, and technologies. Semiconductors and microsystems were part of *Components and*

Subsystems one of the eight domains. Long-term research, integrated manufacturing, software, and the OMI initiative were separated domains. Extra Programs addressed Industrial and Materials Technologies (IMT), Advanced Communication Technologies and Services (ACTS), and Telematics applications. In addition to the research excellence objective, extra emphasis was laid on the exploitation potential, as well as on the international cooperation, dissemination, and training. As a consequence of the larger budget, some of the activities previously integrated were now more separated.

"Components and subsystems" included ICs and multichip modules for wireless and low-power applications, for multifunctional systems, for power control, and for use in harsh environments. R&D was organized in large projects or project clusters and it covered several technology generations in parallel. For example, consolidation in 1996 of a fully economized 0.5 µm CMOS process in several European foundries; availability in industrial pilot lines, also by 1996, of a 0.35 µm CMOS, 3.3 V process based on next-generation lithography, and by the same time, qualification of a state-of-the-art A/D BiCMOS process were achieved. Process modules for 0.25 µm CMOS and below were made available at feasibility level by the research and technology institutes (RTOs). Target for CMOS was by the end of the century five to six layers of metal, 0.18 µm, 50 million transistors on 300 mm wafers. Silicon device research followed the ITRS (More Moore) roadmap. More advanced research topics were part of the separate program for long-term research. Other activities covered system-level design techniques and tools, III–V technologies for millimeter and microwave ICs, SiGe bipolar, optoelectronic devices and new light sources, assessment of prototype equipment, and new concepts for flexible manufacturing.

A separate activity on microsystems (MEMS, MOEMS and Miniaturized Systems) was launched focusing on the transfer of competence from research to industrial use, bringing down the high entry cost, and on the establishment of an industrially oriented microsystem supply. Heterogeneous hybrid as well as monolithic integrated microsystems for medical, environmental, automotive, and smart households were the focus.

The uptake of components, subsystems, and microsystems in advanced applications was stimulated by an action for first users targeting SMEs and access facilitated by a Europractice initiative. "Europractice" provides access for SMEs and academics to design software, prototyping of ASICs, to engineering and low-volume manufacturing of microsystems and multichip components, and to related market intelligence services. Networks were set up to facilitate cooperation and promotion. Europractice is still in existence today with a sole focus on access to silicon technology.

By 1997 more than 80 projects were launched or finalized. Moreover, around 50 projects in the related program dealt with materials and manufacturing for microsystems, 25 projects in the specific SME and innovation activities, and several projects in the biomedical and biotechnology program also dealt with components and microsystems.

25.5
The 5th Framework (1998–2002)

The 5th Framework (€3.6 billion) was conceived to solve problems and respond to major socioeconomic challenges the European Union was facing. FP5 brought the previously separated ACTs, ESPRIT, and Telematics Application Programs together in one single program. A "user-friendly information society" was one of the four thematic programs. The aim was to build a global knowledge, media, and computing space, which is universally and seamlessly accessible to all through interoperable, dependable, and affordable products and services.[3] Key issues of usability, interoperability, dependability, and affordability were addressed ubiquitously throughout the program.

The IST Program (€1.1 billion) included microelectronics, subsystems, and microsystems, The actions aimed to promote excellence in the technologies to accelerate their take-up and broaden their field of application. Cross-programme actions (CPAs) dealt with issues relevant to the entire IST program, such as integrated mobile application platforms, sensor networks, design-for-all, or other generic technology platforms. Future emerging technologies (FETs) including quantum information processing, cognitive sciences, nanotechnology information devices, and research networking were separate activities.

Subsystems and Microsystems targeted research and technology for portable applications, smart network applications; embedded network technologies, and multicomponent assemblies and their integration in products. A total of more than 120 projects were launched for €175 million funding. Resources are split: 64% to R&D, 24% to take-up, and 12% to CPAs and SME stimulation.

(Only for comparison: National funding for 2000: Germany €50 million/year – 35 projects; France €25 million/23 projects; CH €84 million/67 projects?)

The main activities covered were research for components RF MEMS, switches, gyroscopes, accelerometers, MOEMS, and optical sensors. Trials, access, assessment, and best practice actions were given a high importance. As an example, a Europractice microsystem activity (Figure 25.3) at €5 million/year was providing a single-point access for SMEs to a network of 5 foundry services, 12 design services, and 7 thematic competence centers. In 2000 alone, 650 organizations were contacted of which 50% were new to microsystems, resulting in 460 quotations for research and 350 contracts for feasibility studies and prototypes, leveraging another €20.5 million from companies and institutes of which 40% directly from industry, 10% from institutes, and 50% from Member States (public funding). This mechanism allowed bringing microsystems into maturity and preparing for a wider market introduction.

Microelectronics covered microelectronic and optoelectronic research and technology, design and application competencies, process, equipment, and materials, take-up, access, and assessment actions. A basic CMOS process technology

3) Systems and services for the "citizen" (health, administration, environment, transport), "Multimedia content and tools," and "new methods of work and commerce" were the other key areas.

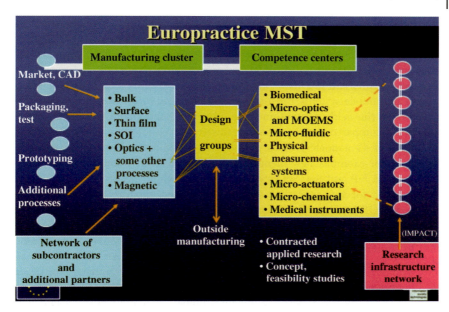

Figure 25.3 Schematic view of the activities covered by EUROPRACTICE Microsystem program.

with front-end below 70 nm and with high-κ gate materials and back-end nodes below 100 nm with low-κ insulators was achieved by industry. Innovative modules addressing ITRS roadmap challenges below 50 nm using next-generation lithography were demonstrated by the institutes. CMOS logic activities were on embedding memory, on-chip integration of RF, and mixed-signal. Further emphasis was on reusable IP blocks, HW/SW codesign, modeling, validation, verification, and test, development of optical and optoelectronic functional components for all-optical networks, interconnection, smart imaging was concentrating on manufacturability, packaging and scalability issues covering information processing and telecommunications, storage, sensing, and imaging applications.

25.6
The 6th Framework (2002–2006)

The 6th Framework (€17.6 billion) introduced the concept of the European Research Area (ERA), namely, the need to create an internal market in research with free movement of researchers, ideas, and technology. The European research fabric was to be restructured avoiding fragmentation and creating more critical mass. At the same time, a European Research Policy was developed to carry forward the Commission's general policy initiatives, in particular the Lisbon Strategy for Europe to become a knowledge-based economy, capable of sustainable growth with better jobs and social cohesion.

Research on components and systems was included as a key ingredient of Ambient Intelligence Systems and Assistive Environments as part of the emerging information Society (IST). European leadership and industrial competitiveness in nanoelectronics and integrated micro/nanosystems was considered a priority. Sensors and microsystems at nanoscale and using new materials, next-generation mobile systems, open-source software, and future and emerging technologies in the field of quantum-scale processing and cognitive sciences were reinforced compared to the previous Framework.

A total of 41 nanoelectronics projects (€190 million) addressed new device concepts, for example, new memory concepts and FinFET and FinFlash structures, mixed SiP and SoC integration, nanoscale RF blocks and new processes, for example, for subwavelength and maskless nanolithography. Developments on the CMOS backbone addressed 45, 32, and 22 nm nodes and below. Also, ultralow power silicon on insulator (SOI), silicon carbide, and platforms for heterogeneous processing and for SoC design were addressed. Specific stimulation actions for SiP design, equipment assessment, and access actions such as Europractice were continued. Photonics became a separate activity. A total of 47 Photonics R&D projects (€132 million funding) were targeting new nanophotonic devices, materials, and technologies: holographic storage, optoelectronic integration for optical networks, new light sources, organic LEDs, and new laser systems, as well as technologies merging electronics and photonics, for example, biophotonic sensors. A European campaign "Fascination of Light" was launched.

A "flying wafer concept" to implement a European distributed 300 mm R&D line, several networks of excellence and roadmap activities targeted the implementation of the ERA concept.

The 79 projects (€300 million funding) in the area of micro- and nanosystems focused on miniaturization, functionalization, and systems integration. MEMS, RF microsystems, plastic and organic microsystems, micro-nano-bio-microsystems, MOEMS, bio, gas, and chemical microsystems, including the integration of microsystems in products and new manufacturing processes, are covered. Health and biomedicine, automotive, food chain management, ambient intelligence applications, and microrobotics have been largely covered. These projects brought together suppliers and users, researchers, and industry from some 500 different organizations from all member states contributing to building an efficient European Research Area.

25.7
The 7th Framework (2007–2013)

During the 7th Framework (€50 billion), the research and development activities continued to be used strategically to respond to Europe's needs in terms of jobs, competitiveness, and quality of life (renewed Lisbon strategy – Europe 2020 objectives) to maintain leadership in the global knowledge economy and to support main political objectives. During this 7-year period, and as a response to the financial crises, the program was several times adapted and gradually more resources were used for, and more emphasis was placed on industrial needs and innovation and to support roadmap-driven research fulfilling the needs of industrially driven public–private partnerships.

In general, FP7 retained the elements of previous programs targeting collaboration across borders, open coordination, and excellence. A stronger focus on major societal research themes, for example, health, ICT, space, and so on, and more intense cooperation with industry using European Technology Platforms providing a direct input to different user-driven research agendas made the program more responsive to the needs of industry and society. Public–Private Partnerships and Joint Technology Initiatives (JTIs)[4] addressed those areas of research where enhanced long-term collaboration and considerable investment are essential to take on objectives that cannot be reached via the fragmented annual calls for proposals. A newly created European Research Council, the first pan-European agency for funding research, aimed to fund longer term, high-risk, high-gain research at the scientific frontiers covering all scientific fields. FP7 also established new Regions of Knowledge, focusing and bringing together various research actors within one region. A new risk-sharing finance facility was established to enhance backing of risk for private investors in research projects, improving access to loans from the European Investment bank.

The different workprograms were updated on a yearly basis. This allowed the 7-year program to adapt its strategy and activities to the different political priorities governed by the 4-year political cycle of the different Commissions and to anticipate shifts in technology trends or in evolving economic situations. For instance, in response to the financial and economic crises, a recovery plan was adopted to help restructure industry and to target policies for innovative key enabling industries (KET). Consequently, the 7th Framework focused more resources in key lead markets setting up contractual public–private partnerships in fields such as advanced manufacturing (factory of the future), advanced automotive (green car initiative), and energy-efficient buildings.

In 2008, five Joint Technology Initiatives (JTIs) were established in order to implement strategic industrial research programs. Two JTIs,[5] ARTEMIS and ENIAC, were in the areas of embedded systems and nanoelectronics.

In 2010, the 7th Framework was again restructured to support the related Commission policy flagships: the Innovation Union,[6] A Digital Agenda for

---

4) Joint Technology Initiatives are public–private partnerships jointly funded by industry, research organizations, some including participating Member States and the Commission's own ICT programme designed to leverage more R&D investments.

5) Other JTIs are Clean Sky for aeronautics and air transport, the "Innovative Medicine Initiative" (IMI) and the "Hydrogen and Fuel Cells" (FCH) initiative.

6) The Innovation Union completes the *European Research Area* installing *Joint Programming with Member States and Regions*, invests in a *single EU Patent, and improves access to capital, making full use of public innovation procurement*, and strengthens partnerships in *knowledge triangle* between education, business, research, and innovation setting up a European Institute for Technology. The Framework supported launching EIP's: 'European Innovation Partnerships' where all EU instruments to support innovation should work together (*Communication COM(2010)245 of 19.05.2010*).

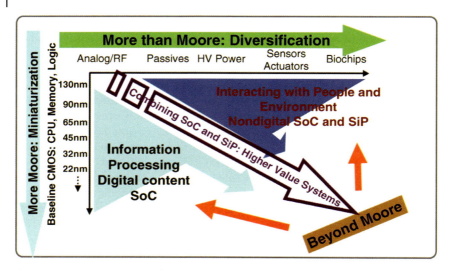

Figure 25.4 European roadmap for nanoelectronics.

Europe,[7] an Industrial Policy for the Globalizing Era', "Agenda for New Skills and Jobs" and a "Resource Efficient Europe."

One of the thematic areas was ICT. The objective of the €9.1 billion ICT program was to ensure the scientific and technology base to improve the competitiveness of the European industry, so that the demands of its society and economy are met, as defined in the Digital Agenda.

In 2010, a core challenge for ICT was micro- and nanoelectronics. The targets were intelligent machines, systems, and processes impacting all sectors. Miniaturization following Moore's law, together with the integration of more functionality on chips (More than Moore), for example, in microsystems and smart systems for health, automotive, and food were focused upon in an integrated manner, as shown in Figure 25.4. These activities were complemented by the closer to industrialization activities in the ENIAC Joint Technology Initiative.

The latest ICT FP7 Workplan 2013 was again restructured. The workplan covered eight "Challenges" of strategic interest to European society,[8] combined

---

7) The DAE "everyone digital" is to create a vibrant digital single market. Part of the ICT research in the 7th Framework was to demonstrate the ICT-enabled benefits for EU society, contribute to interoperability and standards, develop the next-generation ultrafast Internet access, and address digital literacy, skills and inclusion, and trust and security issues (Communication COM(2010)245 of 19.05.2010).

8) Challenge 1: Pervasive and Trusted Network and Service Infrastructures
Challenge 2: Cognitive Systems and Robotics
Challenge 3: Alternative Paths to Components and Systems
Challenge 4: Technologies for Digital Content and Languages
Challenge 5: ICT for Health, Aging Well, Inclusion, and Governance
Challenge 6: ICT for low carbon economy
Challenge 7: ICT for the Enterprise and Manufacturing
Challenge 8: ICT for Creativity and Learning
Future and Emerging Technologies (FET)

with research into *"Future and Emerging Technologies"* (FET) and support for horizontal actions, such as international cooperation and precommercial procurement:

Included in the FET challenge was *the €1 billion (over 10 years)* Graphene Flagship that aimed to bring graphene, and related 2D materials, from academic laboratories to industry, manufacturing, and society. Another FET Flagship, the Human Brain Project Flagship, was set up to develop the most detailed models of the brain to lay the foundation for medical progress and for brain-inspired "neuromorphic" computing systems.

The overall aims for nanoelectronics, micro-nanotechnology, and microsystems were as follows:

- To reinforce European industrial leadership through miniaturization, energy-efficiency, performance increase, and manufacturability for information and communication systems and other applications in 2020 and beyond.
- To enable the convergence of nanoelectronics, nanomaterials, biochemistry, measurement technology, and ICT
- To stimulate the innovation of European industry by well-targeted take-up actions, with special emphasis on SMEs – either as users or as technology suppliers.

The JTIs ENIAC and ARTEMIS did focus on industrial application-driven developments, manufacturability, platforms, and new paradigms, which are applicable across several application domains. ENIAC addressed mainly next-generation technologies in the "More Moore" and 'More than Moore' domains and their integration in new systems and products. These activities closer to the market complemented the nanoelectronics activities under the 7th Framework that essentially covered the "Beyond CMOS" and more advanced "More than Moore" domains preparing Europe for the design and manufacturing of the next-generation components and miniaturized systems with a 10-year time to market.

The ARTEMIS JTI focused on developing industrial platforms for the development and implementation of embedded systems responding to industry requirements in specific application domains.

Related to photonics and to the integration of components and systems, work was aligned with the strategic research agendas of the European Technology Platforms and related associations Photonics21 and EPoSS. In the areas of robotics and photonics, activities in 2013 supported the preparation of Public–Private Partnerships that were to be launched under the next framework called H2020.

25.8
H2020 (2014–2020)

H2020 structure largely follows the Europe 2020 vision putting even more emphasis on jobs and growth and a much larger emphasis on innovation. A

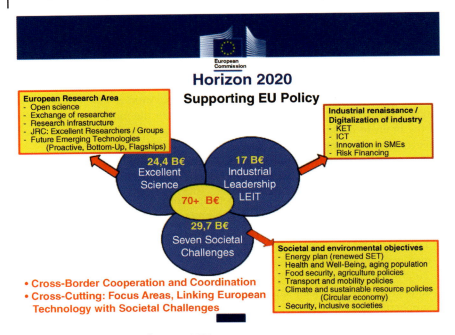

Figure 25.5 Main priorities of Horizon 2020 program.

major objective has been to provide SMEs with adequate support in order to help them grow into world-leading companies.

The activities were regrouped in line with the priorities in three large sets of activities, as shown in Figure 25.5:

- Seven societal challenges
- A program (LEIT) addressing the industrial competitiveness including the PPPs and JTIs and the risk financing
- A program dealing with advanced research including the 2 FET flagships (Graphene and Human Brain Project), the ERC advanced research actions and the activities to stimulate researchers mobility.

Within the field of nanoelectronics, the two JTIs ENIAC on electronic components and ARTEMIS on electronic systems have been merged together, also including the microsystems integration (EPoSS technology platform). Two PPPs have been launched on photonics and robotics. Within the ICT LEIT and next to the KET components and the digitalization of industry, a large emphasis has been given to the emerging data economy, the Internet of Things and the establishing of 5G.

Figure 25.6 shows the holistic approach targeted by micro- and nanoelectronics and embedded systems in H2020 to fulfill the requirements of IoT and the smart connected world driving further the semiconductor and embedded systems industry.

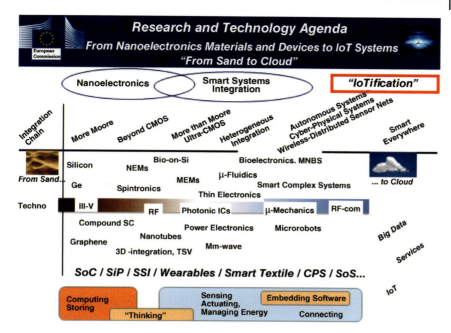

Figure 25.6 Holistic approach targeted by micro- and nanoelectronics and embedded systems in H2020.

25.9
Some Results of FP7 and H2020

25.9.1
At Program Level

At only 3% of total R&D spending in Europe, the €55 billion Framework Programme 7 has had a disproportionate impact on Europe's economy, with each euro spent generating an estimated €11 in return, through new technologies or products. In total, the research and commercial potential of FP7 funding is judged to be worth some €500 billion to the European economy over a period of 25 years.

The scientific output of the FP7 projects is generating more than 165 000 scientific papers, of which 650 were published in the high impact journals Science and Nature. Research projects funded by the European Research Council (ERC) were recognized with a Nobel Prize and two Fields Medals in 2014.

Universities were allocated 44% of total funding, while research and technology organizations obtained 27%, large private companies 11%, and small businesses 13%. Five hundred research institutes obtained 60% of the EU's €55 billion Seventh Framework Programme (FP7) budget. This strong research core includes powerhouses such as the Fraunhofer Gesellschaft and Oxford

University and in particular in the field of nanoelectronics, imec, LETI, EPFL, and Tyndall.

Of the total funding for FP7, 85% went to research facilities and businesses located in Western Europe. Germany put in the strongest performance, closely followed by the United Kingdom. The concentration has led to increased global competitiveness and economies of scale, and fostered the emergence of centers of excellence. Of a total of 139 000 research proposals submitted to FP7, 25 000 were funded

25.9.2
The ICT Research in FP7

Being primarily a precompetitive research program, FP7 ICT has augmented the European knowledge base by creating and disseminating publications, proto-types, patents, expertise, and know-how. The greatest impact of the program has been the knowledge effects for its participants, where the required competencies, resources, scale, and scope could not have been achieved to the same degree at the national level.

A bibliometric analysis for 2007–2013 finds that each project generates 15.7 publications on average, although one project produced 1036 publications. Germany, Italy, the United Kingdom, Spain, and France lead in terms of publica-tion output within the EU-28 with the bulk of output produced by academic institutions. Italy and the United Kingdom are the principal knowledge brokers within the EU-28 group of countries, while Germany is the most favored country to cooperate with. Also, the relative citation rate is well above the world average for the years examined.

The exploration of the distribution of patents across different strategic objec-tives shows that about one-fifth of the patents resulted from projects with the strategic objective of The Network of the Future, with 7% coming from Photonic components and subsystems and 6% from micro/nanosystems. Patents of ICT research in FP7 are associated with very high interorganizational collaboration (more than 50%), much more than what is observed in the industry in general (5%). Patents of ICT research in FP7 also display a unique EU bias with regard to the location of both applicants and inventors.

EU-funded projects in the area of ICT have also led to 125 spin-offs set up to commercialize products and services resulting from EU support.

25.9.3
Micro/Nanoelectronics and Smart Systems

The portfolio in nanoelectronics and smart systems integration in FP7 included 186 projects from the regular Framework Program with more than 1900 partic-ipants as well as 63 projects from the ENIAC JU with 1350 participants.

The 7th Framework targeted the advanced topics or the less mature fields where industry was investing less or where the MS programs were not

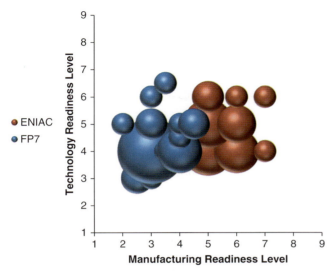

Figure 25.7 Technology and manufacturing readiness levels of FP7 and ENIAC JU programmes.

contributing. Two application fields can be highlighted: power/wireless electronics and electric vehicles with smart grid applications. Access, take-up, and training actions were completing the nanoelectronics and micro/nano-technology research activities to stimulate innovation at SME side. The PPP ENIAC concentrated on the more mature topics in equipment and processes, on power electronics, on MEMS and Sensors, and on main nanoelectronics applications. The main difference with the FP was not (only) in the TRL levels targeted but also in the inclusion of MRL (manufacturing readiness levels) much beyond the manufacturing level reached in the FP7 activities, as shown in Figure 25.7.

There are topics that are finding only marginal support in ENIAC/ECSEL. Among these are, *inter alia*, "Beyond CMOS", the upstream exploration of alternative options for the continuation of computing power improvement.

25.10
Results of the JTI ENIAC and ARTEMIS

Both JUs selected for cofunding 119 projects, with almost 2800 participations engaging €636 million in EU contributions, leveraging €912 million in national contributions and incentivizing research and innovation with total eligible costs approaching €4 billion. The projects represent a total effort of 25 283 person-years or more than 3000 full-time equivalent positions per year. Moreover, each euro contributed by the European Union resulted in €6.4 research and innovation activity in Europe.

At ecosystem level, the majority of the participating entities are SMEs, making up more than 26% of the participants in the program; the large enterprises and the academic or institutional research organizations represent about 40% each.

At the end of 2014, 59 projects reached technical completion; they generated 179 patents, 11 trade secrets, 12 trademarks, 1004 exploitable foreground intellectual property items, and 3099 publications. These numbers are almost double in 2016.

At the end of 2013, ENIAC and ARTEMIS had selected for funding 14 pilot line actions with total eligible research costs of about €1.8 billion. As a result of these activities, Infineon announced building a 300 mm fab in Dresden, the first high-volume fab for power semiconductors worldwide.

More than 50 European manufacturers are working together to establish at IMEC the only facility outside the United States able to drive toward industrial maturity the next generation of semiconductor equipment enabling the transition to sub-10 nm generations 450 mm compatible, used to secure sustainable leadership of the European manufacturers (particularly in lithography).

The fully depleted silicon on insulator (FDSOI) technology enabling up to 40% reduction of the power consumption in portable devices such as smart phones and tablets is developed in a series of "pilot line" projects coordinated by STMicroelectronics.

25.11
An Analysis of Beyond CMOS in FP7 and H2020

Moore's law, underlying all efforts toward miniaturization/integration in the nanoelectronics field, is reaching its limits probably in a decade from now. Further miniaturization/integration of semiconductor transistors will no longer be feasible due to a collection of physical, technological, and economic factors.

The roadmap for "conventional" semiconductor integration along Moore's law probably has three more "nodes to go," as indicated by imec and confirmed by Intel. These are refinement of FinFET, horizontal nanowires, and finally vertical nanowires.

"Beyond CMOS" was coined in the International Technology Roadmap for Semiconductors (ITRS) as the next set of technologies, which may drive the next wave of computing technologies. Beyond CMOS includes quantum computing, spintronics, molecular electronics, or other neuromorphic computing. In essence, the goal is to arrive at integrated computing system where data are represented beyond the usual charge-based binary information.

The topic of "exotic" "Beyond CMOS" technologies is still academic. However, Intel presented an extensive research program on spintronics. Intel also recently awarded a €50 million grant for 10 years to the Technology University of Delft for research on quantum computing. Samsung is very interested in MRAM with several announcements for product release during 2016.

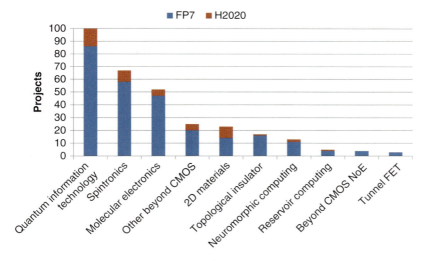

Figure 25.8 Mapping of FP7 and H2020 (until 2015) Beyond CMOS projects by topic.

This area is partly addressed by projects in H2020 LEIT, but as the research is in most cases more at the fundamental level, this portfolio of projects currently represents only a small part of the efforts done on "Beyond CMOS" in the European Union R&I Framework Programme. In fact, about 60% of these projects are supported by the ERC and another quarter by Future Emerging Technologies.

A total of 450 projects in FP7 and 88 projects in H2020 were identified as being related to "Beyond CMOS." A further identification defined which projects have potential for solid-state integration. This resulted in the identification of 311 projects related to "Beyond CMOS" representing a total funding of €408 million in FP7 and €74 million in H2020.

A classification in categories of these 311 projects was done considering their center of gravity resulting in the mapping in Figure 25.8.

The analysis does not contain the Graphene Flagship, but it is clear that 2D materials start gaining importance.

The EU contribution to these projects is substantial: overall €508 million over the 9 years period. Quantum computing attracts the highest funding (Figure 25.9). The field of Beyond CMOS is strongly academic as visible in the figure. The industry accounts for very little participation save for Tunnel FET, which is a topic at the limit between current developments and long-term developments.

The Beyond CMOS topic is still facing many challenges on the fundamental side as only first proof of concept exists for most technology trends. Nevertheless, projects start to appear in higher numbers with more funding focused on industrial cooperation. H2020-ICT-2015 is funding five projects, compared to the 18 funded through the whole FP7 program.

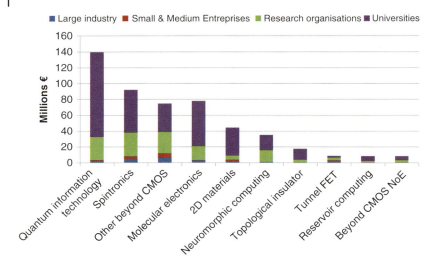

Figure 25.9 European funding for Beyond CMOS topics in FP7 and H2020 (until 2015).

25.12
MEMS, Smart Sensors, and Devices Related to Internet of Things

IoT is expected to be a key growth driver for semiconductors. MEMS and MEMS-based sensors are used in all IoT application segments (wearables, smart home, medical electronics, industrial automation, connected cars, and smart cities) with billions of new devices expected to be made available in the future.

Europe has a leading position in MEMS-based sensors and in low-power electronics, another enabler for IoT. That is why a strategy for electronics in Europe embraces this leadership and embeds it in an extended IoT ecosystem.

In this respect 22 projects were launched under FP7 (2007–2013) (16 projects) and ENIAC (6 projects of which 3 are pilot lines) with 267 participants in total and worth 327 € million of research value. About 70% include the development of a dedicated electronic integrated circuit, an important aspect for functional device demonstration.

Each of the 22 projects includes one or several demonstrator(s). Figure 25.7 is the result of the mapping of these 49 demonstrators along technology and manufacturing readiness levels.

Furthermore, during first 2 years of H2020, the calls for proposals resulted in the support of seven projects, of which four are under ECSEL. The focus is clearly toward system integration for purposes ranging from automotive safety and automation to the Internet of Things. This represents already 40% of the total costs of the 22 projects supported under FP7, ENIAC included. It underlines the alignment between the industry investment directions and the support by the European Union of technologies targeting, *inter alia*, European societal challenges.

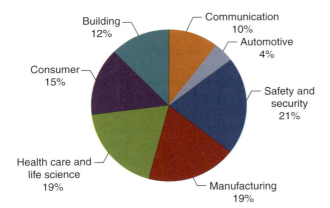

Figure 25.10 Application sectors targeted by IoT projects under FP7 and H2020 (until 2015).

Figure 25.10 presents the application sectors targeted and their intensity in seven main application areas. They cover the main societal challenges put forward at European level.

25.13
From FP6 to FP7: An integrated approach for micro-nanoelectronics and micro-nanosystems

25.13.1
Research cooperation between the Framework and Eureka initiatives

Between 2004 and 2009, the large Integrated Projects NanoCMOS (2004–2006; 17 partners, €24.2 million) and PULLNANO (2006–2008; 40 partners; €25 million) and the accompanying academic Networks of Excellence SINANO (2004–2007; 42 partners); and NANOSIL (2008–2011; 29 partners; €4.3 million) have succeeded in bringing European research and integration efforts to world competitive levels. General-purpose 45 and 32 nm CMOS technologies were developed in a timely manner with integration densities up to 50 million transistors per mm^2. If in the 1970s 1 MB of memory capacity cost more than a house, it costs now – thanks to such an incredible shrinking in size – less than a few cent. This *cooperation* between the academic partners and the industry partners also allowed demonstrating the feasibility of basic technology solutions for the next-generation 22 nm CMOS technology.

The "parallel and complementary processing" between these four projects has led to a new level of European academic integration in the area of very advanced nanoelectronics devices and materials. A *virtual European research institute*, also called SINANO, was founded that was steering European activities in the advanced devices field. Virtual processing, characterization, and simulation platforms were created that allow academic research in the field using common European academic infrastructures. This is an example of the efficiency of

European cooperation using both industrial and academic infrastructures. This integration of formerly scattered research activities in the field also continued in the FP7 project DUALOGIC looking into CMOS solutions beyond the 22 nm technology generation.

Both the 45 and 32 nm CMOS technologies developed by these four IST/ICT projects were *taken up by the EUREKA nanoelectronics clusters* MEDEA+ and CATRENE, respectively, under projects FOREMOST (45 nm) and UTTER-MOST (32 nm). FOREMOST successfully finalized work on an industrial version of the 45 nm technology in mid-2009 that was used in production by the leading European chip manufacturer STMicroelectronics. The project FOREMOST also got the EUREKA innovation award 2010.

Similar examples of the European research food chain, between the Framework ICT programs, the EUREKA nano-electronics clusters (JESSI, MEDEA, MeDEA+, CATRENE), and the ENIAC JTI also exist in the field of semiconductor manufacturing equipment and in particular in the strategic area of lithography and for some application developments.

This dense network of interacting European programs and institutes, guided by industry, also develops the *derivative technologies* such as analog or very high- and very low-power technologies, including extra functions such as embedded memories, integrated sensors, or silicon-integrated MEMS, silicon photonics or biochips delivering a flow of continuing generations of families of devices and components. The latter developments target systems-on-chip (SoC) and system-in-a-package (SiP) to be integrated in smart systems and subsystems for "intelligent" products.

While general-purpose technology R&D is driven by miniaturization objectives in a "technology push" mode, the derivative and enhancement technology R&D were guided by system or application performance considerations in an "application pull" context. This type of research is executed in close collaboration between suppliers and users and often executed in the system- and application-related parts of the Framework program. For instance, high-power technology and related smart components are further demonstrated in the context of the PPP Green Car to deliver advanced smart systems for the electric car of the future, or in combination with subsystems to manage the energy consumption in intelligent housing. Very low-power technology together with sensor integration is demonstrated in the context of future wireless sensor networks for environmental monitoring or for "Internet of Things" applications. It is there – at the end of the high-tech chain – that it becomes obvious that research and funding efforts for components are paying of.

This also delivered *spin-off results* in other areas with a most visible in the fields of photovoltaics, photonics, biochips, microfluidics, micromechanics, microsystems, and microrobotics.

Moreover, *technological borders are blurring* (e.g., silicon photonics) and much *innovation comes from the integration and combination of different technologies* using competencies from different scientific fields, as, for instance, in the emerging multidisciplinary info-bio-nano-cogno field. Such interaction generated totally new products, for example, microrobotics for microsurgery or diagnostic pills.

Most of these developments require very *expensive research infrastructures* and only a few places in Europe offer this capability. European initiatives network the more advanced university laboratories and link them with the most advanced institutes to open the expensive research infrastructures to wider European participation.

Several activities include *support for education and training and to address faster take-up* and innovation for SMEs and an "equipment assessment" activity is helping the semiconductor equipment companies to demonstrate the readiness of their innovative materials, equipments, or products by direct cooperation with some key customers and by using their assessment to promote and introduce their advanced products at global level.

The EUROPRACTICE IC project provides access to industrial CAD tools for learning and research and to advanced IC prototyping technologies. Over 650 European university groups and research institutes are taking part in this initiative under favorable conditions. This enables the next generation of engineers to keep pace with the new developments, allows PhD to use the latest technologies and techniques for their designs and prototypes, and provides SMEs early access to the latest semiconductor knowledge and technology.

Special attention is given to the worldwide recognized European knowledge clusters in nanoelectronics, to network, and make these regions more attractive for further investments and to exploit their infrastructure and competencies in a wider European context.

25.14
Enabling the EU 2050+ Future: Superintelligence, Humanity, and the "Singularity"

EU 2050+ stands for an integrated vision toward a hyperconnected increasingly more global world bridging the virtual and physical worlds instantaneously, where cities become megacities and where people will be more empowered to share knowledge and take informed and responsible artificial intelligence – algorithms-based decisions. Advanced robotics, automation, and new manufacturing techniques (3D printing) will drastically change the world economy, reshoring manufacturing. Technology and ICT will get integrated in the individual and its environment in an invisible manner to improve the human, environmental, and working conditions and to augment the human intellectual, physical, and psychological capacities.

It is a step to a future where entire sectors of economy are data and computing-based giving rise to digitized, dematerialized, and demonetized industries, following new business and new societal models. It is also a step toward a future where human-level intelligence may become integrated with an intelligence cloud and with an intelligent ambient system to govern, where humans, intelligent machines, and a global intelligence ambient will dictate progress. The world will be totally different in 2030–2050–2100. We – as Europeans and human beings – should ensure that this transfer will be managed in a democratic manner using high ethical standards.

25.15
EU 2050+: Driven by a Superintelligence Ambient

Exponential progress in Enabling Technologies and an enhanced mastering of complexity together with a deeper understanding of many aspects of life will drive toward *a smarter, more digitalized, hyperconnected, and more instrumented* world. This in turn will open new business opportunities, increase the efficiency and competiveness of the existing European industry, and allow the grand societal and environmental challenges of the future to be addressed in an often radical new manner bringing direct benefits to the well-being of the European citizens, their living and working, and their environment.

A *hyperconnected world*: Everything is connected and communicates with everything, everywhere, anytime, real-time, using smart, high-performance, and more natural connections. People, things, objects, robots, environments, businesses, business processes, value chains, services, data, cloud, and so on are connected and may explore the cloud to store, exchange, manipulate, and process increasingly larger amounts of data.

A *smarter society*: Nearly all "anythings"[9] are becoming smarter, embedding more intelligence and learning capabilities on or inside them, distributed between them or using the cloud, exploiting big data, HPC (high Performance Computing), simulation, (artificial) intelligence algorithms, processes, and so on to have more informed, evidence-based decisions (*smart anything, everywhere, anytime, real time*).

A more *instrumented world*: (i) Technology get deeply embedded with people and their environments in a very natural invisible manner, for example, using wearable, cognitive technology, or embedding ICT technology with bio or human enhancement. (ii) Living and working processes get a very high degree of automation, for example, using advanced autonomous systems, automated smart sensor-based information, and autonomous decision making.

A more *digitalized society:* (i) The increasingly data/computer-based nature of the economy will disrupt existing businesses models, change jobs, and offer new job opportunities, for example, in the "experience" industry, the economy becomes more dematerialized and demonetized. (ii) The virtual and (augmented) real world are bridging. (iii) (Open) access to data, science, and 3D printing bring a disruptive shift with high impact on education, manufacturing, and the societal, scientific, and economic models of today.

This is made possible by factors shown in Figure 25.11: (i) *an exponential progress in individual Enabling Technologies enhancing each other by their combinations* offering higher performances, better integration, and automation at lower costs and by (ii) *an enhanced mastering of complexity* using large volumes of data together with (iii) *a deeper understanding of many aspects of life, nature, materials, physics, and societal behavior.* This fast exponential progress at all

---

9) "Anythings" is here a terminology for people, things, objects, artifacts, robots, spaces, ways of working, processes, environments, businesses, business processes, value chains, services, data, cloud, and so on.

Hyperconnectivity

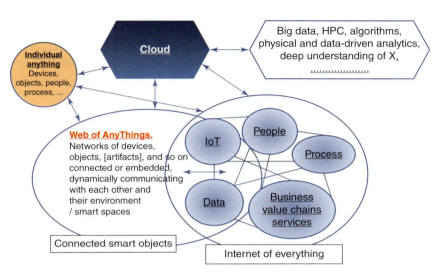

Figure 25.11 Main factors driving toward a hyperconnected world.

fronts, including their multidisciplinary interactions, are the essential means and building blocks.

Investment in emerging disruptive digital technologies to realize EU 2050+ is increasing, leading toward a future that 2050+ will be driven by intelligence beyond today's human capabilities. Progress in artificial intelligence, "superintelligent ambient technologies" and "hyperconnected digital space" may create a global brain with an intelligence (above what humanity can bring) that will allow addressing the large complexity of the "real-world system" and making decisions in a fully evidence-based manner.

Addressing this level of complexity could never happen by an individual human nor a network of individuals. No need to try, human brains are too slow, too imperfect. Predictions could never be done that fast and with such a high precision. This superintelligence global ambient system will manage progress driven by superintelligent machines, build solutions, and implement our future society and our environment. People with augmented capabilities are just one element embedded in this global smart ambient system. This superintelligence system, including superintelligence machines and fully automated approaches, will take over many of the human activities and decisions and will steer progress beyond (but including) humanity-driven progress.

It is expected that at a certain point in time (singularity), artificial intelligence, machine learning, evidence-based algorithms, and so on will not only surpass human capabilities but will also create superintelligent machines operating in this superintelligence environment (system), whose learning and intelligence capabilities will enhance each other, progressing at a speed beyond the imaginable, going beyond the steering capabilities of mankind. It is also shown that it is possible to influence directly the human brain, which makes it not unimaginable that this system also will steer the human activities itself.

Against this background and taking (initially) advantage of this evolution, we see a new role, new behaviors, and new activities emerging for the human being, the human society, and the regions. Against this background and within this emerging "superintelligent ambient world system," people will live longer, a new form of humanism (trans-humanism) will emerge with new ways of integrating technology and the human. We expect new economic and societal models to emerge no longer based on GDP and economic exchanges but based on "human experience" or "behavior" values. Jobs will be drastically different, for example, based on creativity, work changes in nature, and time spend. Economic and scientific powers may shift from big multinationals or institutes dominating one field to continuously changing flexible emerging initiatives and individuals using open methods as crowd funding/crowd sourcing and crowd recognition safeguarded, secured by distributing methods as the block chain. Local manufacturing and local industry, for example, using 3D manufacturing may flourish again. We see new ways of democracy and governance, empowering (flexible groups of) individuals directly in the decision-making processes and directly participating in the governance, steering the different activities with a need for rethinking politics, media and policy making. Science based on open methodologies and education will differ totally. Cities, villages, and manufacturing processes will be redesigned to cope with sustainability, renewables, and the available competencies. The role of the people in this world will be totally different.

Europe should prepare for this future where its human and social values are guiding this progress. Any "superintelligent ambient system" and "intelligent machines" should be intelligent, deliver evidence-based optimal solutions for the future but should also take "European social and human" values as a reference. Many of these issues will need a very broad civil and political debate. Progress is exponentially fast, this future starts now and is totally made possible by progress in nanoelectronics, nanotechnology, nano-biotechnology, new bio and materials technologies, and an emerging digital future incorporating technologies such as the cloud, hypercomputing, the new data economy, artificial intelligence algorithms, hyperconnectivity, and so on.

25.16
Conclusion

With this chapter I want to underline the progress and changing nature of 30 years of consecutive Framework programs and the related ICT and micro-nanoelectronics developments.

Due to the nature of these Frameworks, they have attracted the best European researchers to work together and contributed to structuring the European

Research Area. From an *ad hoc* set of centrally managed research projects, the different frameworks have grown in size with research topics added that were linked with, and contributing to, the different policy initiatives of the different Commissions, which safeguarded political support for increased priority in research.

The programs were gradually managed in a more distributed manner with individual topics spread over several DGs and Institutes, such as the Research Council, and with public-private partnerships such as the Joint Technology initiatives in electronics.

This has allowed focusing large amounts of public funding to the micro and nanoelectronics field and to better link research with industrial initiatives. The consecutive Framework Programs also contributed to better structure the microelectronics landscape and more efficiently organize the research linking the universities, research, and technology organizations and industry together around European priorities and a European oriented roadmap. Emphasis was put on activities contributing to the competiveness of the European semiconductor industry and bringing innovation in the industry at large, covering whole value chains.

This has allowed Europe to stay on par with other regions in the world, in particular in system-oriented competencies, and prepared a good starting position on emerging topics highly relevant for the future of European industry and society, such as the Internet of Things, autonomous vehicles, 5G, and highperformance computing ASO. It could however not prevent in that Europe lost importance in areas where large investments were needed to follow the miniaturization roadmap for high-volume components related to computing functions such as memories and processors. At the same time, the Framework has put Europe in a good position to address the next wave of nanoelectronics devices called "Beyond CMOS."

I also tried to give a rather technocratic engineering view what the next 30 years could bring us if progress continues exponentially. Programs as the Framework can steer to the future we want.

26

The Education Challenge in Nanoelectronics

Susanna M. Thon, Sean L. Evans, and Annastasiah Mudiwa Mhaka

Johns Hopkins University, 3400 N. Charles Street, Baltimore, MD 21218, USA

26.1
Introduction

Educating the next generation of nanoelectronics researchers, engineers, and scientists presents several challenges due to the fast-moving and interdisciplinary nature of the field. This chapter outlines some of the considerations involved with designing modern university-level programs in nanoelectronics. In overviewing traditional programs that confer degrees related to nanoelectronics, specific challenges related to bridging disparate disciplines are addressed. Integrating theory, design, and lab work and enabling new industrial applications are described in detail. A case study involving educating students in the use of nanoelectronics in the health care industry is presented, followed by a discussion of the future of university programs aiming to give students a holistic view of the field.

As overviewed in previous chapters, the field of nanoelectronics encompasses several perspectives that present different challenges in the education sphere. The "More Moore" domain, involving the goal of using nanotechnology to shrink circuit components and increase chip transistor density, requires that students understand traditional chip design as well as new fabrication techniques and complexity issues from a systems perspective. The "More than Moore" domain, involving the integration of different functionality with traditional nanoelectronics devices, requires skills outside of the traditional fabrication and circuit-design spheres, potentially including an understanding of nanoscale chemical, mechanical, biological, and photonic processes. Finally, the "Beyond CMOS" domain, which aims to use nanotechnology to build fundamentally new kinds of electronics to replace scale-limited CMOS components, requires an in-depth knowledge of fundamental semiconductor physics including heat management and quantum processes. These elements will be discussed in the context of traditional curricula, as well as their industrial and societal implications for nanoelectronics education programs.

Nanoelectronics: Materials, Devices, Applications, First Edition. Edited by Robert Puers, Livio Baldi,
Marcel Van de Voorde, and Sebastiaan E. van Nooten.
© 2017 Wiley-VCH Verlag GmbH & Co. KGaA. Published 2017 by Wiley-VCH Verlag GmbH & Co. KGaA.

26.2
Traditional Programs in Nanoelectronics Education

26.2.1
Fields of Study

The nanoelectronics industry, broadly defined, has grown to encompass different competencies at all levels of the engineering process, from traditional circuit design to systems- and applications-level roles for professionals. Therefore, the primary major fields of study at the undergraduate and graduate level have been spread across different traditional disciplines. Electrical and computer engineering and physics have been the most prevalent department homes for nanoelectronics-focused students; however, with the increasing attention on new materials and fabrication techniques, materials science, biomedical engineering, and chemistry departments are also educating students with skills relevant to the nanoelectronics industry. As the field evolves, students desiring a truly comprehensive education in nanoelectronics will need to look to all of these disciplines, along with mathematics and potentially biology, as areas of study, at least at the introductory level.

Figure 26.1 contains a chart with examples of key research categories within the respective traditional disciplines associated with nanoelectronics. The example disciplines listed in the chart could be studied in their respective university departments at an introductory level during the first 2 years of undergraduate education, with further specialization during the 3rd–4th years in a traditional bachelors program. A narrower focus is usually required at the masters and doctorate level (not shown).

Not included in the chart are examples of nontechnical, so-called "soft skills," classes. As the field of nanoelectronics becomes increasingly international, written and oral communications skills, as well as research ethics, management, as well as language and culture, are becoming both more complex and vital for students planning careers in industry and academia. A complete curriculum must include these topics and give students opportunities to practice them in context.

26.2.2
Topics of Study

Independent of the academic departments of interest, there are several broadly defined focus areas associated with an education in nanoelectronics [1]. Different versions of classes that address these areas are often taught in multiple academic departments or cross-listed across several departments. Following is an example of a classification system for nanoelectronics-associated focus areas:

Semiconductor Device Physics
This track focuses on the underlying physics of nanoelectronic devices, starting from semiconductor band theory and progressing through device design. Classes

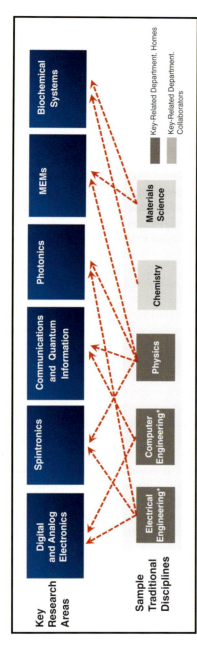

Figure 26.1 Illustrative example of foundational research and disciplines related to nanoelectronics. Electrical engineering*, computer engineering*, physics, chemistry, and materials science are the foundational discipline areas that predominately provide education and skills-set relevant to nanoelectronics. Students typically major in one of these areas, (i.e., obtain a bachelor's degree) with further specialization in these disciplines at the masters and doctorate level. Examples were selected to demonstrate that key research categories are primarily cross-disciplined involving two or more disciplines (two shown). *Electrical and computer engineering are often under a single department in the United States of America.

in quantum mechanics and nanophysics are typically included, and students focusing in this area are traditionally involved in individual device component design.

Circuits

Electronics and circuit theory, as traditionally taught in electrical engineering departments, form the basis for this track. Students learn both analog and digital design principles, as well as circuit modeling and signal processing.

Nanofabrication

This track includes a strong laboratory component in cleanroom processing techniques. These include photo- and electron-beam lithography and associated etching, thin film deposition, and characterization techniques for nanostructured device components. Newer programs will also include self-assembled and scalable fabrication processes, along with computer aided design modules.

"Bottom-Up" and Soft Materials

Classes in nanotechnology with a materials chemistry focus are included in this track, and include an emphasis on organic and soft materials with potential device functionality. Starting with atoms and molecules, students learn how to engineer nanostructures with specific properties that can be integrated with or replace traditional electronic circuit elements.

Systems Design

Systems design is centered around a high-level approach to nanoelectronics and would include courses in control theory and signal processing as well as fundamental nanoelectronic device components. As the field of nanoelectronics approaches the giga-scale, systems-level design principles are becoming more complex and relevant to a modern education.

Beyond the undergraduate level, research in nanoelectronics has grown to encompass many different subfields and systems [2]. Students in Ph.D. programs may be exposed to a variety of different research topics that have broad relevance to nanoelectronics. Specific examples include micromechanical systems, photonics, spintronics, communications, quantum information, hybrid digital–analog and RF systems, networks, and nanosensors, as well as certain biochemical systems. Research in these topics is usually highly interdisciplinary, and students must learn a variety of physical and mathematical concepts and techniques in order to contribute in a significant way.

26.2.3
Example Programs

Throughout the world, many universities currently offer educational and/or research opportunities within the field of nanoelectronics. From an educational perspective, while students can typically find nanoelectronics courses being taught at large or technically focused universities, there are not many universities that offer a formal degree in the field at the bachelors, masters, and/or doctorate levels. While educational programs in nanoelectronics are located on every population-heavy continent, this section focuses on programs found within European and American institutions. With that said, there are notable top universities in this field in Asia, including in Singapore [3], Hong Kong [4], China [5], South Korea [6], and Japan [7]. A discussion of those programs is beyond the scope of this chapter.

While very few academic institutions offer specific degrees in nanoelectronics, the top universities in electrical engineering offer a strong academic experience as well as research opportunities in the field. Some of the top US and European electrical engineering programs based on academic and faculty/researcher reputation as well as impact of their research [8] include the Massachusetts Institute of Technology (MIT) [9], Stanford University [10], the University of California – Berkeley [11], Harvard University [12], ETH Zurich – Swiss Federal Institute of Technology [13], the University of Cambridge (CU) [14], Imperial College London [15], the University of Oxford [16], the California Institute of Technology [17], the University of California – Los Angeles [18], Delft University of Technology, [19] University of Leuven [20], EPFL Lausanne, Politecnico di Milano [21], Politecnico di Torino [22], Dresden University of Technology [23], Technical University of Munich [24], and Grenoble Institute of Technology [25]. There are many others that provide solid academic course-work in nanoelectronic concepts as well as cutting edge research opportunities.

A specific program example is MITs nano-Electrical Engineering and Computer Science program [9], which offers over 15 specific nanoelectronics-focused courses, including *Introduction to Nanoelectronics*, and over 10 faculty who conduct nanoelectronics-related research. A European-based example program can be found at Cambridge University (CU). In addition to course work and research in nanoelectronics, the Department of Physics at CU boasts a number of cross-institutional and academic-private partnerships focused on advancing the field of nanoelectronics [14]. CU has formed a number of research cooperatives with other academic institutions such as the Korea Advanced Institute of Science and Technology as well as private companies, Toshiba and Hitachi. These partnerships provide a strong applied nanoelectronics research environment to its student and faculty researchers [14].

Additionally, there are a few universities that are starting to offer focused degrees in nanoelectronics. European universities have taken the lead in establishing these degree programs. A few examples include the following:

- The University of Oslo offers both a bachelor's and master's degree in nanoelectronics and robotics within the Department of Informatics [26].
- The University of Southhampton offers a master's degree in nanoelectronics and nanotechnology within the Department of Electronics and Computer Science [27].

- The University of Technology in Dresden offers a master's degree in nanoelectronic systems within the Department of Electrical Engineering and Information Technology [28].

For aspiring trainees in this field, the availability and quality of academic and research offerings are critical to success; however, their success may also depend on the availability and scope of support from external initiatives outside of their home university. External support for applied research from private companies is available, but governmental and nongovernmental organizations provide the bulk of the funding and infrastructural support. The European Doctoral Training Support in Micro/Nano-electronics (EURO-DOTS) was one such external initiative supported by a special European Union research funding program [29]. EURO-DOTS provided a central virtual platform that offered quality advanced courses in micro/nanoelectronics to fulfill accreditation criteria [29]. In addition, Ph.D. students taking the courses provided were given scholarships [29]. The funding for this program ceased after the second phase in 2014, despite its successes; however, there are efforts in progress to reinstate the program [29].

While EURO-DOTS provided supplementary coursework and scholarships to help strengthen foundational knowledge in nanoelectronics, the American-based National Nanotechnology Coordinated Infrastructure (formerly National Nanotechnology Infrastructure Network) funded by the National Science Foundation provides an extensive network of nanotechnology laboratories and resources to support nanotechnology research, including in nanoelectronics [30]. There are currently a total of 22 laboratory and education partners located throughout the United States of America that are funded by this program [30]. Additionally, the National Nanotechnology Initiative is a US government program established in 2000 that brings together over 20 federal research agencies to support nanoscale research programs around the country [31].

26.3
Challenges in Nanoelectronics Education

As nanoelectronic devices, concepts, and systems become increasingly more relevant to all areas of technology, addressing the question of how to educate the next generation of engineers and scientists who specialize in nanoelectronics is a pressing matter. The most likely scenario is that a variety of academic programs and philosophies will evolve to meet the needs of specific industries. This section aims to describe some of the main challenges in nanoelectronics education moving forward, and the following sections will propose specific solutions to some of these challenges.

26.3.1
Bridging the Disciplines

The interdisciplinary nature of the field presents the largest challenge in building educational programs in "nanoelectronics." The definition of the field itself is often broken down into three inter-connected subtopics, with intellectual content traditionally located in physics, chemistry, electrical engineering, computer engineering, materials science, and mathematics departments. At the undergraduate level, requiring students to build competency through classwork in all of these areas is a scheduling and time-constraint challenge.

Most bachelors-level engineering degrees include a core science requirement that consists of general chemistry, physics through mechanics and electricity and magnetism, and math through multivariable calculus, differential equations, and linear algebra. One potential solution to the interdisciplinary challenge is for programs to develop tailored versions of these classes with a focus on the topics most relevant to an education in nanoelectronics. For example, a general chemistry class could replace deep dives on certain topics (kinetics, bonding) with an emphasis on physical and nanochemistry including quantum mechanics.

Bridging the gap between different disciplines involves several practical challenges besides the time-constraints involved with requiring students to take many introductory-level classes. Faculty from different departments must make a concerted effort to work together to create an integrated curriculum. The lexicon in different academic disciplines is varied which can create confusion among students. This problem is particularly acute when considering some of the new fields being integrated into the nanoelectronics sphere, such as the biological sciences. Finally, there is the question of the relative value of breadth versus depth in an undergraduate education; depth within a particular discipline is difficult to achieve if students are required to gain competency in a number of engineering and physical science fields.

26.3.2
Theory versus Practice in Classwork

Another challenge facing degree programs in nanoelectronics is the balance between offering courses with a focus on theory, computation/modeling, and practical laboratory work. Ideally, these three types of learning focuses complement each other; for example, a course on integrated circuit design could include theoretical circuit theory, computer-aided simulation and layout, and practical fabrication modules. In practice, however, it is often difficult to include all three facets with enough depth in a single semester-long course. Therefore, students may be required to take classes that focus on only one area at a time, which can impede learning progress and lead to disconnects between the theoretical and practical aspects of a subject. One potential solution is to design year- or multiyear-long courses that integrate theory and practice. A potential drawback to this solution is a loss of flexibility in scheduling.

26.3.3
Resource Availability

It is impractical for most universities to build elite programs across many subspecialties. Therefore, resource availability is a challenge for interdisciplinary subjects such as nanoelectronics. For example, smaller universities may not have the funding or personnel resources to develop both strong schools of engineering and biological sciences.

A comprehensive education in nanoelectronics requires not only expertise across multiple disciplines but also access to a variety of infrastructural resources as well. Primary among these are cleanroom fabrication facilities that include expensive and difficult-to-maintain equipment.

One potential solution to the personnel and infrastructure resource availability challenge is for multiple universities to establish partnership programs for interdisciplinary degree tracks such as nanoelectronics. Pooling resources can be an efficient method of compensating for specific areas of need; however, multiuniversity degree programs come with the drawback of management complexity.

26.3.4
New Applications

Nanoelectronics technology is increasingly being applied in areas outside of the traditional electronics domain, and educating students to apply their skills in these new areas is a challenge. Examples of exciting new nanoelectronics applications include the fields of health care, biological sensing, and energy. These new applications are an extreme example of the interdisciplinary challenge, and the healthcare example will be explored in-depth in the following section.

26.3.5
Industry and Translation

The field of nanoelectronics is currently driven by industrial applications. The rapid progress in technological advancement has been guided by industrial research as well as new applications. Developing educational programs with the ability to adapt to current industrial needs is a challenge. Additionally, universities have struggled to bridge the gap between discovery and innovation to real-life translation, and students graduating from nanoelectronics-related degree programs often lack knowledge in this area.

It is therefore crucial that academic programs maintain contacts with industry in order to educate students with competencies matching the needs of nanoelectronics employers. Integrating internship or co-op programs, often involving university alumni, into degree programs are one potential solution to this challenge. Additionally, classes on translation should complement basic science and engineering classes in nanoelectronics.

Beyond educating students as the future industrial workforce, evaluating the interconnected roles of education, research, and industrial applications in the field of nanoelectronics is a challenge. Industry has undergone a paradigm shift in the past decades in which industrial research centers have been downsized or consolidated [32]. This has shifted the burden of performing "industrial" research to university labs, many of which have traditionally focused on basic scientific questions as opposed to applied projects. This challenge represents a potential opportunity for partnerships between industry, academia, and government to define research collaborations involving students and to innovate new funding models. Cosupervision of students, who could be given both university and industrial personnel mentors, is one option for integrating research and industrial application into a nanoelectronics education.

A further challenge involving the role of industry in a nanoelectronics education is defining the specific skill set that students need to fill jobs in the modern economy. Universities are often primarily equipped to prepare students for a career in research; some of the skills students acquire are applicable to both industrial and academic careers, but there may be some mismatch. Following is an example list of some of the key attributes that students should acquire for successful industrial careers:

- The ability to work in interdisciplinary teams.
- The ability to quickly tackle new problems and develop plans to solve them.
- The ability to communicate with different levels of management and technical personnel.
- "Self-starting" ability; that is, the ability to take independent initiative.
- The ability to stick to a defined timeline and budget.
- The ability to stay current with the latest techniques, tools, and research in the field.
- Literacy in scientific computing and software.
- Strong oral presentation and writing skills.
- The vision to balance between long-term strategic thinking and short-term goals.

26.3.6
Degree Levels

The challenges faced in designing comprehensive bachelors-level degree programs are different from those in designing masters- and Ph.D.-level programs. Bachelors programs are typically more standardized across different universities, and they have the primary goal of preparing students with a broad background across nanoelectronics-relevant disciplines for jobs in industry or for beginning research. Masters programs tend to involve more specialization in one chosen area or subfield, and can focus on systems, design, fabrication, or theory.

Ph.D. programs, on the other hand, are more varied across different universities and often involve small numbers of students. This makes offering a large number of classes at the Ph.D. level difficult, so the summer school model involving students from multiple universities can be used to fill in gaps in Ph.D.-level education. The question of whether nanoelectronics curricula should be standardized for all students or tailored toward specific applications can be answered differently at the different degree levels.

26.3.7
Cultural Challenges

In addition to educational and logistical challenges associated with building educational programs, there are several cultural challenges, both specific to the field of nanoelectronics and more broadly relevant to education in the science, technology, engineering, and mathematics (STEM) fields. As the nanoelectronics industry becomes more globally distributed, the question of how to educate students for an international workplace should be addressed at all degree levels. Study-abroad programs have been slow to gain traction at the bachelors level in the STEM fields, but international collaborations and partnerships are more common at the Ph.D. level where students may have the chance to attend conferences or workshops abroad. An ideal educational program would expose students not just to European and North American perspectives, but would incorporate the emerging nanoelectronics centers in Asia, Africa, and Latin America as well.

Nanoelectronics, and the STEM fields, in general, have been traditionally male-dominated disciplines with low participation from members of under-represented minorities (URMs) [33]. Programs in nanoelectronics education have the potential to remedy and address this challenge by incorporating specific outreach efforts and cultural changes designed to increase participation by women and URMs. Instituting policies that promote work-life balance, forming partnerships with primary and secondary schools and historically black colleges and universities (HBCUs), and aggressively pursuing Title IX [34] cases (in the United States of America) are examples of programs that have demonstrated some success in other areas [35]. Education programs in nanoelectronics should follow suit in order to build a competitive twenty-first century industrial workforce.

26.4
New Cross-Discipline Applications

One example of an exciting new application of nanoelectronics relevant to the twenty-first century economy is in the health care industry. In this section, we take health care applications of nanoelectronics as a case study to illustrate some of the challenges faced in educating students in an increasingly interdisciplinary area.

Nanoelectronics technologies are widely used in the health care and biomedical fields, for applications such as sensing and monitoring, electronic implants, and drug delivery [36]. The interest in applying general engineering techniques and technologies to medical applications led to the creation of the field of biomedical engineering and the rapid expansion in degree programs in that area [37]. An educational program focusing on nanoelectronics for health care applications would face similar challenges, including bridging the gap

Table 26.1 Sample experiment, characterization, and data analysis techniques by primary discipline (biological sciences versus physical sciences and engineering).

Biological sciences	Physical sciences and engineering
• PCR	• Photolithography
• Electrophoresis	• Thin film deposition
• Cell transformation	• Thermal annealing
• Chromatography	• Chemical and physical etching
• Microscopy, for example, optical	• Circuit layout and CAD
• Flow cytometry	• Microscopy
• DNA sequencing	• X-ray diffraction
• Bioinformatics/computational biology	• Profilometry
• RNAseq (expression analysis)	• 2- and 4-point probe, capacitance–voltage,
• Epigenetics	Hall effect profiling

between traditional science and engineering curricula, and teaching nanoelectronics experimental techniques to students with biology training and vice versa.

Assuming most students interested in health care applications of nanoelectronics have a background in either the biological sciences, traditionally associated with degrees in health care, or engineering/physical sciences, traditionally associated with degrees in nanoelectronics, there are specific challenges to bridge the two fields. Table 26.1 summarizes some of the experimental techniques and data analysis tools specific to each discipline.

As Table 26.1 illustrates, bridging fields such as healthcare and nanoelectronics is difficult, not only because of the different jargon and approaches used but also because of the sheer quantity of techniques and methods that may be important for expertise. Students in such programs should understand the important techniques in each field without necessarily having to master all of them. Instead, they are likely to be introduced to these methods through basic lab-focused biology and electronics/materials science courses. Additionally, there are some techniques common to both fields; for example, electron and optical microscopy and fluorescence spectroscopy are often used in both the biological sciences and physical sciences/engineering. As cross-discipline applications become more common, the importance of teaching collaboration and teamwork skills is apparent, given the constraints on the number of techniques individual students can reasonably be expected to master over the course of their degree programs.

26.5
Future Education Programs

As future education programs in nanoelectronics are developed to address twenty-first century needs, it is important to keep in mind the objectives in doing so. First and foremost, educational programs should be designed to facilitate the growth and relevance of the foundational industry sector of nanoelectronics so that it can adequately address the grand challenges of today and tomorrow. This will require extensive partnering among universities, research organizations, industry, government, and policy makers. A robust public–private partnership model will be essential to efficiently educate students and bring to bear the fruits enabled by nanoelectronics across industry and society. With this in mind, the overarching goal of this section is to define a nanoelectronics roadmap to educate future engineers with a holistic, multidiscipline viewpoint. Educational programs should be based around competencies required to generate solutions that can adequately address industry and societal needs. Following are the three essential requirements:

1) A curriculum that extends beyond core nanoelectronics courses to include cross-discipline subjects. The required knowledge base should include competencies in cultural and commercial integration in order to prepare students for real-life utility.

2) Attention to both depth and breadth in a nanoelectronics-related education and research framework. Specific areas of focus and further in-depth specialization can be informed by the applicable industry sector of focus (e.g., Healthcare, Communications, or Energy).

3) In addition to preparing students for a research and development-focused career, more attention should be given to imparting familiarity with IP management, technology transfer, business intelligence, commercial development, and policy. These "soft-skills" elective subjects will give students more flexibility as they enter the workforce.

Building the types of partnerships needed to implement future nanoelectronics programs of study is probably the most challenging aspect of these recommendations. Therefore, the field should initially focus on the types of partnerships necessary to create a robust educational platform that is well-informed by the research and development needs of industry. Public–private partnerships in which all stakeholders come together to share challenges, resources, and infrastructure for mutual benefit must be developed. The potential educational partnerships can take various forms, including bilateral or groups of universities (academic consortia), and similarly among academic, industrial, and government institutions. The major players in such a partnership will include universities, industry, and governments as follows, each of which has a specific role to play in nanoelectronics education:

- *Universities and Research Organizations* should focus on offering educational programs with breadth and depth in nanoelectronics and associated disciplines. Their course offerings can provide basic literacy in nanotechnology and an understanding of industrial grand challenges with competencies in basic science, engineering, and translation skills. In-depth interactions with industry early in the educational cycle will prepare students for problem solving and innovation at a professionally relevant level.

- *Industry* should facilitate these early interactions to achieve its goal of establishing a pipeline of highly talented, sector-relevant research and business leaders. It can do more to influence the shaping of research programs to address under-served needs by direct interactions and financial assistance. This mutually beneficial relationship will provide industry with early access to and direct contact with scientific discoveries and the innovators behind them.
- *Government and Global Society* should be involved in setting priorities in order to maintain an informed population and keep up with evolving societal needs. Notably, a well-trained, multidiscipline workforce can advance innovations for real-life utility and global benefit. Governments can enact education and industry policies that assure a functional and cooperative education ecosystem.

As plans for implementing a roadmap centered on the university as the core educator develop, specific scenarios can be tested. In the final sections of this chapter, we highlight two example plans, as diagramed in Figure 26.2, which could serve as a starting basis for new comprehensive programs in nanoelectronics. Although not specifically included in the figure rubrics, both scenarios should include "soft-skills" (nontechnical) electives as discussed earlier.

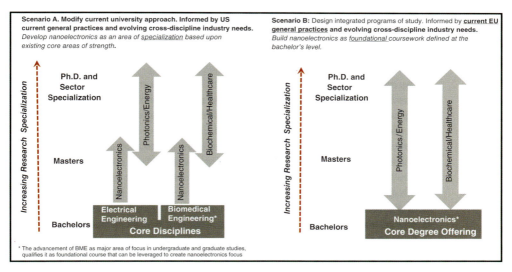

Figure 26.2 Two example scenarios for future comprehensive university programs in nanoelectronics. Scenario A is a step-wise approach to education with focus on specialization in nanoelectronics later in the education timeline starting with some key courses at the master's level. Scenario B integrates key coursework to create a cross-discipline nanoelectronics education foundational platform at the bachelor's level. This provides students with a holistic understanding of relevant disciplines early on, and more time to meaningfully identify areas of specialization that are better suited to address industry's underserved needs. As such, students have more time and knowledge to mature to industry-level expertise. See Figure 26.3 for more details on Scenario B.

26.5.1
Scenario A: Modification of Current University Approach

The first example plan, referred to as "Scenario A," is designed to leverage existing areas of university strength at the bachelor's level and access in-depth area-specific competencies through partnerships beginning at the master's level. This plan has been developed with the US educational system in mind. Bachelor's programs that confer degrees specifically in nanoelectronics are rare in the United States of America; students interested in this field instead obtain more general degrees in the traditional engineering fields, with electrical engineering being the most common.

Scenario A's approach is for universities to identify the programs related to nanoelectronics in which they are strongest and build on these strengths. Undergraduates would continue to major in the traditional disciplines and could then specialize in nanoelectronics at the master's level. This would require cooperation among different universities, with each leveraging their own discipline-specific programs of strength, to establish a common basis to facilitate student exchange and an accrediting system.

Large university systems could offer selections of classes at the bachelor's level between different departments. Included in the program would be some level of industry engagement, such as student exchanges, university–industry cooperative programs, or formal internships. Incorporating experimental lab work with theory classes early on during the Bachelor's degree (year one or two of study) would be a critical element in this plan.

At the master's level, partnerships between different departments, schools, and universities could be formed where needed, including deeper levels of industry engagement. Cooperatively educating master's students would facilitate university and industry alignment on identifying and solving industrial grand challenges while maintaining basic academic research goals and discoveries.

At the Ph.D. level, where students typically engage in a deep and specialized thesis topic, formal partnerships could be formed with industry to identify relevant projects. These partnerships could include shared research and involve a co-supervision element, if desired.

The main advantage of Scenario A is that it offers a short-term and easily-implementable path for building a viable nanoelectronics education program. It can be implemented immediately on top of the existing university educational framework via the formation of academic and industrial partnerships. The main drawback is that it lacks the comprehensiveness of a long-term plan for meeting future educational needs. It assumes maintenance of the current model for US bachelor's programs and classes. Students would not necessarily get the breadth in building blocks early on that would later be required to serve multiple industrial needs.

26.5.2
Scenario B: Comprehensive Nanoelectronics Education System

The second example plan, referred to as "Scenario B," is an integrated program of study, designed to provide a comprehensive, forward-facing education in nanoelectronics starting from the bachelor's level. It is informed by current European general educational practices [38]. The goal is to implement an integrated program of study focused on all of the core disciplines involved in the nanoelectronics space plus a few outside electives.

Figure 26.3 illustrates potential class areas to be included in Scenario B. At the undergraduate level, comprehensive courses could be taken during the first two-to-three years of study, with a focus on further specialization within a specific discipline during the third and fourth years in a traditional bachelors program. A narrower focus is usually required at the masters and Ph.D. level as described for Scenario A.

Scenario B has some of the same basic tenets as Scenario A, including integrating experimental lab work early in the undergraduate program, offering non-technical soft-skills classes in the curriculum, and emphasizing industrial and government partnerships with academia. This scenario may be more immediately applicable to European universities, some of which already have specific degree programs in nanoelectronics or related disciplines. Scenario B is a strategic fit for the European Union's Bologna Model [38] that seeks to standardize education across the entire continuum of degree levels (bachelors, masters, and Ph.D.) and regions.

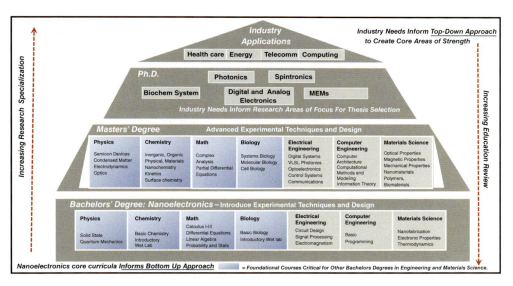

Figure 26.3 Standardizing holistic education in nanoelectronics based on industry needs. Recommended comprehensive courses that would confer a nanoelectronics degree at the bachelors level with a sampling of courses at the master's level and/or Ph.D. level. At master's level and to a lesser extent the Ph.D. level, the area of specialization starts to become more important for course selection followed by a large research effort aimed at generating a thesis. Ultimately, the Ph.D. thesis should focus around a key area of specialization and matching application area of strength combined with industry/academia under-served needs.

The main advantage of Scenario B is that it provides an integrated curriculum from the beginning of the undergraduate education by leveraging traditional classes in different areas. Over time, it provides for further specialization as dictated by industrial and research challenges. The disadvantage is that implementing such a system, especially in the United States of America, will require traditionally disparate departments to work together on an overhaul of the curriculum, which may take some time.

Acknowledgments

The authors have accessed insights and opinions from many people and appreciate their contributions. We would like to thank in particular Marcel Van de Voorde, Livio Baldi, Gilbert Declerk, Johan Van Helleputte, Jo De Boeck, and Stuart Ray for their insight and suggestions.

References

1 Bruun, E. (2008) Provision of a European training infrastructure. EuroTraining Project Report, ET–D2.1.
2 Bruun, E. and Nielsen, I.R. (2013) Trends in university programs in nanoelectronics and microsystems. NORCHIP, 1–6. doi: 10.1109/NORCHIP.2013.6702006
3 NUS Nanoscience and Nanotechnology Institute (NUSNNI). Available at www.nusnni.nus.edu.sg/ (accessed 2016).
4 HKUST. Nano science and technology program. Available at nanoprogram.ust.hk/intro.html (accessed 2016).
5 Institute of Microelectronics, Tsinghua University. Available at http://www.tsinghua.edu.cn/publish/imeen/5943/index.html (accessed 2016).
6 Kaist Institute. Available at kis.kaist.ac.kr/ (accessed 2016).
7 NCRC. Available at http://www.ncrc.iis.u-tokyo.ac.jp/e/ (accessed 2016).
8 Times Higher Education (THE) (2015) Subject Ranking 2013–2014: Engineering & Technology. Available at https://www.timeshighereducation.com/world-university-rankings/2014/subject-ranking/engineering-and-IT.
9 Massachusetts Institute of Technology. nanoEECS index. Available at http://www.eecs.mit.edu/nanoeecs/index.html (accessed 2016).
10 Stanford University. Research and Faculty. Available at http://engineering.stanford.edu/node/36356 (accessed 2016).
11 BNNI. Available at http://nano.berkeley.edu/welcome/welcome.html (accessed 2016).
12 Harvard, John A. Paulson School of Engineering and Applied Sciences. Available at http//www.seas.harvard.edu/ (accessed 2016).
13 ETH Zurich. Available at https//www.ethz.ch/en.html (accessed 2016).
14 The Nanoscience Centre. Available at http//www.nanoscience.cam.ac.uk/ (accessed 2016).
15 Imperial College London. Available at http//www.imperial.ac.uk/nano-at-imperial (accessed 2016).
16 Nano Science & Technology. Available at http//www.eng.ox.ac.uk/nst (accessed 2016).
17 Kavli Nanoscience Institute at Caltech. Available at http//kni.caltech.edu/ (accessed 2016).
18 California NanoSystems Institute. Available at http//www.cnsi.ucla.edu/ (accessed 2016).
19 Kavli Institute of Nanoscience Delft. Available at kavli.tudelft.nl/ (accessed 2016).
20 KU Leuven, Master of Nanoscience, Nanotechnology and Nanoengineering – Groep Wetenschap & Technologie.

Avaialble at http://set.kuleuven.be/ nanotechnologie/eng (accessed 2016).

21 Politecnico di Milano, NanoLab. Avaialble at http://www.nanolab.polimi.it/index.html (accessed 2016).

22 Politecnico di Torino, DET Department of Electronics and Telecommunications. Available at http://www.det.polito.it/ (accessed 2016).

23 TUD, Forschung. Available at https://tu-dresden.de/forschung/ (accessed 2016).

24 nanoTUM. Available at http://www.nano.tum.de/index.php?id=5 (accessed 2016).

25 INP, G. Grenoble INP - Master Micro and Nano Technologies for Integrated Systems - Nanotech. *Grenoble INP.* Available at http://www.grenoble-inp.fr/engineering-degrees/master-micro-and-nano-technologies-for-integrated-systems-nanotech-482448.kjsp (accessed 2016).

26 University of Oslo, Informatics: Nanoelectronics and Robotics (master's 2 years). Available at http://www.uio.no/english/studies/programmes/inf-nor-master/ (accessed 2016).

27 University of Southampton, MSc Nanoelectronics and Nanotechnology, Electronics and Computer Science (ECS). Available at http://www.ecs.soton.ac.uk/programmes/msc-nanoelectronics-and-nanotechnology (accessed 2016).

28 TUD, Master's Program Nanoelectronic Systems. Available at http://www.et.tu-dresden.de/etit/index.php?id=master-nes (accessed 2016).

29 EURO-DOTS. Available at http://www.eurodots.org/ (accessed 2016).

30 NNCI. Available at http://www.nnci.net/ (accessed 2016).

31 Nano. Available at http://www.nano.gov/ (accessed 2016).

32 Usselman, Steven (2013) Research and development in the United States since 1900: an interpretive history. Economic History Workshop, Yale Univ.

33 National Science Foundation (2015). Women, Minorities, and Persons with Disabilities in Science and Engineering. Available at http://www.nsf.gov/statistics/wmpd/ (accessed 2016).

34 TitleIX.info. Available at http://www.titleix.info/ (accessed 2016).

35 Estrada, M. Ingredients for improving the culture of stem degree attainment with co-curricular supports for underrepresented minority students. Available at http://sites.nationalacademies.org/cs/groups/dbassesite/documents/webpage/dbasse_088832.pdf (accessed 2016).

36 Cohen-Karni, T., Langer, R., and Kohane, D.S. (2012). The smartest materials: the future of nanoelectronics in medicine. *ACS Nano*, **6**, 6541–6545.

37 Biomedical Engineering Society. Available at bmes.org/history (accessed 2016).

38 The European Higher Education Area (2015) Bologna Process Implementation Report. Education, Audiovisual and Culture Executive Agency.

27
Conclusions

Robert Puers,[1] Livio Baldi,[2] and Marcel Van de Voorde[3]

[1]*Catholic University Leuven, Elektrotechniek - ESAT-MICAS, De Croylaan, 3001 Leuven, Belgium*
[2]*Formerly technology development manager at STMicroelectronics, Milano – Independent consultant on Nanoelectronics*
[3]*University of Technology Delft, Rue du Rhodania, 5, BRISTOL A, Appartement 31, 3963 Crans-Montana, Switzerland*

In the previous chapters, we have tried to give an overview of the present status of nanoelectronics, its challenges, its foreseen evolution, and its wide field of applications. However, in spite of all the care given to the choice of topics and of the competence of the authors, the picture given in this book cannot claim to be complete, because of the fast evolution of the sector. Claiming to say a final word in a field, where a new technology generation appears every two years, is unrealistic, and perhaps even arrogant.

Fifty years of continuous evolution of microelectronics, as well described in the contribution of Paolo Gargini, have introduced products that have completely changed our life. The move from microelectronics to nanoelectronics has simply accelerated and extended the process, drastically reducing cost of logic devices and making it possible to incorporate intelligence in a variety of applications.

The rate of progress in this period has been astonishing. In 1959, at the annual meeting of the American Physical Society, Nobel Prize winner Richard Feynman gave a talk that is considered as the foundation act of nanotechnology (and nanoelectronics). He asked "Why cannot we write the entire 24 volumes of the Encyclopedia Brittanica on the head of a pin?," and went on, giving an explanation of how it could be done. Only 6 years later, Gordon Moore, looking back at the first 3 years of history of integrated circuits, made the observation that complexity had been growing at a rate of a factor two per year, and stated that "there is no reason to believe it will not remain nearly constant for at least 10 years." It was the seed of the so-called Moore's Law. Fifty years later, integrated memory modules are available that squeeze 256 GB of flash memory in a footprint of about 1 cm^2, which means about 4.5 GB, or more than 20 000 pages on the head of a pin "a sixteenth of an inch across," as indicated by Feynman. What was looking as if a provocation has become reality.

Nanoelectronics: Materials, Devices, Applications, First Edition. Edited by Robert Puers, Livio Baldi, Marcel Van de Voorde, and Sebastiaan E. van Nooten.
© 2017 Wiley-VCH Verlag GmbH & Co. KGaA. Published 2017 by Wiley-VCH Verlag GmbH & Co. KGaA.

In the same paper of 1965, Gordon Moore made the half-joking remark that one day "handy home computers" could be sold in general stores along with cosmetics. A quite daring forecast, since the IBM360, a typical computer of that time was filling two large rooms. Today we have in our smartphones many times the computing power and the memory of those mainframe computers, at a small fraction of the cost, and with added functionalities, like connection capability, high resolution cameras, music and video reproduction, navigation, and plenty of services.

This example shows also how the semiconductor technology, originally developed for processor and memories, has been capable to extend to other fields, allowing the low cost realization of sensors, microelectromechanical systems (MEMS) and solid-state power devices.

In parallel, the waves created by the continuous advance of microelectronics and nanoelectronics have extended to a variety of sectors, far removed from the original computing and communication fields where they have started.

In this book, several chapters have been devoted to the application of nanoelectronics in different domains of our everyday life. In some areas, like automotive, nanoelectronics is now a consolidated presence. The semiconductor value in cars is constantly increasing, and could reach 50% for full electrical cars, while it is evident from advertisement that the added value of new models is essentially based on nanoelectronics-enabled features. In other areas, like environment, food, smart cities, the role of nanoelectronics is less prominent, yet constantly gaining importance, mostly mediated by data collection, communication, analysis, and modeling.

However, as it is clear from the chapters dedicated to applications, we are still at the beginning of the wave of change that nanoelectronics is bringing about. What has been achieved until now, except perhaps in the fields of communication and social life, has been mainly replacement innovation, substituting older mechanical or electrical control systems and sensors with more efficient and lower cost Nanoelectronics-based ones. The full exploitation of the potential of nanoelectronics will be felt when systems will be designed around the capabilities made available by it. The situation is similar to what was happening at the beginning of the car industry, when the first cars were just carriages, with an engine replacing the horse. Progress really started when cars were designed around the engine.

A similar picture can be drawn for health applications: nanotechnology currently enables the realization of very complex and computational intensive medical systems (e.g., MRI tomography, etc.) and extreme device miniaturization leads to new applications with a plethora of ultralow power wearable devices that will improve our quality of life and health. We saw a glimpse of such solutions in the health chapter where the focus was on wearables. However, further innovation driven multidisciplinary research will create also here – similar to what has happened in the automotive sector – a completely new family of – currently nonexisting – devices and systems. As only one of many possible examples, complex clinical lab functionality available at the GP office or even at home is becoming already reality step by step. In this way technology integration can enable a better and affordable future health care both in developed and developing countries.

Of course, most of the considerations about the potential use of nanoelectronics have been based on the present status of the technology, as extensively illustrated in dedicated chapters. But there is no reason to think that the impressive rate of progress that has characterized nanoelectronics in the last 50 years will stop now. In all chapters, experts have indicated the present challenges, and the directions in which research is moving to overcome them. The end of the Moore's Law has been announced several times in the past, but engineering solutions have always been found around it. Even if now fundamental limits appear to be near for conventional semiconductor technology, probably more economic than physical, we have only to think about the performance in terms of computing power, computing density and low power that is offered by the human brain and by the nervous system of animals, to understand that there is still much space for progress and innovation in front of us.

Probably a more disruptive change of approach is needed. New materials, new physical properties, new system architectures will be required, but the world of nanoelectronics industry and research is well equipped to match the challenges. When CMOS technology started, it made use only of five–six elements, now more than half of the periodic table is finding use in the industry. The innovation potential of industry is huge and is supported by investment rates in R&D that exceed by far those of other sectors, except perhaps biotechnology.

New cross-functional education paradigms will be needed to form the next generation of innovators in the field of nanoelectronics, and Europe can play an important role in it, thanks to the presence of a network of advanced research centers that can be considered as world leaders in their fields. The importance of nanoelectronics has been recognized by the European Commission that has listed it among the key enabling technologies (KET) for future economic and social development. Dedicated research initiatives have been started, trying to involve European Commission, member states, and industry in a common effort toward the focusing of resources on a common innovation program, with the ambition of covering the full development chain, to transform research into innovation and social benefits. Like for all pioneering initiatives, constant monitoring and revision will be needed, to learn from mistakes, to remove bottlenecks, and improve efficiency. International cooperation initiatives are also taking form between different regions of the world, because the challenges that nanoelectronics is called to solve, such as green energy, environment protection, support to an aging population, are also global and require the broader possible cooperation on research.

In this book, we have tried to give an overall picture of the status of nanoelectronics and of its applications. It is by necessity a moving picture: it will be an excellent source of information not only for students but also for managers and policy makers. We hope it will stimulate creativity and new approaches for research and innovation, on technology and more importantly on the potential applications of nanoelectronics in a variety of fields, to solve social and economic challenges.

Index

a
acceleration energy 272
acceleration sensors 493, 510
accelerometer structure 209
AC–DC-converter 482
acid etching 234
actinic blank inspection 296
active matrix (AMOLED) 516
actuators 55, 214, 402
– artificial muscles 216
– electrostatic, electromagnetic, and piezoelectric 214
– pneumatic, phase change, and thermal actuators 216
adaptive cruise control (ACC) 493
adaptive power converters 483
ad hoc process 342, 369
– algorithms 352
– disposable sensors 544
Advanced Communication Technologies and Service (ACTS) 573
advanced driver-assistance system (ADAS) 155
advance plasma doping approaches 275
aging-aware logic synthesis 348
agrifood sector 451
Airbus of Chips 554
air quality 540, 541
air traffic management 383
aka Moore's law 317
ALD benefits 278
– composition control 278
– conformality 278
– thermal budget 279
AlGaN/GaN system 187
– buffer 187, 194
– interface 186
algebraic transformations 338

aluminum
– diffusion of 9
– diode formation 11
aluminum-doped zinc oxide (AZO) 518
aluminum penetration 30
ambient assisted living (AAL) 540
Ambient Intelligence Systems and Assistive Environments 576
ambipolar silicon-nanowires 334, 350
amplified resists 298, 313
analog integrated systems 375
analog-to-digital converter (ADC) 148
anamorphic imaging techniques 301
ancillary activities 434
angular rate errors 422
angular rate sensors 422, 423
anisotropic etching techniques 208
annealing 191
anodic bonding 416
anomalous diffusion 274
apnea-hypopnea index events (AHI) 463
Apple A8 chip 73
Apple II 24
Apple iPhone 6, 510, 516
application-specific integrated circuit (ASIC) 398, 572
approximate logic synthesis (ALS) 352
– SALSA, template uses 352
APT. *see* atom probe tomography
architectural decomposition 378
architecture analysis 385
architecture design 387
Ar ions beam processes 282
artificially constrain 371
artificial molecular motors 218
artificial systems 205
aspect ratio trapping (ART) 104
assisted living 460

Nanoelectronics: Materials, Devices, Applications, First Edition. Edited by Robert Puers, Livio Baldi, Marcel Van de Voorde, and Sebastiaan E. van Nooten.
© 2017 Wiley-VCH Verlag GmbH & Co. KGaA. Published 2017 by Wiley-VCH Verlag GmbH & Co. KGaA.

Atomic force microscopy
– nanoprobe sweeping 270
atomic layer deposition 277
atomic layer epitaxy 277
atomic layer etching (ALE) 284, 285
atom probe tomography (APT) 246, 250
Auger electron spectroscopy (AES)
– depth profiling 258
autarkic sensor nodes 539
autofocus camera modules 399
autofocus sensing 464
automata-based formalism 381
automated cyber-physical systems 489, 490
– societal challenges 490
automated theorem 383
automatic control 364
automatic identification and data capture
    (AIDC) 532
automatic logic synthesis 352
automatic test equipment (ATE) 433
automatic test pattern generation (ATPG) 431
automation 373
– automotive electronics 57
– automotive embedded systems 385
autonomous driving 57
autonomous vehicles 593

b
bachelors programs 603, 608
back end of line (BEOL) 126
back-end processing 267, 489
back metallization 175
backside analysis 258
ballistic transistors 101
band-to-band tunneling (BTBT) 143
BASIC language 25, 26
batteries
– energy management of 164
– operated devices 459
beam line implanters 274
behavioral modeling 370
– SiNW and graphene p–n junction 356
Beyond CMOS 583
bi-directional energy flow 483
Big Data 452, 494
Bill Gates 26
binary decision diagram (BDD) 356
biochemical signals 205
biochips 588
biomedical chips 57
biophotonic sensors 576
biosensors 148, 213
biosignal sensors 459

bipolar junction transistor (BJT) 7, 16, 131,
    176, 182
bipolar memories 131
bipolar technology 19
birefringence 290
bisimulation 388
32-bit adder, switching energy 44
bit cell 88
bit-cost scalable (BiCS) 119
BJT. *see* bipolar junction transistor (BJT)
blech effect 322
block copolymers 302, 309
blood cell analysis 463
bluetooth 503
body fluid assessment 461
body sensor network 458
bolometer detector 211
Boltzmann's constant 158
Boolean function 338
Boolean gates 438
Boolean network 338
– algebraic expressions of logic function 338
– model 338
– optimization 339
Boolean relations (BR) 352
Boolean techniques 341
boron Coulomb effect 236
boron diffusion 274
boron-implanted region 273
Bosch BMX055, 510
Bosch Sensor Swarm approach 532
BOX SOI substrate 94
BOX thickness 92
Bragg geometry 246
Bragg reflectors 295
braking energy 164
bridge fault model 430
British health care think tank 457
broadband terahertz detectors 161
BTI. *see* bias-temperature instabilities (BTI)
buffer-induced dispersion 197
built-in self-test (BIST) 432
– architecture 432
– logic 433
bulk transistor 89
buried-channel-array-transistor (BCAT) 117
buried oxide (BOX) 91

c
CAD. *see* Computer-aided design (CAD)
CAGR. *see* compound annual growth rate
    (CAGR)
call emergency services 461

cancer
– point of care 462
capacitances, in MOSFET structure 83
capacitive actuation 215
capacitive sensing 214
capacitors
– banks 472
– quality factor 159
carbon 232
carbon based RRAM 129
carbon dioxide emissions 541
carbon nanotube (CNT) 442, 504
carrier mobility 327
carrier transport 89
carrying gas 269
car-to-car links 57
car-2-x communication 542
cascode 192
cassettes 236
CATRENE organization 549
CAVET. *see* current aperture vertical electron
    transistor (CAVET)
CBD. *see* contract- based design (CBD)
CBL. *see* current-blocking layer (CBL)
CD. *see* critical dimension (CD)
cell displacement defect 441
cell footprint 119
cell library 338
cell misalignment defect 441
cell omission defect 441
cell organization 130
cell pitches 167, 169, 170
cell-to-cell interference 120, 123
cell-to-cell variability 121
center gate (CG) 354
central energy management 473
chalcogenide-based electrolyte (CBRAM) 127
channel bandwidth 154
channel capacity 154
charge carrier layer 75
charge coupled device (CCD) 513
– sensors 513
charge/discharge cycle 506
charge incorporation 191
charge-trapping layers 119
chemical and biological MtM devices 59
chemical instability 145
chemical–mechanical polishing 345
chemical microsystems 576
chemical reactive sputtering 282
chemical sensors 451
chemical vapor deposition (CVD) 147, 196,
    228, 275

chemisorption 285
chemoepitaxy 302, 303
chronic disease management 457
circuits 536
– densities 318
– design 317
– performance 317
– – driven requirement 321
cloud computing paradigm 467
clustered sleep transistor insertion 344
CMOS. *see* complementary metal-oxide-
    semiconductor (CMOS)
CMOSFET. *see* complementary metal oxide
    semiconductor (CMOS)
CNTFETs. *see* CNT field effect transistors
    (CNTFETs)
CNT field effect transistors (CNTFETs) 442
CO_2 emissions 482
cognitive technology 590
cold filament 4
colloidal silica 234
color electrochromic displays 518
commercial motion processing unit chips 511
Commission's Research Framework
    Programme 570
communication 56, 449, 452, 453, 503, 536
– devices 161, 567
– means 55
– network 449
– technologies 473
– with traffic lights 57
complementary metal-oxide-semiconductor
    (CMOS) 28, 39, 84, 102, 118, 122, 164, 165,
    167, 271, 351, 397, 510, 595
– based circuits and systems, types of variability
    in 345
– blocks 344
– challenges and opportunities 156
– device scaling prediction 76, 88
– image sensor technology 513
– logic
– – activities 575
– – device, low-power mobile and multimedia
    applications 88
– – speed performance 80
– nodes 78, 100, 159
– pixel 149
– projects, mapping of FP7 and H2020, 585
– sensor cleaning 270
– technologies 28, 39, 53, 77, 85, 98, 103, 292,
    334, 615
– – area and delay optimization 335
– – gate to channel capacitance 78

– – limitations of 437
– – multilevel optimization 337
– – sequential synthesis 339
– – speed performance metrics 78
– – switching delay formulation 78
– – two-level logic minimization 336
– terahertz circuit blocks 160
– transistors 50
complex sensor fusion algorithms 494
component compatibility 377
component consumption
– by market segment 560
compositional analysis 257, 386, 389, 390
compound annual growth rate (CAGR) 489
computational complexity 384
computational imaging 211
computational power 22
computation tree logic (CTL) 383
compute clusters 364
computer-aided design (CAD) 333, 572
– commercial tools 343
computer engineering 597
concurrency 367
conductive bridge (CBRAM) 127, 128
conductive filament (CF) 126
congestive heart failure (CHF) 461
connected vehicle 491
consumer applications 57, 501, 502, 508
– D materials/flexible electronics 502
contact lens
– integrated LCD display 465
contamination reduction 268
continuous abstraction 382
continuous optimization engine 387
contour metrology 300
contract-based design (CBD) 373
contract composition 377
control algorithms 365
control design 388
controllable-polarity (CP) 443
controller equations 388
controller synthesis 388
control theory 598
control units, additional requirements 495
convergence applications 115
cooling filament 4
CoolSiC 180
co-op programs 602
copolymers 302
Core Partners Program 38
Coriolis effect 210
correctness of connections 369
cosimulation 389

cosmic ray 182, 325
cost-effective ICs 18
Coulomb interaction 77
coupling-faults, definition 442
covalent bonding 238
CPS. *see* cyber physical system (CPS)
CPU 74, 113
critical dimension control, plasma etching
    limitation 281
crossbar array architectures 441
cross-cutting paradigm 373
cross-discipline applications 605
cross-programme action (CPA) 574
cryogenic 269
crystallization process kinetics 254
crystal-originated particles (COPs) defects
    236
CTL. *see* computation tree logic (CTL)
Cu conductors
– reliability 320
Curie temperature 251
current aperture vertical electron transistor
    (CAVET) 188, 189
current-blocking layer (CBL) 188
current-carrying capability 318
current density 321
current fitness trackers 465
current-induced magnetic field 123
current-limiting design rules 321
CVD. *see* chemical vapor deposition (CVD)
cyber physical system (CPS) 363, 364, 390, 473,
    489, 532, 533
Czochralski growth method (CZ) 228,
    229, 230

d

damascene trench barrier layers 318
3D analysis techniques 246
dangling bonds 5, 44
dark-field holography (DFH) 251, 252
dark silicon 50, 139
Dash neck method 231, 232
data analysis techniques 605
data dependencies 379
data/information chain 545
data processing 55
data security 546
3D chip stacking 423
DC power consumption 158
DE. *see* discrete-event (DE)
decentralized microgrids 471
deep trenches 172
defect density 306

defect engineering 273
defect level (DL) 429
defects 427, 443
defects per million (DPM) 429
Defense Advanced Research Projects Agency
    (DARPA) 32
2-DEG. *see* two-dimensional electron gas
    (2-DEG)
delay fault models 430, 433
demographic evolution 455
Dennard's equations 18
Dennard's scaling 34, 36, 39
– equations 241
– laws 18
– rules 20, 139
Department of Energy (US-DOE) 449
design for manufacturability (DFM) 346
design for testability 428
device under test (DUT) 433
DFH. *see* dark-field holography
DG/FinFET devices 94
3D horizontal cross-point™ memory 133
3D horizontal ReRAM
diborane 236
dielectrics 280
– constant materials 318
– electrostatic charging of 407
– etching chemicals 280
– parasitic charging of 405, 407
– porosity 319
– refill 171
– reliability 319
digital factories 545
digital functionality 54
digital information processing 58
digital integrated circuits 371
digitalized society 590
digital lifestyle 56
digital memory 114
digital mirror device 215
digital MOS circuit 53
digital power management 485
digital products 568
digital signal 55
digital technology 141
digitization 531, 537
dim chip 140
directed self-assembly (DSA) 292, 294, 302,
    303, 304, 307
discrete abstractions 388
discrete-event (DE) 382
discrete power devices 164, 165
disease monitoring 462

dislocation density 187
dispersion phenomena 195
display technologies 515
DMOS 165, 166, 169
3D NAND architectures 119
domain-specific languages (DSL) 382
dopants 147, 175, 232, 236, 271–273, 275
double-disk grinding (DDG) 234
double-gate devices 93, 443
double patterning 281, 321
– flow 291, 292
double-side polishing (DSP) 234
3D power scaling 48
3D printing 589
drain-induced barrier lowering (DIBL)
    80, 97
– *versus* Leff 94
– reduction 80
– TBOX, simulation of 92
drain outreach 38
DRAM 17, 26, 116, 131
– architectural methodology 18
– basic schematic 116
– cell capacitance 118
– cell capacitor 118
– demise 31
– devices 117
– die size trend 22
– unit volume production 25
drift resistances 165, 169
dry etching 280
– pattern transfer 314
drying techniques 269
DSA. *see* directed self-assembly (DSA)
2D scaling, physical limits 45
DTCO. *see* design-technology co-optimization
    (DTCO)
dual epitaxy 172
dummy-gate device, STEM image
    253
3D vertical RRAM structure 132
dynamic energy demand 483
dynamic or switching power 340
dynamic power 340
– for CMOS gate 340
– dissipation 85
dynamic toll systems 541

e

e-beam maskless lithography 309
echocardiography 461
EDA. *see* electronic Design design
    Automationautomation (EDA)

educational partnerships 606
EELS. *see* electron energy loss spectroscopy
E-ink, eReader products of 517
elastomer 399
electrical and thermal energies 539
electrical defects 271
electrical energy 472, 483, 535
electrical engineering 597
electrically sensitive hydrogel 217
electrical resistance 212
equivalent oxide thickness (EOT) 37
electric cars 541
electric/electronic systems 473
electric field failure time model 319
electric lighting 3
electric vehicles 163
– electrified trucks 489
electrochromic materials 518
electromagnetic forces 207
electromagnetic interference (EMI) 402,
    409
– risk of 407
electromagnetic wave 59
electromechanical switching 144
electromigration 322
electron-beam lithography 598
electron diffraction 252
electron energy loss spectroscopy (EELS)
    246
electron holography 259
electronic circuits 537
Electronic Components and System for
    European Leadership (ECSEL) 554
electronic control unit (ECU) 436
electronic design automation (EDA) 333, 365
– tools 427
electronic devices
– continued scaling of 370
– manufacturing 437
electronic grade polysilicon
– production of 228
electronic grade silicon 226
electronic industry 568
electronic readers (eReaders) 517
electronic security systems 546
electronic skin 459
electron tomography (ET) 246, 250
electro-optic lenses 464
electrorheological fluids 216
electrostatic actuators 214
electrostatic discharges (ESD) 408
– risk of 407
electrostatic doping 355

electrowetting 518
embedded automation systems 489
embedded integrated circuits 409
embedded memories 133
– in microcontrollers and digital signal
    processors 124
embedded sensor fusion 363
emerging devices, silicon NWFET 355
emerging memories 127
– architectures 130
– devices 245
– storage class memory 133
– technologies 120
emerging wearables 459
EMI. *see* electromagnetic interference (EMI)
emitter pipes 19
E-mode 191, 192
EM simulations 161
encapsulations 409
end-of-manufacturing 436
energy bands 4
energy consumption 163, 471, 532, 536
energy dispersive X-ray spectroscopy
    (EDS) 246
energy efficiency 141, 149, 451, 520
– algorithms 536
– computing/sensing technologies,
    convergence 148
– trajectories 491
energy flow 483
energy generation 471, 473, 538
energy grid 472, 473
energy harvesting 459, 508, 537
– applications 481, 484
energy management 364, 483
– systems 471, 480
energy-saving aspect 538
energy-saving lighting technology 164
energy scaling requirements 142
energy storage 506
– system 503
energy supply 471, 534
energy systems 473, 483
– of electric vehicle 482
environmental gas analyzers 514
environmental issues 471
environmental monitoring 567
epitaxial growth process 165, 170, 177, 184
EPoSS technology platform 580
error coding 435
– correction codes 118
ESD. *see* electrostatic discharges (ESD)
e-society 57

etching 267, 345
– existing solutions 284
Eureka 557, 569, 570, 588
European budget, for research/innovation
   activities 571
European Commission 449, 554, 615
European consumption 559
– of electronic components 560
European Fund for Strategic Investment
   (EFSI) 558
European funding of research and innovation
   programs 567, 569
– ESPRIT 572
– FP7 and H2020 581
– H2020 (2014-2020) 579
– ICT Research in FP 582
– micro/nanoelectronics and smart
   systems 582
– nanoelectronics integrated hardware
   perspective 571
– nanoelectronics/micro-nanotechnology
   570
European general educational practices 609
European Industry, overview of 553
European Investment Bank manages 558
European leaders 561
European manufacturing capacities 563
European market share 561
– consumption 559
– supply 560
European parliament 569
European policy makers 563, 565
European private sector 565
European Regional Development Fund
   (ERDF) 558
European Research, Development and
   Innovation Programme 449
European Research Policy 575
European semiconductor companies and
   institutes 572
European strategic initiatives 554
– combining instruments 555
– ECSEL Joint Undertaking 554
– European Commission 554
European Structural and Investment Fund
   (ESIF) 558
European Technology Platform (ETP) 554, 577
European Union (EU) 553
– budget of Member States in 2014, 559
Europe's market position 558
EUROPRACTICE IC project 589
EUROPRACTICE microsystem program
   574, 575

eutectic bonding 416
EUV. *see* extreme ultraviolet (EUV)
EUVL. *see* extreme ultraviolet lithography
   (EUVL)
evidence-based algorithms 591
EXNOR 357
EXOR, boolean comparator 357
exponential complexity 388
exposure wavelength 290
extended X-ray absorption fine structure
   (EXAFS) 256
extremely thin SOI (ETSOI) 92
extreme ultraviolet (EUV) 119, 297
– anamorphic imaging 302
– insertion into N7, 298
– lithography 119, 309
extreme ultraviolet lithography (EUVL) 294,
   299, 300, 301
– extendibility 301
– insertion 294, 301
– patterning 299

f

Facebook 449
FACTS/HVDC-systems 486
failure mechanisms 410
failure times 319, 321
fast access cache memory 114
fault collapsing 440
fault coverage FC 429
fault-free products 428
fault models 427, 429
– definition 439
fault simulation 439
FDSOI. *see* fully depleted silicon on insulator
   (FDSOI)
ferroelectric material 121
ferroelectric memories 120, 121
ferroelectric polarization RRAM 129
ferroelectric RAM (FeRAM) 120–122
ferroelectric thin films 251
fetal health 466
fiber extension 155
fiber optic gyroscopes 402
field-assisted superlinear threshold
   (FAST) 131
field-effect transistor 176
– at Bell Labs 6
– sensors 148
field emission display (FED) 517
field plate (FP) 196
field-stop (FS) 173
fine-tuning production techniques 508

FinFET 90, 93, 94, 96, 99
FinFET sensor characteristics, full-scale pH
    influence 149
finite state machine (FSM)
– decomposition 342
– model 339
– and sequential circuit implementation 339
fin pitch (FP)
– *versus* technological node 96
flash anneal 274
flash memory producers 46
flash NAND 132
flip chip assembly 416
flip-flop configuration 48, 117, 431
flip-type IGBT 183
floating gate memories 253
floating gate transistor 17
float zone (FZ) technique 228
fluorine codoping 273
fluorine doped tin oxide (FTO) 518
Focus Center Research Program (FCRP) 36
food chain 451
fossil fuels 471
Fourier transform infrared spectroscopy
    (FTIR) 256
FP7/H2020, beyond CMOS 586
FP7 project 585
– displaying EU bias 582
– DUALOGIC 588
– and ENIAC JU programmes 583
– funding 581
FP88 project 585
Framework Programs 582
– European Research 569
– FP7 budget 581
free space path loss (FSPL) 154
free-standing IGBT 183
free-wheeling diodes 179, 180
front collision warning (FCW) 493
front-end process 267
– cleaning 268–270
– deposition 275–279
– etching 279–285
– silicon oxidation 271
frozen gas particles 269
FS. *see* field-stop (FS)
full-adders (FA) 353
fully depleted silicon on insulator (FDSOI) 90,
    272, 561, 564, 584
– capacitance breakout for 98
– device 92
– planar 281
– structure 91

– technology 93
Future and Emerging Technologies (FET) 574,
    579, 585
– negative capacitance 150
future technology nodes 322
FZ growth method 232
FZ techniques 229

g
1G 153
2G 153
3G 153
4G 153
5G 153, 154
GAA. *see* gate all around
galium nitrate (GaN)
– bandgap 195
– bulk substrate 200
– device 188
– – material and physics 184
– HEMT 195
– layer 195
– on-Si technology 193, 197
– power device 184
– power semiconductors 199
– power transistors 193, 194
– technologies 157, 197
– transistors 191, 192
– trench 189
– vertical devices 189
– wafers 188
gallium nitride (GaN) 198, 536
GaN. *see* galium nitrate (GaN); gallium nitride
    (GaN)
5G, and high-performance computing
    ASO 593
GaSb 104
gas-sensing technology 212
gate all around (GAA) 249, 273
– transistors 13, 96
gate charge 169
gate density requirement 100
gate dielectric layer 77
gate dielectric stack 322
gate electrode architecture 98
gate field plate 196
gate first approach 98
gate leakage current 322
gate mask, inevitable misalignment 14
gate oxide
– partial nitridation 271
– scaling 1997 NTRS 35
– thickness 87

gate replacement 342
gate–stack deposition
– on Ge channel 104
gate-to-epitaxy capacitances 98, 99
– parasitic capacitance effect 98
gate voltage 75, 76, 141
Gbitp MTJ arrays 124
GDP/economic exchanges 592
GdTiO$_3$ 145
generic modeling 385
geometrical phase analysis (GPA) 251
geometrical scaling 291
germanium 103
– transistor effect 225
germanium-antimony-tellurium
(GeSbTe) 124, 125
gestational diabetes 466
Ge, transport parameters for 103
GIXRD configuration 255
glassy silicon dioxide 9
Global Foundries 564
global positioning satellite (GPS) 504
– receiver 460
– sensors 496
Glucodoc 463
glue dispensing process 420
gold nanoparticles 512
gold-standard PSG testing 463
gold wires 6, 11
goniometer 253
Gordon E. Moore 13
GPA. *see* geometrical phase analysis (GPA)
graphene 502, 519
– nanoribbons 334
– p–n junctions 350, 354, 355
graphene nanoplatelet paper (GNP) 505
grapho-epitaxy 302, 303
grazing incidence 255
green economy 452
greenhouse gas 451
grid interface 164
GSM telephony 565
GW. *see* gigawatt (GW)
gyroscope 210, 459

h

Hamming distance 342
hard disk drive (HDD) 115
hard-switching circuits 173
hardware description language (HDL) 371
hardware-in-the-loop (HIL) 497
H$_2$ bake cycle 236
HC. *see* hot-carrier (HC)

HDL. *see* hardware description language (HDL)
HDMI services 155
health care expenditure 456
health care system 455
health monitoring 219
HEMT. *see* high electron mobility transistor
(HEMT)
heteroepitaxy 200
heterogeneous system 382, 397
– calibration of 420
– design 400, 401, 402, 403
– – analysis 401, 402
– – assembly and testing 403, 412
– failure mechanisms in 411
– integration 48, 414
– packaging 414, 418
– processing 576
hexagonal boron nitride (hBN) 502
HfO$_2$–TiN interface 258
HF vapor 269
– removal 271
high electron mobility transistor (HEMT)
187, 195
– FET hybrid 192
– transistors
– – fabrication of 191
high energy efficiency 538
high-energy self-ion implantation 273
high-temperature annealing process 272
high-voltage blocking 178
high-voltage power semiconductors 542
high voltage transmission 479
high-volume manufacturing (HVM) 31,
33, 298
HIL. *see* hardware-in-the-loop (HIL)
H2020 LEIT 585
HMOS II generation 30
homoepitaxy 200
honeycomb structure (HCS) 119
Hook's law 238
Horizon 2020 program 449, 580
– as societal challenges for Europe 450
horizontally stacked 3D RRAM (HRRAM)
architecture 132
hot-carrier (HC) 322
HR-TEM corresponding image 259
4H-SiC layers 183
4H-SiC wafers 177
humanity 589
humidity 540
HVM. *see* high-volume manufacturing (HVM)
HW/SW codesign 575
hybrid CBRAM devices 128

hybrid linear temporal logic with
  expressions 382
hybrid solutions 437
hybrid vehicles 163
hydrofluoric acid 280
hydrogen contamination
– in back-end process 121
hydrogen-to-oxygen ratio 29
hyperconnected world 590, 591
hyperpure silicon 227, 228
hypertension 463
hysteretic characteristics 145

i

Ibbetson theory 186
IBM 26, 32
IBM360, 614
IC. *see* integrated circuit (IC)
idealstacked nanowire transistor
– schematic of 96
IEEE RC 51, 52
IEEE Rebooting Computing Organization 50
IEEE 1500 standard 433
IGBT. *see* insulated gate bipolar technology
  (IGBT)
illumination-related effects 274
image sensors 211, 463
immersion liquid 290
immersion lithography 291, 293, 295, 306, 309
immersion multiple patterning 299, 302, 307,
  309, 311
IMOS devices 143
impact ionization 143
IMU 421
incentive programmes 553
in-depth electrical characterization 252
indium tin oxide (ITO) 518
indoor communication 540
indoor lighting 540
Industrial and Materials Technologies
  (IMT) 573
industrial automation 57
industrialization 471
Industry Strategic Symposium (ISS) 37
inert gas flash 278
inertial measurement unit (IMU) 398
in-field test 436
info-bio-nano-cogno field 588
information and communication technology
  (ICT) 531
– infrastructure 535
– program 572
– related technologies 545

– support for logistics 546
Information Society (IST)
– ICT projects 588
– program 574
information society tools 452
Infrared SNOM 256
infrared spectrometry 212
InGaAs alloys 104
InGaN quantum well blue LED 515
innovation capacity 449
innovative logistics strategy 544
inorganic LED point source 69
inorganic resist 298
in-plane switching (IPS) 516
InP technologies 157
in'situ thermal annealing 255
insulated gate bipolar technology (IGBT)
  165
insulated gate bipolar transistor (IGBT) 173
insulators 271
integrated algorithm framework 442
integrated circuit (IC) 11, 15, 164, 205, 225,
  289, 290, 333, 397, 409, 427, 473,
  483, 537
– degradation mechanisms 317
– economical advantages 13
– fabrication, doping technique 272
– failure mechanisms 320
– fundamental building blocks 47
– manufacturing 268, 271
– packaging technology 397
– Robert N. Noyce 12
– transistor effect 51
integrated power devices 164
integration density 53
Intel 8080, 24
intellectual property (IP) 353, 371, 433
intelligent activity tracker 461
intelligent energy management 473
intelligent environment 219
Intel's DRAM business 28
interconnect capacitance 318
interconnection technology 219
interconnect reliability issues 318
inter-diffusion phenomena 256
interlayer (IL) 256
intermodal transportation 490
internal gettering (IG) 240
internal transduction 143
internal voltage amplification 143
International Electron Devices Meeting
  (IEDM) 20, 21
International Energy Outlook 472

International 300 mm Initiative (I300I) 33
International Nanotechnology Conference on
   Communication and Cooperation (INC) 42
International Roadmap for Devices and
   Systems (IRDS) 49, 51, 534
International Technology Roadmap for
   Semiconductor (ITRS) 34, 76, 534, 584
International Technology Roadmap for
   Semiconductors (ITRS)
– ITRS 2.0 49, 51
– roadmap 324
Internet of Everything (IoE) 140, 148, 154
Internet-of-Nano-Things concept 515
Internet of Thing (IoT) 55, 57, 154, 334, 370,
   450, 453, 459, 514, 532, 534, 586, 588, 593
– driving force for smart cities and
   semiconductors 535
– under FP7 and H2020, 587
internship programs 602
interpoly dielectrics (IPD) 118
interstitial adsorption 274
interstitial fluids 466
interstitials annihilation 273
intraocular lens 464
invasive catheters 461
in-vehicle software 496
inversion coefficient (IC) 156
inverted temperature dependence (ITD) 347
inverter
– delay 79
– parasitic capacitances of MOSFET
   devices 79
– switch, parasitics capacitance in logic
   devices 81
inverter chain 82, 84
Io-E requirements 150
ion beam curing 313
ion bombardment 283, 284
ion-hosting capacity 507
ionic gating liquids 145
ionic polymer metal composite 216
ion implantation 174, 179, 182, 272
ion-induced perturbation thickness 284
IP. *see* intellectual property (IP)
iPhone 6 smartphones 73
irredundant threshold networks 440
ISFET sensor 148
isothermal annealing 254

j
Johnson–Nyquist theorem 141
Joint Technology Initiative (JTI) 577
– JTI ECSEL 570

Joint Undertaking (JU) 555, 558
junction barrier Schottky (JBS) 179
junction field-effect transistor (JFET) 180
junction termination extension (JTE) 179

k
key enabling technologies (KET) 449, 554, 577,
   615
Kirk effect 182

l
lane change assistant (LCA) 492
lane departure warning (LDW) 493
lane keeping assistant (LKA) 493
Langmuir-Blodgett deposition 279
Laplace force 270
lapping 234
Lars' brain registers stress 465
Lars' prostate problem 463
laser
– annealing 175, 274
– beam scans 274
– irradiation techniques 270
– melting 274
– produced plasma (LPP) 294
– pulses 270
lateral power devices 167
Laue microdiffraction 251
layout optimization solution 118
leakage power 343
– of CMOS 139
– in CMOS gate 343
leakage problems 38
LEIT addressing industrial
   competitiveness 580
LG G-series 516
Li batteries 507
LiDAR sensors 493
light-emitting diodes (LED) 69
light illumination 19
lighting 539
Li-ion/nanostructured battery systems 509
linear temporal logic (LTL) 382
line edge roughness (LER) 297, 311, 312, 320
LiNe process flow 304
linewidth roughness (LWR) 311
liquid crystal display (LCD) 516
liquid gate technology 148
– FinFET ion sensor, cross section of 149
liquid phase epitaxial 274
Lisbon European Council 554
lithium-ion cells 482
litho-etch (LE) 291

litho-etch-litho-etch (LELE) 291
lithography 267, 345
– imaging 292
– insertion 314
– pitch scaling in 292
– projection lenses used in 290
– technologies 205, 320
local atomic density 250
local energy management systems 473
local field enhancement 320
local magnification effect 250
local oxidation of silicon (LOCOS) 19
logic equivalence checking (LEC) 442
logic function 439
logic gates 23
logic standard cell 292, 300
logic storage elements 326
logic synthesis 334, 347
– for approximate computing 351
– for manufacturability and PV
  cmpensation 346
– and optimization, emerging trends in 350
– strategies, taxonomy of 335
– tools 334
lower electron temperature 284
low-noise amplifier (LNA) 156
low-power (LP) market 74

m

macromolecular memory 129
magnetic core memories 23, 120, 122, 230
magnetic memory (MRAM) 122
magnetic tunnel junctions 350
maneuvers lane changing 494
manually driven vehicles 496
manufacturing readiness levels (MRL) 583
manufacturing technologies 205
MapperLithography 310
mapping 373
– specifications to implementations 386
marangoni/rotagoni drying 270
March algorithms 431
maritime traffic flow 490
market share, of leading companies 562
Mark-8, microcomputer design 24
mask handling 296
MASTAR model 76
Masters programs 603
material phase transition 143
material properties 363
material removal system 279
maternal/fetal parameters monitoring 467
mathematical optimization software 382

Mbit Ag/GeS$_2$ CBRAM memory core 129
64 Mbit DRAM memory devices 33
mechanical coupling 407
mechanical forces 205
mechanical noise 207
mechanical parameters 59
mechanical sensors 207
– gyroscopes and accelerometers 209
– pressure sensors and microphones 208
– resonators 210
medical imaging 211
medium energy ion scattering (MEIS) 252
megabit memory generation 46
memory-centric architecture 114
memory devices 27
memory hierarchy, evolution 114
memory market 115
memory producers 46
memory products 27
memory reliability 119
memory technologies 113, 114
– computing systems 113
– data-intensive applications 113
merged PiN Schottky (MPS) 179
metal capping layers 321
metal gate 148
– stack 249
metal-insulator-metal (MIM) 126
metal-insulator transition (MIT) 143, 145
– FETs 150
metallic CNTs 442
metallic mobile charge 145
metallurgical silicon
– production 226
– purification of 227
metal nano-wire based transparent
  conductor 519
metal oxide nanoparticle resist 298
metal oxide semiconductor field effect
  transistor (MOSFET) 166, 169, 172, 502
– device 84
– – parasitic capacitances 79
– electrostatics 80
– leakages 85
– optimization 88
– planar bulk MOSFET, TEM cross section 75
– structure
– – capacitances of 84
– structures 89
– technology performance evaluation 79
– threshold voltage 76
– transistor 73, 75, 80, 81, 87
metal resistance 322

metronomy 378, 389
mHealth 460
microactuators 207
micro- and nanoelectromechanical systems
(MEMS/NEMS) 49, 59, 205, 363, 397, 404,
451, 503, 586, 614
– actuator 400
– advantages 206
– based sensors 586
– holistic approach 581
– market trends 207
– microreflector technologies 515
– MOEMS and Miniaturized Systems 573
– processing 269
– silicon-integrated 588
– tuning fork resonator 504
microcomputers 24
microdevices 215
microelectronic products 54, 397
microelectronics 271, 450, 536, 613
– industry 53
– structures 247
microenergy grid 473, 480
microfluidics 206, 462, 588
– devices 55
– platform 213
microhydraulic actuators 216
micromechanics 588
micro-nanoelectronics 537, 539, 568, 592
micro-nanosystems 568
micro-nanotechnology 568
microprocessors 50
micropumps 206
microrobotics. 588
microscale energy transducers 484
microscale sensors 219
microscopic sensor 364
microsystems 207, 397, 588
microvalves 206
mid-wavelength infrared (MWIR) 506
miniature gas sensors 212
miniaturization 53, 54, 409, 486, 537, 567
– capabilities 280
– program 568
miniaturized system 533
minimum voltage supply, limitation of 87
misaligned carbon nanotubes 442
MISHEMT 188
mixed integer linear constraints 384, 387
mixed-ionic-electronic conduction (MIEC) 131
mixed valence oxide RAM 129
mobile devices 460
mobile health 460

mobile Internet device (MID) 74
mobility and transport, in smart city 541
mobility applications, of smart energy
systems 482
model- based design (MBD) 365
model for advanced semiconductor roadmap
(MASTAR) 76
modeling languages 385
model-in-the-loop (MIL) 497
molecular beam epitaxy 147
molecular doping 274
molecular layer deposition (MLD) 275
molecular monolayers doping 275
molecular motors 217
molecule bond breaking 275
molybdenum disulfide (MoS_2) 502
monitor communication 540
monitoring simulation 389
monocrystalline silicon 237
– crystal structure 237
– defect kinetic behavior 240
– doping 237
– ingots 228
– intrinsic defect categories 239
– production 229
– silicon dioxide 238
monolayer deposition techniques 279
monolithic integrated microsystems 573
monolithic microwave integrated circuit
(MMIC) 504
monostable–bistable transition element
(MOBILE) 439
Moore's law 20, 33, 34, 40, 45, 51, 53, 54, 73,
118, 156, 229, 241, 280, 291, 333, 397, 428,
458, 502, 584
– for micro-nanoelectronics 70
– revolutionary and scalable technologies 51
– vision, validation of 229
More Moore domain 595
More-than-Moore 54
– devices 58
– – application domains 61
– – industrial applications 67
– – powering 59
– – safety and security 65
– photodetector 56
– technologies 55, 60, 70
morphing 52
MOS channel noise factor 158
MOS circuits 16
MOS device 5, 117, 131
– of Shockley's dreams 15
MOS process 29

MOS transistor 8, 11, 14, 16, 18, 34, 54
– p- and n-channel 28
MOS yields 20
motion processing unit (MPU) 510
Mott transition RAM 130
MP3 players 49
MRI tomography 614
MtM devices/technologies for sensing/
  actuating
– potential taxonomy of 60
MtM technologies 58
mulitenergy smart grid 477
multibillion transistor chips 334
multibit errors 326
multichannel vital sign monitoring system 463
multienergy smart grid 476
multifunctional heterogeneous subsystems 56
multipatterning 293, 298
multiple-epitaxy growth 171
multiple patterning techniques 291
multiple p-type doping 171
multiple sensor systems, on-chip integration
  of 510
multirobot systems 389
multi-sensor system-on-chip (MUSEIC)
  463, 464
multiwire sawing 233

n
NAND 116
– capability 115
– flash 125
– flash memory 118
– gate 78
– memories 117
– type array organization 122
nanocharacterization 245
nanodevices 258, 363
nano-electro-mechanical memory
  devices 130
nanoelectromechanical system (NEMS) 59,
  504
nanoelectronics 141, 205, 212, 219, 245, 449,
  450, 453, 530, 536, 539, 541, 568, 595, 599,
  613, 614, 615
– based hardware 529
– based solutions 541
– components 245, 317, 537, 567, 568, 600
– degree programs, challenge facing 601
– education 604, 606
– sector 471
– in Europe 564, 578
– fabrication technology 205

– foundational research and disciplines 597
– future comprehensive university
  programs 607
– hardware challenges 546
– manufacturing processes 212
– related degree programs 602
– research clusters 570
– as smart energy systems 483, 533, 545
– standardizing holistic education 609
– technologies 602, 604
Nanoelectronics Research Initiative (NRI) 42
nanofabrication technologies 205, 208
nanoimprint lithography 309, 311, 313
– challenges 19, 311
nanometer era 345
nano-MOSFETs 323
nanoparticles 424
– based electrospray deposition methods 517
nanopillars 518
nanopirani-based sensors 209
nanoscale manufacturing 277
nanosensors 148
nanostructuring 509
– supercapacitor systems 508
nanotechnologies 208, 363
– enhanced glucose sensors 512
– plasma etch challenges 285
nanotweezers 270
nanowire 96, 245, 281
– conduction lines 519
– fabrication 400, 404, 424
– – defects 441
– – methods 424
– – techniques 66
– grid polarizers 516
– parasitic capacitances of 97
National Institute of Standards and Technology
  (NIST) 42
National Nanotechnology Initiative (NNI) 41
National Network for Manufacturing
  Innovation (NNMI) 556
National Science Foundation (NSF) 42, 449
National Strategic Computing Initiative
  (NSCI) 51
National Technology Roadmap for
  Semiconductor (NTRS) 33
N_2-based aerosols 270
n-Doped MOS (NMOS) 18, 19, 43, 78, 259,
  260, 344
near-adiabatic logic circuits 141
near-field optical techniques 251
negative bias temperature instability
  (NBTI) 323, 348, 350

neuristors 147
neurosynaptic chips (SyNAPS) 147, 148
nFET transistor 104
n-IGBT 184
nitride spacer etching 281
nitrogen aerosols 270
nitrogen exodiffusion 258
nitrogen plasma treatment 271
noise figure (NF) 157
nonaqueous cleaning steps 269
nondestructive testing 247
nonlinear behaviors 387
non-Newtonian fluids 269
nonpunch-through (NPT) 173
non-silicon channel 102
nonstandard oxidation techniques 271
nonthermal activation techniques 274
nonthermal illumination effects 274
nonvolatile memory (NVM) 20
– applications 125
NOR Flash 131
NOR gate 78
NP junction 43
n/p-MOS transistors 58
N-push VCOs 161
nuclear magnetic resonance (NMR) 210
nucleotides 214
numerical aperture 291

o

object oriented programming 375
obstructive sleep apnea (OSA) 463
off-axis electron holography 260
off-axis illumination 290, 301
ohmic contacts 190, 198
on-body sensors 458
on-button-cameras 48
one-dimensional nanowires 275
one-pass synthesis (OPS) 356
ON–OFF switching process 79
on-resistance 177, 178, 180, 182
open-circuit failures 19
open fault model 430
Open Microprocessor Systems Initiative
    (OMI) 572, 573
operating temperatures 347
optical lithography 289, 290, 291, 295
optical projection lithography 294
optical proximity effect correction 290
optical pulse oximetry 459
optical transducers 206
optical wavelength 291
optimal safety 495

optimal trajectories 494
optimization-based design methods 373
optoelectronic integration 576
optoelectronics 251
ordinary differential equation 383
organic-based areal light emitters range 69
organic contamination 268
organic films 279
organic light-emitting diode (OLED) 516
– commercial OLED flat panel display 517
– displays 516
organic thin films dep 279
original equipment manufacturer (OEM) 437
OSA. *see* obstructive sleep apnea (OSA)
ovonic threshold switch 131
oxide–electrolyte interface 145
oxide electronics 143
OxRAM, oxide based resistive RAM 127, 128
oxygen plasma 268
oxygen scavenging 258

p

Packaged TLens 399, 400, 417, 418
– assembly of 413
package parasitic elements 408
Pan-European partnership in micro- and
    nanoelectronic technologies and
    applications (PENTA) 557
parallel e-beam lithography 309
parallel testing systems 325
parasitic capacitances 405
– calculation method 83
parasitic current paths, in cross-point
    arrays 130
parasitic inductances 405
Pareto optimal 389
park distance control (PDC) 493
parking assistant (PA) 493
particle removal efficacy 268, 270
pass-diagram (PD) 357
pass-transistor-logic (PTL) 356
pass-XNOR logic (PXL) 357
patched flows 371
path distribution at different temperatures 347
pattern collapse 270
pedestals 404, 406
people-to-thing 153
performance modeling, analysis 370
permanent defects 438
personal health, factors determining 457
personalized stress measurement
    techniques 466
personal rapid transport (PRT) 541

pFET device 93
pFET mobility on Ge channel 104
p-GaN devices 192
P-GaN/P-AlGaN gate structure 189, 192
phase change memories (PCM) 120, 124, 125
phase noise (PN) 158
phase separations, morphology 304
phase-shifting masks 290
phosphorous, as dopants 30
photo-beam lithography 598
photoemission electron microscopy
    (PEEM) 256
photon flux 247
photonics 588
photoresist mask 281
photoresist removal 270
photovoltaic (PV) 179, 588
physically unclonable functions (PUF)
    technologies 534
physical vapor deposition (PVD) 275, 276
physiological sensors, medical-grade 457
piezoelectrics 509
– actuation 215
– behavior 185
– polarization 185
piezoresistive 209
p-IGBT 184
pigment-silica composite nanoparticles 517
pilot lines projects 555
PiN diodes 181, 183
pin reordering 349
piranha strip 268
planar-bulk transistor architecture 89
planar double-gate devices 90, 93
planar process 10
planar transistors 317, 327
plasma-based native oxide removal 269
plasma cleaning procedures 283
plasma doping 274, 275
plasma-enhanced ALD (PE-ALD) 279
plasma etching 279, 280
plasma-induced damage 284
plasma oxidation 271
plasma processing 282, 283
plasma smoothing 313, 314
plasma–surface interactions 282
platform-based design (PBD) 365, 372, 373,
    380
PMOS/NMOS transistors 18, 28, 78, 340, 349
POC devices 462
polarity gate (PG) 354
policy implementation instruments
– European Union 557, 558

– intergovernmental programmes 557
– joint undertaking 558
– national programmes 557
– public announcements 556
– world 556
pollution 452
polychromatic X-ray beam 251
polymethylmetacrylate (PMMA) 302
polysilicon 30, 38
– channel 120
– chunks 229
– etch mechanism 283
– floating gates 119
– gate 283
– lines 30
– rods 228
polysomnography (PSG) 463
polystyrene (PS) 302
pooling resources 602
porous inter-metal-level dielectrics (ILD) 318
porous low dielectrics failure times 319, 320
positioning sensors 493
positron annihilation spectroscopy (PAS) 256
post CMOS 49
– and beyond silicon 354
post-Dennard or leakage-limited scaling 139
postlitho smoothing techniques 313
post-mapping transformations 342
power-added efficiency (PAE) 157
power amplifier (PA) 156
power budget 459
power combiners 158
power consumption 4, 486
– per sensor 458
power converters 540
power density 486
power devices 164, 176
power dissipation
– in CMOS circuit, sources of 340
– in transistor devices 84
power electronics 163, 164, 195, 451, 483,
    536, 583
– applications 163
– components 484
– converters 472
power factor correction circuit (PFC) 179
power grids
– stabilization 163
power management 459
– electronics 537
power MOSFET structures 168
power-optimized power supplies
– *versus* gate length 87

power performance trade-off 75
power semiconductors 164
power spectra density (PSD) 312
power transistors 187
power wall 340
PPP Green Car 588
preamorphization 273
precession electron diffraction (PED) 251
predictive medicine 567
preeclampsia 466
pregnancy, monitoring 466
presbyopia 464
preterm birth 466
printed circuit board (PCB) 54
process architecture 99
process engineering 317
processor–memory performance gap 114
processor power consumption 50
process variations 435
product design 544
programmable logic array (PLA) 441
projection lens 290
projection systems 289
protein identification 463
PSA test 462
p-type dopant 171, 272
p-type layers 192
Public-Private Partnership 570, 577, 606
public support 564
puller furnace 232
pulmonary pressures 461
pulse-arrival time (PAT) 464
pulsed laser deposition 147
pulsed plasma technologies 284
pulse sensor 461
punch-through (PT) 173

q
quality constraint circuit (QCC) 352
quality of information 544
quality of life 457
quantum confinement effects 104
quantum-dot cellular automata 438, 439
quantum effects 424
quantum mechanics 4, 143
quasi ballistic 101

r
racetrack memory 124
RADAR 542
radar/lidar 212
radar sensors 493
radio communication 542

radio frequency (RF) 505
– performance 157
– RF CMOS 153
radio frequency identification (RFID) 515
– tags 532
radio transmitter circuits 503
random-access memory (RAM) 16, 25
random telegraph noise (RTN) 324
– noise 129
rapid escalation
– in energy/operation 41
rate of soft errors (SER) 325
Rayleigh equation 289, 301
razor 435
RC. *see* reverse-conducting (RC)
reactive radicals 282
reactive synthesis 388
read-only memory 15
realistic simulation 542
real-time computing power 493
real-time sensing 458
real-time software capabilities 499
recess-channel-array transistor (RCAT) 117
recess phenomena 284
recognition, mining, and synthesis (RMS) 351
redeposition 269
reduce surface states 17
refractive index 290
register transfer level (RTL) 334, 371
remote monitoring 503
renewable energies 163, 451, 471, 472, 537
ReRAM *see* Resistive RAM cell
Research and Innovation for societal
    welfare 450
research and technology institutes 573
Research Technology Organization (RTO) 561
resist removal processes 269
resist trimming processes 281
resistive RAM (RRAM or ReRAM) 126, 129,
    134
resonance frequency 210, 404
resonant tunneling diodes 438, 439
restricted fin geometry 325
RESURF 166
reverse-blocking IGBT 175
reverse-conducting (RC) 175
risk of inconsistency 369
Robert H. Dennard scaling laws 21
robotics 451
root mean square (RMS) 234
– applications 351
RRAM 131, 132
rural health care 462

S

sacrificial thin film layers 268
safety-critical scenarios 497
sapphire substrate 200
SAR500 398, 399, 402, 404, 406, 407, 409,
   410, 419
– ceramic package of 411
satisfiability modulo theory (SMT) 382
scaffolding 373
scaling existing technologies 120
scaling in Dennardian and post-Dennardian
   nanoelectronics era and exploitable silicon
   area 140
scaling laws 208
scalpel conductive atomic force 246
scan chain 431
scan in signals 431
scanners 289, 290
scanning electron microscope 247
scanning near-field optical microscopy
   (SNOM) 251
scanning transmission electron microscopy
   high-angle annular dark-field imaging
   (STEM-HAADF) 249
scan out signals 431
scan test 431, 432
Schottky diode 176, 179
science, technology, engineering, and
   mathematics (STEM) fields 604
secondary ion mass spectrometry (SIMS)
   256
security measures and technologies 68
segment insertion bit (SIB) 435
self-aligned deposition techniques 267
self-aligned double patterning (SADP) 291
self-aligned quadruple patterning (SAQP) 291
self-aligned silicon gate MOS process 15
self-faults, definition 441
self-sustaining energy systems 471
Sematech consortium 32
SEM, electron beam 248
semiconductor 8, 15, 115, 167, 193, 333
– atom probe tomography, analysis 249
– based LASERs 69
– business environment 27
– chips 313
– companies 48
– components 333, 537
– enhanced CNTs 505
– industry 11, 22, 48, 279, 317, 431
– – history of 3
– memory
– – lowest-cost and highest-density 116

– packaging technologies 397, 416
– processing 140, 193, 225, 267, 473,
   483, 486
– units 225
– value 614
– wafers 9
Semiconductor Industry Association (SIA) 36
SEMI standard 33
sensing 532
sensors 55, 208, 509, 533, 539
– and communication networks 533
– die 404, 412
– equipped with receiver/transmitter unit 548
– gas sensors 212
– mechanical sensors 208
– motion processing units 510
– nanosensors for biomedical applications 511
– nodes 471
– optical sensors 513
– for smart home, and building energy
   management 539
– terahertz imaging 211
– vision 210
SER. *see* rate of soft errors (SER)
shape memory alloy 216
shift-register memory 124
shock absorber 404, 407
shock margins 415
shock resistance 405
short-channel device threshold voltage 76
short-channel effect (SCE) 73, 322
– DIBL component 91
– immune MOSFET architectures 89
SiC
– BJT 182
– devices 178
– diodes/rectifiers 179
– power devices 183
– substrate 200
– wafers 177, 183
SiCOH-based porous dielectrics 318
Siemens process 228
SiGe-HBT 157
signal critical path 85
signal mapping 401
signal processing 598
signal temporal logic (STL) 382
signal-to-noise ratio 154
silicic acid residues 269
silicon 225
– availability and technologies 226
– based sensors 513
– channel 89, 92

– crystal orientation 102
– metallurgical production 226
– purity of 226
– transport parameters 103
– worldwide production of 226
silicon boule 237
silicon carbide (SiC) 176, 199, 536
silicon circuits
– radiation-induced soft-errors 325
silicon coimplantation 273
silicon crystal, growth 231
silicon dioxide 31, 38
silicon emerging technologies 335
silicon fabrication 397
silicon food chain 226
silicon fusion bonding 416
silicon gate devices 19
silicon gate process 15, 30
silicon ingot 231
silicon lattice 230
silicon microelectronics 370
silicon nanowire 354, 438
– FET 443
– transistors 249
silicon nitride layer 19
silicon nitride spacer 284
silicon on insulator 38, 90, 576
silicon oxidation 271
silicon oxide
– consumption 268
– deposition 279
– as gate material 271
– growth kinetics 271
silicon oxynitride (SiON) 271
silicon photonics 588
silicon platform 458
silicon recess 284
Silicon Research Corporation (SRC) 36
silicon-silicon dioxide interface 16, 271
– properties 271
silicon substrate 200
silicon surface 275
silicon (Si) technology 536
silicon transistors 8
Silicon Valley 18
silicon wafers 188, 194, 242
– production 232
– – chemical-mechanical polishing
    (CMP) 232, 234
– – chemical treatments 232, 234
– – cleaning methods 233
– – epitaxy 236
– – final cleaning and packaging 235

– – mechanical treatments 232, 233
– – wafer manufacturing costs 233
silver nanoink 515
silver nanowire 519
SIMS. *see* secondary ion mass spectrometry
    (SIMS)
simulation techniques 370
simultaneous iterative reconstruction
    technique (SIRT) 248
Si nanowire channel 250
SiN, chemical vapor deposition of 196
Si_3N_4 film 252
SiO_2/HfO_2 gate dielectrics 323
SiO_2 interface layer 318
SiP design 576
Si power devices 164
SiP–System-in-Package 54
Si wafers 289
skin conductance 466
skipping methodology 31
sleep apnea 463, 464
sleep monitoring 463
slower access mass-storage memory 114
smart autonomous systems 148
smart building 537, 538
– cockpit 539
smart cities 451, 452, 453, 529, 531, 532, 540,
    543
– areas to define 530
– challenges to development of
    nanoelectronics 542
– data security 546
– environment 547
– flow of information 531
– information infrastructure 534
– infrastructure of 543
– mobility and transport 540
– production and logistics 543
– safety and security 546
– by semiconductor business, future for 530
– urban innovation 533
smart contact lens 464, 465
smartdust 532
smart energy 473–475, 486
– grid 536
– key products for 483
– management 472
– and societal challenges, applications of 476
– systems 471, 473, 474, 483–486
smarter society 590
smart factories 545
smart grid 164, 471, 539
– energy revolution 472

smart mobility 542
– and transport 537
smartphone 49, 449, 460, 501
– based point of care biomedical sensors 512
– camera modules of 402
– glucose sensor 463
smart production and logistics 537
smart reservation and booking 541
smart sensor 424, 533, 586
– based information 590
– and sensor systems 542
smart system 53, 530, 539
smart watch 467
SME businesses 571
sneak paths 442
social welfare system 452
societal challenges 449
societal demands 472
societal needs, digital domain 58
soft error
– effects 318
– in SRAMs, technology scaling trends 326
soft skills 596, 607
software architecture 499
software-based self-test (SBST) 437
software-related aspects of authorization 534
software solutions 547
solar energy 163, 538
solid-phase epitaxial growth (SPEG) 273
solid-state drives (SSD) 115
solid-state imaging 513
solid-state lighting 538, 539
solid-state technology 145
solution space of ITD-aware synthesis tool 347
sonic pulsed spray 269
source–drain epitaxy 97
source–drain fringing capacitance 79
source–drain structure 90
source field plate 196
source-to-drain current 38
SpaceEx 383
spacer-defined patterning 291
spacer formation 279
S-parameter 97
spatial frequency 312
spectrometer 213
speculative execution 50
speed sensors 493
spike rapid thermal annealing 274
spiking neural network (SNN) 134
spin coating 275
spin-transfer torque (STTM or STT-RAM or
   STT-MRAM) 44, 123, 124, 131

spontaneous polarization 185
sputtering 147
sputter PVD techniques 277
stable power source 537
stacked nanowires (SNW) 96
– structure, capacitance breakout for 98
standard dielectric cap 321
standard low surface tension approaches 270
standard megasonic cleaning 269
standard MOS
– FEOL process flow 267
– transistors 351
state encoding for low power 343
state-of-the-art (SOA)
– A/D BiCMOS process 573
– energy per conversion 148
– finFET CMOS technology 514
– terahertz CMOS detectors 161
– tunnel FETs from various international
   groups and 144
static random access memory (SRAM) 87
– cells 292
– circuits 85
– yield 77
statistical process control (SPC) 230
steep-slope switches address 140
steering wheel 493
steppers 289
STIM300, 398, 402, 420
ST LIS2DS12 microchips 510
storage class memory (SCM) 115, 126
strain analysis 250
– precession electron diffraction 252
– techniques 251
strained Si 86
strained-SOI 102
strain field distribution 251
stress 465
– decoupling 405, 406
– – substrate 404, 420
– evaluation, single sensor 465
– free polysilicon rods 232
structured silicon beams 404
stuck-at fault model 431
study-abroad programs 604
substrate implantation 194
substrate isolation 165
substrate removal 194
substrate resistance 169
subtractive processing 267
sulfuric acid-based chemistries 268
supercapacitors 506, 507
supercritical CO_2 270

supercritical nitrogen fluid 270
superintelligence 589–591
superjunction 176
– devices 170
– research and development activities 172
superjunction MOSFET 199
– HV-MOSFET 171
superjunctions 166, 199
– technology 165
surface-induced dispersion 195
surface modification agent 271
surface passivation 198
surface plasmon resonance (SPR) 213
surface preparation 268
surface treatment 273
sustainable energy 164
switch devices 180
switching circuits 169
switching ON–OFF 84
switching PAs 158
switching polarity 126
synchrotron 246
synthesis tools 369
synthetic biology 373
system architectures 385
– design 380
system engineers 390
system in package (SiP) 588
– structures 246
system modeling 386
system of systems 531
system on chip (SoC) 54, 363,
    433, 537
– embedded instruments 433
– test architecture 434
system properties 375
system requirements 48

t
tablets 49
1T1C CBRAM memory structure 128
1T/1C ferroelectric memory 121
tear analysis sensors 464
tear fluid glucose levels 464
technology-dependent transformations 342
technology mapping 349
– for low power 341
technology readiness level (TRL) 554, 571
telecommunication 536, 575
telematics application programs 574
temperature 540
terahertz circuits 161
terahertz CMOS 160

terahertz imaging 211
terahertz or submillimeter wave spectrum 159
terahertz products 160
terahertz radiation 211
terahertz waves 159
test access mechanism (TAM) 434
test access port (TAP) 434
test data register (TDR) 435
testing role 427
test pattern generation 440
test process 429
test vector generation circuitry 433
textile-integrated health patch 465
therapy monitoring 460
thermal actuators 216
thermal annealing process 275
thermal-aware logic synthesis 347
thermal budget 271
– techniques 274
thermal conductivity 200, 238
thermally controlled artificial muscles 216
thermal voltage noise limit 141
thermionic limit 141
thermo-, piezo-, or electrodynamic
    generators 537
thin AlGaN gate barrier 191
thin film deposition 275
thin film EUV lithography 279
thin film transistor (TFT) 505
thin oxide electrical quality 271
three-dimensional analysis 246
threshold logic gates (TLG) 439
threshold logic synthesizer (TELS) 439
through silicon via (TSV) 246
thyristors 484
time-dependent dielectric breakdown
    (TDDB) 318, 322
time redundancy 436
timing language 386
tip-enhanced Raman spectroscopy
    (TERS) 251
topology manipulation 349
traditional beam line implanter 274
traffic density 540
traffic management 490, 540, 541
transducers 205
– integration and connectivity 218
transfer characteristics of tunnel FET pixel, at
    light intensity 150
trans-humanism 592
transient defects 438
transient-enhanced diffusion (TED) 273
transistors 141, 206, 280

– architecture 318
– Bardeen, John 7
– Brattain, W.H. 7
– channel, mass of carriers 102
– concepts 53
– current 15, 100
– degradation 323
– – mechanisms, voltage dependence 323
– devices 87
– drain current changes 324
– effect 5, 51
– electrical width 94
– fabrication 43
– innovation
– – 1998 ITRS, equivalent scaling vision 35
– reliability issues 322–325
– self-mixing behavior of 161
– threshold voltage 73
transistor-to-transistor isolation 18
transmission gate logic,and one-pass
    synthesis 356
transmission lines 407
transnational collaboration 569
transportation systems
– CO_2 and toxic emissions reduction 491
transport enhancement, material
    engineering 101
1T1R architecture 131
trench access transistors 275
trench MOSFET 189, 190
trichlorosilane 227, 228
TriGate 90, 99
triple-level cell (TLC) 120
true fabrication technology 464
truth table, algebraic expressions, and PLA
    implementations of logic function 337
tungsten silicide 30
– polysilicon gate MOS process 30
tunnel FET 43, 143, 145, 148, 150
– optical sensors 149
tunnel transistor. *see* tunnel FET
turing machine 113
two-dimensional electron gas (2-DEG)
    185, 186

u
ultrahigh voltage–high-injection devices 183
ultralow power (ULP) 133
– electronics 536
– wristband 467
ultrashallow junctions 273
ultrashort channel 100
ultrashort devices 100

ultrasonic sensors 493, 496
ultrasonic transducers 215
ultrathin body (UTB) 92
ultrathin body and BOX (UTBB) 92
– device 92, 93
– parasitic capacitances of 97
– three-dimensional schematics 90
ultrathin oxides 271
under-represented minoritie (URM) 604
unified modeling language (UML) 368, 385
unipolar memories 131
United States
– built equipment 27
– health care cost 455
– owners 563
– US National Nanotechnology Institute 502
university-industry cooperative programs 608
unmanned automated vehicles 490
Uppaal-Tiga 389
urban energy generation 538
urban manufacturing locations 543
urban mobility 541
urban processes 531
– infrastructure 541
UV light 17

v
vanadium dioxide (VO_2) 145
Van der Waal (vdW) 502
Van Neumann architecture 52
vapor phase doping 172
variability-reliability interactions 325
vehicle communication systems
– security breaches 497
vehicle-to-infrastructure (V2I) 491
vehicle-to-vehicle (V2V) 491
velocity limitation 101
ventilation 538
vertical alignment (VA) 516
vertical buffer 193
vertical-channel CAT (VCAT) 117
vertical devices 188
vertical DMOS 171
vertical 3D RRAM (VRRAM) architecture 132
vertically stacked double-gate SiNW
    transistors 354
vertically stacked nanowires 90
Vertical NAND (V-NAND) Flash 120
vertical pillar transistor (VPT) 117
vertical ReRAM 132
video cameras 493
video sensors 543
virtual European research institute 587

virtual factories 546
virtual integration 371, 372
virtual world models 499
vision zerotarget for road safety 490
VLSI Symposium 32, 33
V-model, pictorial representations for 372
volatile etch products 282
voltage-controlled oscillator (VCO) 156
voltage converters 537
voltage doping transformation (VDT) 76
voltage rating 170, 177
volume-dependent forces 208
Von Neumann architecture 39, 113
Voronkov theory 239

w
wafer
– cost of producing 27
– fabrication 184
– heat 543
– size conversions 31, 32
– specification, for key parameters 242
– wafer-to-wafer assembly 416
water-based cleanings 269
water causing pattern collapse 269
watermarks 269
wave radar applications 212
wax-mounting process 234, 235
wearable bioimpedance sensor patch
    system 461
wearable devices 363, 460
– medical-grade 457
wearable health care application domains,
    technologies for 64
wearable sensors 363, 459
wearable wireless bioimpedance patch 461
weather conditions 541

Weibull failure distribution 319
welding 418
Western Institute of Nanoelectronics (WIN) 42
wet-ambient diffusion 9
wet cleaning 268
wet etching 279
wide-bandgap semiconductors 164, 177
WiFi 155
wind energy 163
wire bonding 415, 416
wireless communications 154, 536
– for consumer electronics 503
– systems 540
– technologies 536
wireless sensors 461
– nodes 211
World Energy Outlook 472
World Semiconductor Council (WSC) 34
WPAN 155

x
1975-203X, ages of scaling 47
X-ray
– detection chain 248
– micro-Laue diffraction 246
– for strain measurements 253
X-ray absorption near-edge structure
    (XANES) 256
X-ray computed tomography (XCT) 246
X-ray diffraction (XRD) 253
X-ray fluorescence (XRF) 256
X-ray focusing devices 251

z
Zeno behaviors 370
zero-power technologies 534
zone technique (FZ) 232